Emerging Trends in Mechatronics

This book series helps understand how Mechatronics will impact the practice and research of developing advanced techniques to model, control and optimize complex systems. It presents recent state of advances in mechatronics and related technologies including: Automatic Control, Robotics, Agent Based systems, Smart Manufacturing and Industry 4.0. The selected topics give an overview of the state-of-the-art and present new research results and prospects of the future development in the interdisciplinary field of mechatronic systems. The book series will provide up-to-date and useful knowledge for researchers and engineers involved in mechatronics and related fields.

Within the above dimensions, the scope of the book series welcomes high quality original research and review articles that cover a broad range of topics related to theoretical and applied aspects of modern mechatronics. Potential topics include, but are not limited to the following:

Controller design

Many mechatronic systems have challenging control difficulties due to the high nonlinear structures and time-varying dynamic behaviors. In addition to these, there are external disturbances, which cannot be predicted and change according to uncontrolled working environments. Conventional controllers are insufficient to solve the problems and to compensate the environmental disturbances, therefore designing efficient controllers for complex systems has a great importance in the current book series.

Sensor Design & Data Collection Approaches

Essential part of a mechatronics system is the measurement system that senses the variations in the physical parameters, such as temperature, pressure, displacement, and so on, and converts it to voltage or current. In mechatronic systems, accurate measurement of the dynamic variables plays a vital role for the actuators to function properly, therefore on of the aims of this book series, to focus on understanding what should be done to design accurate sensors and efficient data collection methods to improve the performance of mechatronic systems.

Optimization Techniques

Designing a complex mechatronic system involves multiple variables, objectives, constraints, and evaluation criteria as well as their nonlinearly coupled relationships. One of the challenging research areas of this book series is utilizing optimization techniques to design an optimal mechatronic system.

Artificial Intelligence

From the point of view of dynamics, mechatronic systems can be characterized by model uncertainties, high nonlinearities, complicated coupling, and stringent performance requirements, among others. To deal with these facts, this book series would like to focus on investigating and utilizing Artificial Intelligence techniques such as Neural Networks, Fuzzy logic, Neuro-fuzzy, evolutionary techniques etc. to model, design and optimize complex mechatronic systems.

Robotics

In the field of Robotics, there is a need of focusing on kinematic, dynamic, control and optimization of innovative robotic systems. This book series can introduce new and novel robot designs such as hybrid (serial-parallel) redundant manipulators that have a high potential of application in special fields like space robots. State of the art researches on selection of optimum kinematic solutions and path planning based on such an optimization, dynamic analysis and new control methods developments and application of artificial intelligence in this field can be published in this book series.

Smart manufacturing

Application of networked sensors and machinery equipment using internet and intranet in order to monitor and automation of the manufacturing process and improve the performance efficiency and is a point of interest. Advanced robotics, multi input-multi output control approaches combined with the artificial intelligence and optimization techniques, big data processing approaches, multi-scale modelling and simulation of dynamic sub-systems etc. are the hot topics that will be focused in this book series.

AGVs

Automated Guided Vehicle (AGVs) are going to be very popular is industry logistics, surveillance, defense and social applications in order to make processes safer and more cost efficient. Innovative AGVs designs for various missions, Robot Operating System usage on AGV control, Indoor and outdoor navigation and control algorithms, intelligent path planning techniques, sensor fusion, simultaneous application of GPS, INS and offline maps are the topics that are planned to be published in this book series.

Masoomeh Mirrashid · Danial Jahed Armaghani ·
Aydin Azizi
Editors

Advanced Decision-Making Under Uncertainty

Springer

Editors
Masoomeh Mirrashid
College of Engineering
Abu Dhabi University
Al Ain, United Arab Emirates

Danial Jahed Armaghani
School of Civil and Environmental Engineering
University of Technology Sydney
Sydney, NSW, Australia

Aydin Azizi
School of Engineering
Computing and Mathematics
Oxford Brookes University
Oxford, UK

ISSN 2731-4855 ISSN 2731-4863 (electronic)
Emerging Trends in Mechatronics
ISBN 978-981-95-8695-0 ISBN 978-981-95-8696-7 (eBook)
https://doi.org/10.1007/978-981-95-8696-7

This Springer imprint is published by the registered company Springer Nature Singapore Pte Ltd.
The registered company address is: 152 Beach Road, #21-01/04 Gateway East, Singapore 189721, Singapore

Preface

The reason for writing *Advanced Decision-Making Under Uncertainty* stems from a need across both the scientific and industrial landscapes: how can one choose among optimal alternatives when faced with incomplete information, nonlinear systems, and multiple, often-clashing, objectives? The answer does not rest on acquiring perfect, often unattainable, information, but rather on constructing intelligent systems that operate successfully within a given range of information and a defined system environment. This collection offers a series of studies that illustrate this approach. The chapters constitute an argument on the ability of cutting-edge technology to quantify and manage uncertainty. The book starts with methods for the control of uncertainty, including the similarity measure for Fuzzy Multi-Objective Linear Programming, which solves for efficient solutions by maximizing the closeness to ideal solutions. This is complemented by a Quartic Fuzzy framework for selecting Solar PV Recycling Technologies that demonstrates how complex, opinion-driven uncertainties related to a group decision can be structured and solved. Next, the book shows the relevance of AI and Machine Learning. The theme of interpretability is further extended with a Complex Linear Diophantine Fuzzy model for the selection of e-waste recycling siting which provides a mathematically implementable tool for achieving sustainable urban development. The following chapters demonstrate the use of these tools for advanced decision making in the layers of the infrastructure. Scan-to-BIM processes for structures create reliable digital twins from uncertain real-world data, while the development of a aerial ropeway network presents a visionary solution for sustainable urban mobility, optimized through intelligent control systems. In the domain of applied deep learning, XSS-Shield demonstrates how uncertainty created by covert cyber-attacks may be mitigated by employing dual-branch deep learning models. To conclude, the review on AI for fiber-reinforced soils mentions how the predictive capabilities of data-driven models highlight the disparity between the behavior of materials, thus the value that these methods possess in advancing geotechnical engineering is undeniable. Appreciation is extended to the contributing authors for the research that is foundational to this book. Their research not only pushes the boundaries of each discipline but also, in unison, paves the way for a future whereby systems are managed with greater intelligence and flexibility.

Al Ain, United Arab Emirates — Masoomeh Mirrashid
Sydney, Australia — Danial Jahed Armaghani
Oxford, UK — Aydin Azizi

Contents

Optimal Mathematical Design of Enhanced Volunteer Activities in a Monastery, Human Potentials-Based Search, Poker Strategy-Based Optimization and Conservation of Parrot-Based Algorithm

Lenin Kanagasabai

Abstract Volunteer activities in a monastery characteristically encompass compassionate, administrative, and maintenance responsibilities, such as spiritual based actions, food preparation, gardening, and supporting the events. Humans utilize imagination and expectations to search for the potentials, resolve problems, and forecasting the future, along with the drawing imagination from internal familiarity and understanding, whereas expectations are views about forthcoming consequences. Playing optimal poker necessitates an amalgamation of cognitive and emotional features, comprising judgment, hand appraisal, game theory, and mathematical abilities to comprehend varieties and possibilities. The Torah, Tanakh, and Talmud are the fundamentals of Jewish life, delivering a structure for religious exercise, beliefs, and communal distinctiveness. Jewish populations have traditionally established an outline of migration, frequently compelled by a prerequisite for endurance due to oppression, struggle, and monetary destitution. This has ensued in substantial migration groups and unstable population dispersals crossways universally. Jewish populations endure to acclimatize and transfer in rejoinder to numerous factors, comprising financial prospects, safety apprehensions, community and political variations in their nations of origin. The Holocaust stemmed in the demises of a substantial percentage of the Jewish population and had a permanent influence on the Jewish demographics and relocation outlines. Parrot preservation encompasses numerous significant features: habitation protection, combating illegitimate trade, public commitment, and enslaved breeding schemes. Shielding parrot habitations is decisive, as environment loss is a foremost danger.

Keywords Monastery · Volunteer · Poker · Migration · Parrot

L. Kanagasabai (✉)
Prasad V. Potluri Siddhartha Institute of Technology, Kanuru, Vijayawada, Andhra Pradesh, India
e-mail: gklenin@gmail.com

A. Azizi et al. (eds.), *Advanced Decision-Making Under Uncertainty*, Emerging Trends in Mechatronics, https://doi.org/10.1007/978-981-95-8696-7_1

1 Introduction

Volunteer activities in a monastery characteristically encompass compassionate, administrative, and maintenance responsibilities, such as spiritual based actions, food preparation, gardening, and supporting the events. Noble volunteer qualities comprise trustworthiness, accountability, empathy, consistency, flexibility, resilient communication, and co-ordination in all type of activities, all directed by the monasteries precise virtuous, traditional, and religious procedures. Humans utilize imagination and expectations to search for the potentials, resolve problems, and forecasting the future, along with the drawing imagination from internal familiarity and understanding, whereas expectations are views about forthcoming consequences. Both are vital for creativeness, personal development, and human evolution, but they are also restricted to the prevailing knowledge and can lead to dissatisfaction as soon as reality deviates from the expectations. The association amongst imagination and opinion is robust, as the imagination can modify the real experience of the realm. Imagination and expectations assist the humans to comprehend others' perceptions, attain novel knowledge, and generate connotation. Imagination is restricted by the prevailing knowledge and understandings, creating it challenging to imagine things out of the human's present understanding. Imagination will be effective at antedating obstructions and problems as it is at evolving novel notions, leading to cynicism or apprehension. What the humans imagine can modify the perception about the world, distorting the lines amongst belief and reality. Poker game requisite cognitive and emotional features, comprising judgement, hand appraisal, game theory, and mathematical abilities to comprehend varieties and possibilities. Important psychological characteristics are strength of mind, emotive control, concentration, and self-awareness to dodge tilt and make coherent judgements. The mental forte to preserve a noble balance between reviewing and playing, and to twig to noble decision-making even after trailing, is vivacious. The capability to dodge the negative emotion which influences the playing activity and manage the feelings like annoyance or obstruction is precarious for preserving concentration and building coherent decisions. Remaining attentive and focused all the way through a session, even as soon as cognitive functions obviously weakening over time, is essential to evade mistakes. Identifying one's individual forte, feebleness, and emotive state permits for superior self-regulation and modification to the varying crescendos of the game. Virtuous sleep, appropriate exercise, and a strong diet can expressively enhance the player's corporeal and intellectual state, primary to improved performance in the game. The Torah, Tanakh, and Talmud are the fundamentals of Jewish life, delivering a structure for religious exercise, beliefs, and communal distinctiveness. The Torah, the principally five books of the Hebrew Bible, is deliberated the verbatim conversation of God and a controller for existence, whereas the Talmud delivers explanations and solicitations of the Torah's commandments and preachings [1]. The Torah assists as a directive for all features of Jewish life, from spiritual performances to moral actions. The Torah's descriptions and commandments outline Jewish antiquity, distinctiveness, and association with God.The Tanakh, also entitled as Hebrew Bible, comprehends the Torah, the Prophets

(Nevi'im), and the Texts (Ketuvim), delivering a comprehensive ancient and literary perspective for Jewish credence. The Nevi'im suggests antique interpretations and prophetic communications that promote the Jewish understanding of God's role in ancient times. The Ketuvim (Writings) deliver knowledge literature, rhyme, and devotions that offer additional moral and spiritual supervision. The Tanakh stimulates replication on Jewish antiquity, distinctiveness, and the association between God and humankind. Jews be certain of the Torah comprehend the conversation of God and revealed to Moses on Mount Sinai. The Torah summarizes 613 commandments (mitzvot) and administrates numerous features of Jewish life, from nutritional regulations (kashrut) to prayer and moral actions. The Ten Commandments, a fundamental part of the Torah, deliver the base for moral behavior and associations. Studying and understanding the Torah and Tanakh assistances maintain the Jewish ethnicities, civilizations, and morals crossways of the generations. Optimal influence of Torah: Willie wagtail optimization algorithm is added with the procedure to expand the search. Willie wagtail [2] teaches the younger ones about predation and procreant abilities by relocation, and young grownups might turn out to be leaders of the fresh flock and should take care of the flock against the predators. Jewish population is highly intellectual and possesses the robust survival ability. Jewish populations have traditionally established an outline of migration, frequently compelled by a prerequisite for endurance due to oppression, struggle, and monetary destitution. This has ensued in substantial migration groups and unstable population dispersals crossways universally. The annihilation of the second shrine in Jerusalem steered to the principal foremost Jewish migration, with Jew's dispersal all over the Roman Empire and elsewhere. All the way in the history, Jews have met evictions, exterminations, and various forms of oppression in numerous parts of the globe, compelling them to travel in quest of protection and shelter. Samples comprise evictions from France, Spain, and other parts of Europe.

Financial adversity and the longing for superior opportunities have also played a key role in Jewish relocation outlines. The upsurge of Zionism and the formation of the State of Israel in the year 1948 prompted large-scale migration of Jews around the world, predominantly from Middle East countries. At present, the preponderance of Jews lives in the Israel and United States of America. In the Middle East and North Africa, a largely Israel base has seen an upsurge in Jewish population share, whereas North America's portion of Jews population has been decreased. Jewish populations endure to acclimatize and transfer in rejoinder to numerous factors, comprising financial prospects, safety apprehensions, community and political variations in their nations of origin. The Holocaust stemmed in the demises of a substantial percentage of the Jewish population and had a permanent influence on the Jewish demographics and relocation outlines. Jewish populations in the migration are facing encounters such as decreasing of birth rate and elderly populations, whereas Israel has a comparatively greater fertility rate and abundant quantity of younger population. Immigration of Jewish population continues from around the world to Israel remains, however at a slower speed than in the preceding years. Jewish populations around the globe are engaged to preserve the Jewish distinctiveness and beliefs in the façade of integration and other contests. Jewish relocation is a multifaceted and complicated phenomenon molded by ancient, political, financial, and community factors. Endurance and the protection of Jewish distinctiveness have been foremost drivers of these relocations all the way through the history and endure to be appropriate in the current era.

Parrot preservation encompasses numerous significant features: habitation protection, combating illegitimate trade, public commitment, and enslaved breeding schemes. Shielding parrot habitations is decisive, as environment loss is a foremost danger [3]. Sustainable exercise, such as ecological based tourism and public headed ingenuities, also play a vibrant role in preservation. Furthermore, addressing the illegitimate trade of parrots and endorsing liable pet proprietorship are imperative steps. Enslaved breeding schemes, predominantly those concentrating on reinstatement, can aid in strengthening the populations. Parrots are frequently forest reliant, so preserving forests is vital. A Parrot Park (SGS parrot park) established by Pontiff Dr. Sri Ganapathy Sachchidananda, Founder of Avadhoota Datta Peetham, Mysore, India. Rare and endangered species of Parrots are preserved in healthy habitation in the SGS parrot park. An ecological park is created which centered on the natural habitation of the parrots. Pontiff safeguards various types of parrot's especially endangered species. Pontiff allocates funds to organize medical camps for the parrots frequently to safeguard the parrots from various diseases. Ms. P. Siva Sindhu Bramarambika is acting as one among key donor years to years. This attribute is scientifically defined in the exploitation segment of the algorithm. Romanticism-based optimization is added with the procedure to boost the local search. Romanticism is a literary and creative effort is regarded as concentration on sensation, individuality, and the exaltation of nature, frequently divergent with the purpose grounded epitomes of the illumination. It accentuates idiosyncratic understanding, imaginings, and the supremacy of the specific entity, whereas also reconnoitring leitmotifs of the paranormal, the historical, and the transcendent [4]. Romanticism was a fictional and creative effort accentuating sensation, imaginings, and uniqueness, flourishing in Europe and the Americas in the late 18th and initial nineteenth centuries.

Optimal mathematical design of enhanced volunteer activities in a Monastery (OMVM), Human potentials-based optimization algorithm (HPOA), Poker strategy-based optimization algorithm (PSOA), optimal influence of Torah, Tanakh, and Talmud in Jewish life inspired algorithm (OTTJ), Sustainable conditions of Jewish population-based optimization algorithm (SCPO), and Optimal conservation of parrot-based algorithm (OCPO) are substantiated in 7 benchmark functions and IEEE 30 bus systems.

2 Optimal Mathematical Design of Enhanced Volunteer Activities in a Monastery

Volunteer activities in a monastery characteristically encompass compassionate, administrative, and maintenance responsibilities, such as spiritual-based actions, food preparation, gardening, and supporting the events. Noble volunteer qualities comprise trustworthiness, accountability, empathy, consistency, flexibility, resilient communication, and co-ordination in all type of activities, all directed by the monasteries' precise virtuous, traditional, and religious procedures. SGS monastery headed by Pontiff Dr. Sri Ganapathy Sachchidananda [5] has been considered for the analysis and design. Activities of disciple and senior volunteer Ms. Siva Sindhu Bramarambika and other junior volunteers in SGS monastery have been studied years after years. Population is generated randomly as,

$$X = \begin{bmatrix} X_1 \\ \vdots \\ X_i \\ \vdots \\ X_N \end{bmatrix}_{N \times m} = \begin{bmatrix} x_{1,1} & \cdots & x_{1,m} \\ \vdots & \ddots & \vdots \\ x_{N,1} & \cdots & x_{N,m} \end{bmatrix} \quad (1)$$

$$x_{i,j} = min_j + G\left(max_j - min_j\right)$$

Value of objective function is calculated as,

$$F = \begin{bmatrix} F_1 \\ \vdots \\ F_i \\ \vdots \\ F_N \end{bmatrix}_{N \times 1} = \begin{bmatrix} F(X_1) \\ \vdots \\ F(X_i) \\ \vdots \\ F(X_N) \end{bmatrix}_{N \times 1} \quad (2)$$

A scientifically designed methodology to augmenting volunteer activities in a monastery encompasses an organized, needs-based procedure from forecasting to maintenance, centering on bringing into line of volunteer services with monastery requirements, providing opportunities for reciprocal development and support, instituting vibrant prospects, progressions, and resilient communications. This comprises evaluating the monastery's requirements, employing appropriate volunteers, planning the tasks that provide a balance of employing prevailing abilities and novel learning for the volunteer's development, and applying systems for on-going commitment, backing, and well-being to guarantee volunteer fulfillment and long-standing commitment.

$$X_{i,j}^{L1} = X_{i,j} + G \cdot \left(C_j - G_1 \cdot X_{i,j}\right) \quad (3)$$

$$X_{i,j}^{L1} \rightarrow new\,postion\,of\,\,ith\,volunteer$$

$$C_j \rightarrow prevaling\,condition\,in\,monastery$$

$$G \in [0,1]$$

$$G_2 \in \{1,2\}$$

$$X_i = \begin{cases} X_i^{L1}, F_i^{L1} \le F_i \\ X_i, otherwise \end{cases}$$

Volunteers in the monastery contribute to knowledge gaining activities, especially to the learners belonging to deprived families and to certain rural area population who are unable to afford the cost of private education. Senior volunteer Ms. Siva Sindhu Bramarambika worked as role model, providing direction and backing to the junior volunteers who joined newly in the service. Volunteers engage in the creation of major knowledge resource center inside the monastery, which aids the disciples and general public to assess various types of books, e-content rendering to religion, science, and social needs.

$$W_i = \{X_k, F_k > F_i\} \tag{4}$$

$$W_i \rightarrow \text{creation of knowledge resource center}$$

$$X_{i,j}^{L2} = X_{i,j} + G \cdot \left(X_{i,j} - G_2 \cdot P_{i,j}\right) \tag{5}$$

$$X_{i,j}^{L2} \rightarrow \text{new postion of ith volunteer}$$

$$G_2 \in \{1,2\}$$

$$P_{i,j} \rightarrow \text{utilization factor}$$

$$X_i = \begin{cases} X_i^{L2}, F_i^{L2} \le F_i \\ X_i, otherwise \end{cases}$$

Volunteers assist with the routine tasks like food preparation, cleaning, and preserving the tradition of monastery. Monasteries depend on volunteer assistance for maintenance, comprising responsibilities like gardening and building certain assemblies. Volunteers aid the visitors and pilgrims, guaranteeing a smooth experience for all disciples and visitors.

$$X_{i,j}^{L3} = X_{i,j} + G \cdot (1 - 2 \cdot G) \cdot \frac{max_j - min_j}{t} \tag{6}$$

$$X_{i,j}^{L3} \rightarrow \textit{new location of the ith member}$$

$$X_i = \begin{cases} X_i^{L3}, F_i^{L3} \leq F_i \\ X_i, \textit{otherwise} \end{cases}$$

In the period of festivals, additional volunteer support is needed by the monastery towards event planning, ceremonial assistance, and the common escalation in visitor supervision, intensifying the opportunities beyond characteristic daily responsibilities. Volunteers actively aid with establishing and expediting distinctive rituals, managing enlarged crowds and logistics, and supporting with communal outreach. This enhanced participation permits the volunteers to gain deeper insight into hermetic life and tradition, contributing a vibrant and immersive experience beyond the regular volunteer day to day work. Certain situations during large festival days controlling the disordered crowd inside the monastery is uphill task by the volunteers and this activity is scientifically defined through chaotic tent map [6].

$$Y_{t+1} = Y_t + CTM * \beta * (Y_{new} - Y_t * \alpha) \tag{7}$$

$$\alpha = \frac{2z}{T}$$

$$CTM \rightarrow \textit{Chaotic tent map}$$

$$f(x) = \mu x, x < 0.5$$

$$f(x) = (1 - x), x \geq 0.5$$

The additional efforts of the volunteers during the festival period is defined as,

$$\varphi_i = \varphi_i + b(\varphi_{best} - \varphi_i) \tag{8}$$

Since festivals draw more people to the monastery, it is necessary to aggregate the requirement towards logistics and hospitality. This activity encompasses providing specific information, guiding the people, and guaranteeing a diplomatic environment inside the monastery which acts as the significant task of the volunteers.

$$\varphi_i = \varphi_i + c(H - 0.5) \tag{9}$$

All the volunteers will play an active role in the organization and implementation of festival events. The probability of engagement by the volunteers in last-minute arrangements, handling dissimilar activities, and guaranteeing smooth coordination amongst different features of the festival is defined as,

$$PBOM = e^{\frac{\Delta f}{r(t)}} \tag{10}$$

$$PBOM \rightarrow = probability\ of\ engagement$$

$$\Delta f \rightarrow difference\ of\ fitness\ value$$

$$among\ volunteers$$

Rapid actions of the volunteers during festival period is defined as,

$$r(t) \rightarrow factor\ of\ rapidness$$

$$r(t) = \beta \cdot r(t-1) \tag{11}$$

$$\beta = 0.65$$

The energy level of the volunteer during the extreme level of services is defined as,

$$e(t) = \beta \cdot e(t-1) \tag{12}$$

$$e \rightarrow energy\ level$$

Behavioral activity of the volunteers is defined by Kurt Lewin's Behavior Equation [7] as follows,

$$B = f(P, E) \tag{13}$$

$$B \rightarrow behaviour$$

$$P \rightarrow individual$$

$$E \rightarrow environment$$

Optimal scientific design of enhanced volunteer activities in the Monastery approach is integrated with white whale optimization algorithm. White whales utilize the sound waves to discover prey in the habitation of shady and cluttered atmospheres.

$$P_i^0 = min + (max - min) \cdot B \tag{14}$$

$$P_i^0 \rightarrow Preliminary\ location\ of\ each\ entity$$

$$min, max \rightarrow limits$$

$$B \in [0,1]$$

$$P^* = \text{argmin} f\left(P_i^0\right) \tag{15}$$

$$P^* \rightarrow Premium\ solution$$

White whales create sounds through clacks, which become more louder when passing close to the prey. Occurrence of the clacks is greater than that of communiqué noises and differs amongst the other species.

$$e\left(P_i^t, P^*\right) = 1 - \frac{P_i^t \times P^*}{\|P_i^t\| * \|P^*\|} \tag{16}$$

$$e \rightarrow distance$$

$$P_i^t \rightarrow Present\ location\ of\ the\ entity$$

Sound wave concentration is defined through sine function,

$$SWC\left(P_i^t, t\right) = L \cdot |sin(a \cdot e_i - \varphi \cdot t)| \cdot e^{-\delta t} \tag{17}$$

$$SWC \rightarrow Sound\ wave\ concentration$$

$$L \rightarrow sound\ wave\ amplitude$$

$$L = 1$$

$$a = 2\pi$$

$$\delta = 0.01$$

$$\varphi = 2\pi$$

When the vitality of the prey is high, it owns high chances of fleeing away, sequentially white whales search for fresh prey and renew their location.

$$P_i^{t+1} = P_i^t + H_V * \left(P^* - P_i^t\right) + SWC\left(P_i^t, t\right) * B \tag{18}$$

$$H_V = 2 * g * b_1 - g$$

$$b_1 \in [0,1]$$

$$B \in [0,1]$$

$$H_V \rightarrow concentration\ of\ exploration$$

Extraordinary prey vitality levels have a significant consequence on impacting the predator's examining behavior.

$$when b_1 < examination_r \rightarrow entity\ in\ the\ stage\ of\ exploration$$

$$r \rightarrow ratio$$

$$when b_1 > examination_r \rightarrow entity\ in\ the\ stage\ of\ exploitation$$

$$r \rightarrow ratio$$

The vitality of the prey is defined as,

$$V_{prey}(t+1) = V_{prey}(t) * e^{(-\tau \cdot t)} \tag{19}$$

$$V \rightarrow vitality\ of\ the\ prey$$

$$\tau = 0.001$$

$$when\ V \geq 0,\ then$$

$$V_{prey}(t+1) = maximum\left(V_{prey}(t), 0\right) \tag{20}$$

$$when\ V < 1,\ then$$

Depending on the prey location white whales arbitrarily alter their position and initiate the attack toward the prey. This random position is defined through Lévy flight distribution [8] as follows,

$$P_i^{t+1} = P_i^t - H_V * \left(P^* - P_i^t\right) + D_w * S_f\left(P_i^t, t\right) * SWC\left(P_i^t, t\right) * B \cdot L(d) \quad (21)$$

$$SWC\left(P_i^t, t\right) = \frac{V_{prey}(t)}{1 + \|P^* - P_i^t\|}$$

Changeover from exploration to exploitation is defined as

$$L(t) = 2 - \left(\frac{2t}{maximum\ iteration}\right) \quad (22)$$

$$Prey_{Vitality} < 0.5 \rightarrow procedure\ will\ transfer\ to\ exploitation$$

3 Human Potential-Based Optimization Algorithm

Humans utilize imagination and expectations to search for the potentials, resolve problems, and forecast the future, along with the drawing imagination from internal familiarity and understanding, whereas expectations are views about forthcoming consequences.

$$h_{i,j} = min_j + Q\left(max_j - min_j\right) \quad (23)$$

Imagination is the base of human invention, permitting to apprehend of novel notions and answers, leading to the evolution of civilization.

$$H_{i,j}^{L1} = H_{i,j} + Q \cdot \left(E_j - Q_1 \cdot H_{i,j}\right) \quad (24)$$

$$H_{i,j}^{L1} \rightarrow humans\ in\ the\ phase\ of\ imagination$$

$$E_j \rightarrow evolving\ novel\ notions$$

$$H_i = \begin{cases} H_i^{L1}, F_i^{L1} \le F_i \\ H_i, otherwise \end{cases}$$

Imagination is a perceptive ability that engenders thoughts and metaphors from inside, rather than depending directly on exterior sensory input.

$$A_i = \{H_k, F_k > F_i\} \quad (25)$$

$$A_i \rightarrow engendering\ thoughts\ and\ metaphors$$

$$H_{i,j}^{L2} = H_{i,j} + Q \cdot \left(E_j - Q_2 \cdot C_{i,j}\right) \tag{26}$$

$$H_{i,j}^{L2} \rightarrow perceptions\ of\ humans$$

$$C_{i,j} \rightarrow \text{human intellect}$$

$$H_i = \begin{cases} H_i^{p2}, F_i^{p2} \leq F_i \\ H_i, Otherwise \end{cases}$$

Expectations can range from antedating achievement to apprehensive of failure and are frequently entrenched in previous familiarities or needs.

$$H_{i,j}^{L3} = H_{i,j} + Q \cdot (1 - 2 \cdot Q) \cdot \frac{max_j - min_j}{t} \tag{27}$$

$$H_{i,j}^{L3} \rightarrow expectations\ of\ humans$$

$$H_i = \begin{cases} H_i^{L3}, F_i^{L3} \leq F_i \\ H_i, otherwise \end{cases}$$

Human potentials-based algorithm is combined with Blesmol optimization algorithm. Blesmol Queens engender the mating process, through vocalizing and then squatting in front of a selected male and espousing the lordosis position, in which the male Blesmol does the mating.

$$A_{ij} = lb_j + U\left(ub_j - lb_j\right) \tag{28}$$

Worker Blesmol action is scientifically defined as follows,

$$A_i^{t+1} = A_i^t + \varphi\left(A_y^t - A_z^t\right) \tag{29}$$

$$\varphi = \delta_{min} + (\delta_{max} - \delta_{min}) * Y^{(k-1)}$$

$$A_i^{t+1} \rightarrow renewed\ candidate\ solution$$

$$A_i^{t+1} = \begin{cases} A_{best} + g_1\left(\left(ub_j - lb_j\right)g_2 + lb_j\right)x \geq 0 \\ A_{best} - g_1\left(\left(ub_j - lb_j\right)g_2 + lb_j\right)x \leq 0 \end{cases} \tag{30}$$

$$A_i^{t+1} = (A_d \cdot b_1 \cdot b_2 \cdot b_3) + A_{best} \tag{31}$$

$$t, T \rightarrow \textit{current and maximum iteration}$$

$$A_l = A_n \cdot \left(A_{best} - A_i^t\right)$$

Specific male Blesmol mate with Queen and this action is demarcated in the breeder section of thc algorithm.

$$A_i^{t+1} = (1 - \varphi)A_i^t + \varphi\left(A_{best} - A_i^t\right) \tag{32}$$

Queen Blesmol selects the male for mating and the selection process is defined as,

$$A_{new}^{t+1} = \begin{cases} A_{new} iff \, (A_{new}) < f\left(A_i^t\right) \\ A_i^t Else \end{cases} \tag{33}$$

In order to evade the stagnation, the process is updated as follows,

$$A_1 = A_i - Y_1\left(E_1 \cdot A_{new} - A_i^t\right) \tag{34}$$

$$A_2 = A_i - Y_2\left(E_2 \cdot A_{new} - A_i^t\right) \tag{35}$$

$$A_3 = A_i - Y_3\left(E_3 \cdot A_{new} - A_i^t\right) \tag{36}$$

$$A_{new}^{t+1} = \frac{A_1 + A_2 + A_3}{3}$$

Through Cauchy distribution function fresh solutions are generated as follows,

$$f_{cd,(o,c)}(\delta) = \frac{1}{\pi} \frac{cd}{\left(cd^2 + \delta^2\right)} \tag{37}$$

$$cd \rightarrow \textit{Cauchy distribution function}$$

$$\omega = 1/2 + 1/\pi \, arctan\left(\frac{\delta}{cd}\right)$$

$$\delta = tan\left(\pi\left(\omega - \frac{1}{2}\right)\right)$$

The final examination for obtaining new solutions is defined as,

$$A_i^{t+1} = A_i^t + \alpha * cd(\delta)\left(A_{best} - A_i^t\right) \tag{38}$$

Striped hyena optimization algorithm is integrated with the procedure to advance the local search. Enclosing the prey by the cluster of striped hyena is scientifically delineated as,

$$E_{SH} = |U \cdot Y_P(t) - Y(t)| \tag{39}$$

$$Y(t+1) = Y_P(t) - Q \cdot E_{SH} \tag{40}$$

Procedure augments the exploitation as iteration increases. Striped hyena modernizes its location in the direction of the position of the prey and this activity is mathematically defined through "U" and "Q." Striped hyena arbitrarily appraises its location rendering to the position of the prey.

$$E_{SH} = |U \cdot Y_{SH} - Y_K| \tag{41}$$

$$D_{SH} = Y_K + Y_{K+1} + Y_{K+2}, \ldots, +Y_{K+SH} \tag{42}$$

$$N = Sum(Y_{SH} + Y_{SH+1} + Y_{SH+2}, \ldots, (Y_{SH} + A)) \tag{43}$$

In the attacking phase, striped hyenas cluster will attack the enclosed prey collectively and it defined as.

$$Y(t+1) = \frac{D_{SH}}{N} \tag{44}$$

4 Poker Strategy-Based Optimization Algorithm

Playing optimal poker necessitates an amalgamation of cognitive and emotional features, comprising judgement, hand appraisal, game theory, and mathematical abilities to comprehend varieties and possibilities.

$$p_{i,j} = lb_j + U\left(ub_j - lb_j\right) \tag{45}$$

Cognitive and strategic attributes are considered as imperative proficiencies, encompassing the capability to guess a challenger's prospective range of hands centered on their activities and outlines, for this will progress the timing and decision-making in the duration of the game.

$$P_{i,j}^{L1} = P_{i,j} + U \cdot \left(M_j - U_1 \cdot P_{i,j}\right) \tag{46}$$

$$P_{i,j}^{L1} \rightarrow new\,position\,of\,players\,in\,the\,game$$

$$M_j \rightarrow timing\,and\,decision-making$$

$$P_i = \begin{cases} P_i^{L1}, F_i^{L1} \leq F_i \\ P_i, Otherwise \end{cases}$$

Understanding conceptions of game theory, possibility, bet sizing, and the equity in pot is vital for practical arithmetic and creating sound tactical choices. Strategic knowledge is highly needed to deeply understand the game's rules, approach, and the explicit distinctions of dissimilar hands and gambling assemblies.

$$B_i = \{P_k, F_k > F_i\} \tag{47}$$

$$B_i \rightarrow tactful\,playing$$

$$P_{i,j}^{L2} = P_{i,j} + U \cdot \left(M_j - U_2 \cdot E_{i,j}\right) \tag{48}$$

$$P_{i,j}^{L2} \rightarrow new\,position\,of\,players\,in\,the\,game$$

$$E_{i,j} \rightarrow gambling\,assemblies$$

$$P_i = \begin{cases} P_i^{L2}, F_i^{L2} \leq F_i \\ P_i, Otherwise \end{cases}$$

Identifying one's individual forte, feebleness, and emotive state permits for superior self-regulation and modification to the varying crescendos of the game.

$$P_{i,j}^{L3} = P_{i,j} + U \cdot (1 - 2 \cdot U) \cdot \frac{ub_j - lb_j}{t} \tag{49}$$

$$P_{i,j}^{L3} \rightarrow new\,position\,of\,players\,in\,the\,game$$

$$P_i = \begin{cases} P_i^{L3}, F_i^{L3} \leq F_i \\ P_i, Otherwise \end{cases}$$

Progressive poker strategy-based optimization is integrated with Platylophidae Platylophus optimization. The procedure is centered on the swarm behaviors of the Platylophidae platypus.

$$\Delta E_j = (M - F) + a \tag{50}$$

$$Y_j = F + \Delta E_j \tag{51}$$

$$a = e^{-\frac{2bt}{T_{max}}} \tag{52}$$

Both the male and female Platylophidae platypus will feed the food, and this aspect is defined as,

$$\Delta F_j = Z_j + a \tag{53}$$

Younger Platylophidae platypus will be fed by the male Platylophidae platypus, and this aspect is mathematically defined as,

$$\Delta F_j = a + \left(Z_j - M\right) + M \tag{54}$$

If the male is unable to feed the younger Platylophidae platypus, then the female will feed the younger ones, and it defined as,

$$\Delta F_j = a + \left(Z_j - F\right) + sin(\gamma) \tag{55}$$

At this condition, the position of the Platylophidae platypus is defined as,

$$PP_j^{t+1} = PP_j^t + \Delta F_j \tag{56}$$

$$PP_j^{t+1} = PP_j^t + \left(a \cdot PP_j + sin(\gamma)\right) \tag{57}$$

The density of the endurance of the individual function is defined as,

$$EOI(t) = P(T > t) = \int_t^{\infty} P(T)dt, t > 0 \tag{58}$$

Risk function with reference to the endurance of the individual function is defined as,

$$\varphi(t) = \lim_{\Delta t = 0} \frac{P(t < T < t + \Delta t | T > t)}{\Delta t} = \frac{P(t)}{EOI(t)} \tag{59}$$

$$EOI(t) = e^{-\int_0^t \varphi(\sigma)d\sigma}$$

Each individual may face road accidents on the road, and the probability function with reference to the occurrence of the accidents is denoted as,

$$PF\left(\theta;\{z_i,t_i,\alpha_i\}_{i=1}^{n}\right)=\prod_{i\in D}PF(t_i|z_i,\theta)\prod_{j\in O}EOI\left(t_j|z_j,\theta\right) \tag{60}$$

The proportional risk function is defined with reference to the occurrence of the road accident as follows,

$$\varphi(t|z,\theta)=\varphi_0(t)exp\left\{z^T\theta\right\} \tag{61}$$

During the phase of the accident, the fractional probability of endurance of the individual is defined as,

$$PF_F=\left(\theta;\{z_i,t_i,\alpha_i\}_{i=1}^{n}\right)=\prod_{i\in D}\frac{exp\left\{\theta^T z_i\right\}}{\sum_{j:t_j\geq t_i}exp\left\{\theta^T z_j\right\}} \tag{62}$$

$$PF_F\rightarrow fractional\,probability$$

The concordance directory finds the enactment as a measure of the proposed method,

$$CD=\frac{1}{|\mathfrak{M}|}\sum_{(i,j)\in\mathfrak{M}}PI\left(F(Z_i)<F\left(Z_j\right)\right) \tag{63}$$

Collaborative learning is incorporated into the process to enhance the learning procedure. In the common functional mapping,

$$x=F(z;\gamma) \tag{64}$$

$$\gamma\rightarrow parameters\,of\;F$$

$$\left(\mu_q,\sigma_q\right)=\underset{\mu,\sigma}{argmin}\sum_{i=1}^{n}\mathsf{C}\left(x_i,F_{q-1}(z_i)+\mu f(z_i;\sigma)\right) \tag{65}$$

$$\mu_q=\underset{\mu}{argmin}\sum_{i=1}^{n}\mathsf{C}\left(x_i,F_{q-1}(z_i)+\mu f\left(z_i;\mu_q\right)\right)$$

$$\sigma_q=\underset{\sigma}{argmin}\sum_{i=1}^{n}\left(L_{iq}-f(z_i;\sigma)\right)$$

$$L_{iq}=-\left[\frac{\partial\mathsf{C}(x_i,F(z_i))}{\partial F(z_i)}\right]$$

$$F_q(z)=F_{q-1}(z)+\mu_q f\left(z_i;\sigma_q\right) \tag{66}$$

To control exceeded time analysis parameter "ω" has been introduced as follows,

$$F_q(z) = F_{q-1}(z) + \omega\mu_q f\left(z_i; \sigma_q\right) \tag{67}$$

The cost function of the distribution is defined as,

$$C(x, F) = -\sum_{i=1}^{n} \alpha_i \left\{ F(Z_i) - log\left(\sum_{j:t_j \geq t_i} eF\left(z_j\right) \right) \right\} \tag{68}$$

Concordance directory learning through the collaborative learning method is defined as,

$$ICD = \frac{1}{|\mathfrak{M}|} \sum_{(i,j)\in\mathfrak{M}} \frac{1}{1 + e^{\beta\left(F(z_i) - F\left(z_j\right)\right)}} \tag{69}$$

$$L_{iq} = \left[\frac{\partial ICD}{\partial F(z_i)} \right]_{F(z)=F_{q-1}(z)} \tag{70}$$

$$\mu_q = arg\min_{\mu} \frac{1}{|\mathfrak{M}|} \sum_{(i,j)\in\mathfrak{M}}^{n} \frac{1}{1 - e^{\beta\begin{pmatrix} F_{q-1}(z_i) + \mu f\left(z_i; \sigma_q\right) - \\ F_{q-1}\left(z_j\right) + \mu f\left(z_j; \sigma_q\right) \end{pmatrix}}} \tag{71}$$

5 Optimal Influence of Torah, Tanakh, and Talmud in Jewish Life Inspired Algorithm

The Torah, Tanakh, and Talmud are the fundamentals of Jewish life, delivering a structure for religious exercise, beliefs, and communal distinctiveness. The Torah, the principal five books of the Hebrew Bible, is deliberated the verbatim conversation of God and a controller for existence, whereas the Talmud delivers explanations and solicitations of the Torah's commandments and preachings. The Tanakh, comprising of the Torah and added books of the Hebrew Bible, compromises ancient and prophetic perceptions that further shape the Jewish uniqueness and principles. The Torah, also entitled as the Pentateuch, encompasses the basic commandments and decrees of Judaism, comprising the Ten Commandments. Jews have confidence in the Torah as it was what God revealed to Moses on Mount Sinai and is the verbatim conversation of God.

$$G = \begin{bmatrix} G_1 \\ \vdots \\ G_i \\ \vdots \\ G_N \end{bmatrix}_{N \times m} = \begin{bmatrix} g_{1,1} & \cdots & g_{1,m} \\ \vdots & \ddots & \vdots \\ g_{N,1} & \cdots & g_{N,m} \end{bmatrix} \tag{72}$$

$$G \rightarrow population$$

$$N \rightarrow members$$

$$m \rightarrow variables$$

$$g_{i,j} = min_j + V\left(max_j - min_j\right)$$

$$max_j, min_j \rightarrow limits$$

$$V \in [0,1]$$

$$F = \begin{bmatrix} F_1 \\ \vdots \\ F_i \\ \vdots \\ F_N \end{bmatrix}_{N \times 1} = \begin{bmatrix} F(G_1) \\ \vdots \\ F(G_i) \\ \vdots \\ F(G_N) \end{bmatrix}_{N \times 1} \tag{73}$$

The Torah assists as a directive for all features of Jewish life, from spiritual performances to moral actions. The Torah's descriptions and commandments outline Jewish antiquity, distinctiveness, and association with God.The Tanakh, also entitled as Hebrew Bible, comprehends the Torah, the Prophets (Nevi'im), and the Texts (Ketuvim), delivering a comprehensive ancient and literary perspective for Jewish credence. The Nevi'im suggests antique interpretations and prophetic communications that promote the Jewish understanding of the role of God in ancient times. The Ketuvim (Writings) deliver knowledge literature, rhyme, and devotions that offer additional moral and spiritual supervision. The Tanakh stimulates replication on Jewish antiquity, distinctiveness, and the association between God and mankind. The Talmud is a gigantic assortment of rabbinic deliberations, explanations, and authorized decisions that enlighten and relate the Torah's preachings. It signifies the verbalized Torah, the explanations of the inscribed Torah delivered down over and done with generations. The Talmud encompasses Halakha, a Jewish rule, which delivers practical direction for living a Jewish lifespan, comprising compliance of mitzvot (commandments), nutritional laws, and domestic life. The Talmud is a subject of on-going reading and explanation, replicating the vibrant nature of Jewish custom. The Torah, Tanakh, and Talmud collectively shape the Jewish distinctiveness, credence, and rituals. The Torah delivers the basic text, the Tanakh proposes ancient and prophetic framework, and the Talmud delivers on-going explanation and application, guaranteeing that Jewish decree and custom endure as relevant and significant for each generation.

$$G_{i,j}^{p1} = G_{i,j} + V \cdot \left(L_j - V_1 \cdot G_{i,j}\right) \tag{74}$$

$$G_{i,j}^{p1} \rightarrow understanding\ the\ commandments$$

$$L_j \rightarrow fundamental\ laws$$

$$V \in [0, -1]$$

$$V_1 \in \{1,2\}$$

$$G_i = \begin{cases} G_i^{p1}, F_i^{p1} \leq F_i \\ G_i, Else \end{cases}$$

Jews are certain that the Torah comprehends the conversation of God and Moses revealed on Mount Sinai. The Torah summarizes 613 commandments (mitzvot) and administrates numerous features of Jewish life, from nutritional regulations (kashrut) to prayer and moral actions. The Ten Commandments, a fundamental part of the Torah, deliver base for moral behavior and associations. The Torah assists as a spring of knowledge and directive for circumnavigating life's encounters and creation of moral decisions.

$$Y_i = \{G_k, F_k > F_i\} \tag{75}$$

$$k = 1,2,3,..,Q$$

$$i = 1,2,..,N$$

$$Y_i \rightarrow attributes\ of\ Torah,\ Tanakh,\ and\ Talmud$$

$$G_{i,j}^{p2} = G_{i,j} + V \cdot \left(G_{i,j} - V_2 \cdot B_{i,j}\right) \tag{76}$$

$$G_{i,j}^{p2} \rightarrow understanding\ the\ principles$$

$$Q_2 \in \{1,2\}$$

$$B_{i,j} \rightarrow behaviour\ and\ associations$$

$$G_i = \begin{cases} G_i^{p2}, F_i^{p2} \leq F_i \\ G_i, Else \end{cases}$$

The Torah signifies the bond between God and Jewish societies, which emphasizes the unique relationship and responsibilities. The Tanakh comprehends prophets who spoke God's conversation, contribution of cautions, direction, and confidence to the Jewish people all over the history. The Torah narrates the antiquity of the Jewish people, from formation to the Israelites' passage to the guaranteed region; it enhances their understandings about their history. Studying and understanding the Torah and Tanakh assistances maintain the Jewish ethnicities, civilizations, and morals crossways of the generations. Common descriptions and preachings inside the Torah generate a sagacity of unison and belonging amongst Jews in global level. The commandments of the Torah attach the Jews to their dynasties and their understandings, nurturing the sense of endurance. The Torah is recited in public on a weekly basis, and its reading is deliberated as a fundamental mitzvah. Torah reading is seen as a track to individual progress and spiritual progression, assisting entities to associate with God and expand their understanding of Jewish preachings.

$$G_{i,j}^{p3} = G_{i,j} + V \cdot (1 - 2 \cdot V) \cdot \frac{max_j - min_j}{t} \tag{77}$$

$$G_{i,j}^{p3} \rightarrow transformation\ of\ moral\ values\ to\ subsequent\ generations$$

$$V \in [0,1]$$

$$G_i = \begin{cases} G_i^{p3}, F_i^{p3} \leq F_i \\ G_i, Else \end{cases}$$

Optimal influence of Torah, Tanakh, and Talmud in Jewish life inspired algorithm is integrated with Willie wagtail optimization algorithm. This integration advances the local search competence of the algorithm. The growth of young ones in the colonies, performance of the Willie wagtail to feed their younger ones subsequent to breeding, and unique strategies are utilized to evade predators are defined in the Willie wagtail optimization algorithm. Willie wagtail younger ones' growth, breeding, nourishing, and evading the predators are mathematically defined in the algorithm. Willie wagtails stalk with an elongated, slender beak, comprehensive underbelly, and are nonetheless susceptible to predators. Each member in the population is the candidate solution and defines the decision variable value in the exploration region.

$$Z = \begin{bmatrix} Z_1 \\ \vdots \\ Z_i \\ \vdots \\ Z_N \end{bmatrix}_{N\times m} = \begin{bmatrix} z_{1,1} & \cdots & z_{1,m} \\ \vdots & \ddots & \vdots \\ z_{N,1} & \cdots & z_{N,m} \end{bmatrix} \tag{78}$$

$$Z \rightarrow population$$

$$N \rightarrow members$$

$$m \rightarrow variables$$

$$z_{i,j} = lb_j + U\left(ub_j - lb_j\right) \tag{79}$$

$$ub_j, lb_j \rightarrow \ upper\ and\ lower\ bounds$$

$$U \in [0, 1]$$

Willie wagtail espouse an accommodating breeding stratagem, with clusters of four to six members protecting their terrain year-round, comprising a pair of female Willie wagtails and supplementary male assistants. The sex ratio is deranged, with smaller quantity of female Willie wagtail, prominent to certain male Willie wagtail to acquire health assistances by assisting close relations.

$$F = \begin{bmatrix} F_1 \\ \vdots \\ F_i \\ \vdots \\ F_N \end{bmatrix}_{N\times 1} = \begin{bmatrix} F(Z_1) \\ \vdots \\ F(Z_i) \\ \vdots \\ F(Z_N) \end{bmatrix}_{N\times 1} \tag{80}$$

In Willie wagtail optimization algorithm, the location of population associates in the examination region is rationalized and grounded on the replication of regular performance amongst young and adult Willie wagtails. The procedure of continually learning lead to gaining experiences and the location will be modernized rendering to the growth of young Willie wagtail.

$$Z_{n_{ij}} = Z^t_{ij} + lb_j + U\left(ub_j - lb_j\right), a > 0.5 \tag{81}$$

$$Z_{n_{ij}} \rightarrow new\ location$$

$$U \in [0,1]$$

Once the danger threshold is small, the Willie wagtail will pass to the breeding phase and utilize exceptional parenthood challenging deeds wherein gestating the eggs to avert invasion by predators. Threat threshold is computed as,

$$X = a_1 * 20 + a_2 * 20 \tag{82}$$

$$a_1, a_2 \rightarrow random\, constants$$

The rationalized location of Willie wagtail during the breeding period is defined as,

$$Z_{n_{ij}} = Z_W + \left(Z_b - Z_{ij}^t\right) \cdot \tau \tag{83}$$

$$Z_W = Z_b * K$$

$$a < 0.5$$

$$X < 20$$

$$Z_b \rightarrow present\, optimal\, location$$

$$K = 0.8$$

$$\tau = sin\left(\left(ub_j - lb_j\right) * 2 + \left(ub_j - lb_j\right) * n\right)$$

$$n = \left(\frac{t}{T}\right) * 2$$

During this period Willie wagtails besieged by a predator will rapidly escape, instigating a trivial change in the associate's location. The other associates will fly in the mid-air to evade predators, ensuing in a great alteration in location and it is specified through Lévy Flight [8].

$$Z_{n_{ij}} = Z_b + Z_{ij} * Lévy * S \tag{84}$$

$$a < 0.5$$

$$X > 20$$

$$L(s) \sim |s| - 1 - \beta$$

$$0 < ß < 2$$

$$\mathrm{F(k)} = \exp\left[-\alpha|\mathrm{k}|^{\beta}\right], 0 < \mu < \mathrm{s} < \infty$$

$$S = 0.2 * \left(\frac{\pi}{2} - v\right) sin$$

$$v = \frac{\pi}{2} * \frac{t}{T}$$

Predator evading activity is computed as,

$$Z_{n_{ij}} = \begin{cases} Z_{ij}^{t} + lb_j + U\left(ub_j - lb_j\right), a > 0.5 \\ Z_W + \left(Z_b - Z_{ij}^{t}\right) \cdot \tau a < 0.5 \, and \, X < 20 \\ Z_b + Z_{ij} * Lévy * Sa < 0.5 \, and \, X > 20 \end{cases} \tag{85}$$

6 Sustainable Conditions of Jewish Population-Based Optimization Algorithm

Jewish populations have experienced miscellaneous histories, comprising epochs of prosperous integration and eras of oppression and segregation. Jewish communities in diverse countries or areas have fluctuating degrees of communication with their non-Jewish nationals. Israeli Jews, having a robust national distinctiveness, also contend with the interrogation of whether their Jewish distinctiveness or Israeli uniqueness takes antecedence. Based on the grouping population of Jewish community, the preliminary population of the projected algorithm is defined as,

$$Z = \begin{bmatrix} Z_1 \\ Z_2 \\ Z_3 \\ Z_4 \\ \vdots \\ Z_n \end{bmatrix} \in \mathfrak{A}^{JC \times d} \tag{86}$$

$$Z_n = \begin{bmatrix} Z_{n,1,1} & \cdots & Z_{n,1,d} \\ \vdots & \ddots & \vdots \\ Z_{n,m,1} & \cdots & Z_{n,m,d} \end{bmatrix} \in \mathfrak{A}^{m\times d} \tag{87}$$

$$JC \rightarrow Jewish\,community$$

$$m \rightarrow members\,of\,\,the\,population$$

$$d \rightarrow dimension$$

Fitness value is calculated as,

$$F = \begin{bmatrix} F_1 \\ F_2 \\ F_3 \\ F_4 \\ \vdots \\ F_n \end{bmatrix} \in \mathfrak{A}^{JC\times 1} \tag{88}$$

$$F_n = \begin{bmatrix} F_{n,1} \\ F_{n,2} \\ F_{n,3} \\ F_{n,4} \\ \vdots \\ F_{n,m} \end{bmatrix} \in \mathfrak{A}^{m\times 1} \tag{89}$$

Migration rapidness of the Jewish population is defined as,

$$M = \begin{bmatrix} M_1 \\ M_2 \\ M_3 \\ M_4 \\ \vdots \\ M_n \end{bmatrix} \in \mathfrak{A}^{JC\times 1} \tag{90}$$

$$M_n = \begin{bmatrix} M_{n,1,1} & \cdots & M_{n,1,d} \\ \vdots & \ddots & \vdots \\ M_{n,m,1} & \cdots & M_{n,m,d} \end{bmatrix} \in \mathfrak{A}^{m\times d} \tag{91}$$

$$M \rightarrow Migration\,Rapidness$$

When assemblages of Jewish community migrate to a certain region there will be cooperative actions among the members of the Jewish population. Certain times there will be chaos condition when population clusters upsurge in sharing the resources and various internal conflicts. So the cooperation factor among the members of the Jewish population is defined through sine chaotic mapping.

$$a_t = \begin{cases} X, ift = k \cdot m, k = W \\ sin\left(\frac{\pi}{2}a_{t-1}\right) otherwise \end{cases} \quad (92)$$

Through the sine chaotic mapping the impulsive nature of the individual members is mathematically defined and these members in the Jewish population may shift their loyalties in the assemblage rendering to various internal conflicts at various intervals of time period. Each individual of the Jewish population has altered mind-set despite being in the assemblage and the course of actions in the phase of survival is defined as,

$$M^{t+1}_{n,m,d} = 0.6 \cdot M^{t}_{n,m,d} + b_1 a^{t}_{1,n,m,d}\left(Z^{best}_{n,m,d} - Z^{t}_{n,m,d}\right)$$

$$+ b_2 a^{t}_{2,n,m,d}\left(Z^{best}_{JS} - Z^{t}_{n,m,d}\right) \quad (93)$$

$$M^{t+1}_{n,m,d} \rightarrow Rapidness\ of\ the\ mth\ member\ in\ the\ population$$

$$JS \rightarrow Jewish\ population$$

$$0.6 \rightarrow inertia\ factor$$

$$t \rightarrow Time$$

$$b_1, b_2 \rightarrow actions\ of\ individual\ members$$

The conflict may rise exterior to the assemblage of the Jewish population. Certain times Jewish population members may face the problem for the other population in the region while sharing the resources and internal conflicts increases. During that period the actions of the members of the Jewish population actions may be in diverse mode.

$$E^{t}_{n,m,d} = b_3 a^{t}_{3,n,m,d}\left(Z^{t}_{n,m,d} - Z^{best}_{C}\right) \cdot L(D) \quad (94)$$

$$C \rightarrow competition$$

$$L(D) \rightarrow Lévyflight$$

$$M_{n,m,d}^{t+1} = \begin{cases} M_{n,m,d}^{t+1} - E_{n,m,d}^{t} \; if \; F_n^{best} < F_C^{best} \\ M_{n,m,d}^{t+1}, \; otherwise \end{cases} \tag{95}$$

Position of the Jewish population members in the competitive conditions are updated as,

$$Z_{n,m,d}^{t+1} = Z_{n,m,d}^{t} + M_{n,m,d}^{t+1} \tag{96}$$

7 Optimal Conservation of Parrot-Based Algorithm

Parrot preservation encompasses numerous significant features: habitation protection, combating illegitimate trade, public commitment, and enslaved breeding schemes. Shielding parrot habitations is decisive, as environment loss is a foremost danger. Parrots are frequently forest reliant, so preserving forests is vital. This comprises creating endangered regions, averting deforestation, and endorsing ecological environment. A Parrot Park (SGS Parrot Park) was established by Pontiff Dr. Sri Ganapathy Sachchidananda [5] Founder of Avadhoota Datta Peetham, Mysore, India. Rare and endangered species of parrots are preserved in healthy habitation in the SGS parrot park. In this paper, safeguarding and protecting the parrots is scientifically defined in the optimal conservation of parrot-based algorithm. Population arbitrarily is created as,

$$P = \begin{bmatrix} P_1 \\ \vdots \\ P_i \\ \vdots \\ P_N \end{bmatrix}_{N \times m} = \begin{bmatrix} p_{1,1} & \cdots & p_{1,m} \\ \vdots & \ddots & \vdots \\ p_{N,1} & \cdots & p_{N,m} \end{bmatrix} \tag{97}$$

$$p_{i,j} = min_j + D\left(max_j - min_j\right)$$

$$max_j, min_j \rightarrow maximumandminimumlimits$$

$$D \in [0,1]$$

$$F = \begin{bmatrix} F_1 \\ \vdots \\ F_i \\ \vdots \\ F_N \end{bmatrix}_{N\times 1} = \begin{bmatrix} F(P_1) \\ \vdots \\ F(P_i) \\ \vdots \\ F(P_N) \end{bmatrix}_{N\times 1} \tag{98}$$

An ecological park is created which centered on the natural habitation of the parrots. Pontiff safeguards various types of parrots especially the endangered species. A country with the thickest population, practices deforestation for the livelihood and advancement of the infrastructure, including increasing the pollution and these activities push the birds, flora and fauna to the endangered condition. The safeguarding attributes of the parrots are scientifically defined in the exploration section as follows,

$$P_{i,j}^{L1} = P_{i,j} + D \cdot \left(L_j - D_1 \cdot P_{i,j}\right) \tag{99}$$

$$P_{i,j}^{L1} \rightarrow new\ locations\ of\ the\ parrot$$

$$L_j \rightarrow preservation$$

$$D \in [0, -1]$$

$$D_1 \in \{1,2\}$$

$$P_i = \begin{cases} P_i^{L1}, F_i^{L1} \leq F_i \\ P_i, Else \end{cases}$$

Different species of parrots are safeguarded and it is a noteworthy aspect that an appropriated condition has been created by the pontiff in SGS Parrot Park to protect almost major types of parrots. This specific action is scientifically defined as,

$$O_i = \{P_k, F_k > F_i\} \tag{100}$$

$$k = 1,2,3,..,Q$$

$$i = 1,2,..,N$$

$$O_i \rightarrow creation\ of\ protecting\ environment$$

$$P_{i,j}^{L2} = P_{i,j} + D \cdot \left(P_{i,j} - D_2 \cdot P_{i,j}\right) \tag{101}$$

$$P_{i,j}^{C2} \rightarrow renewed\ location\ of\ the\ parrot$$

$$D_2 \in \{1,2\}$$

$$P_{i,j} \rightarrow zone\ of\ protection$$

$$P_i = \begin{cases} P_i^{L2}, F_i^{L2} \leq F_i \\ P_i, Else \end{cases}$$

SGS park authorities concentrate much on maintaining good health condition of the parrots. Throughout the year various activities are implemented to safeguard the parrots from diseases. Virus, fungus and bacteria, and parasitic infections, frequently will be found in the parrots naturally. Psittacine beak, feather disease, Pacheco's disease, psittacosis, and aspergillosis are commonly found in parrots. Pontiff allocates funds to organize medical camps for the parrots frequently to safeguard the parrots from various diseases. This attribute is scientifically defined in the exploitation segment of the algorithm.

$$P_{i,j}^{L3} = P_{i,j} + D \cdot (1 - 2 \cdot D) \cdot \frac{ub_j - lb_j}{t} \tag{102}$$

$$P_{i,j}^{L3} \rightarrow renewed\ location\ of\ the\ parrot$$

$$D \in [0,1]$$

$$P_i = \begin{cases} P_i^{L3}, F_i^{L3} \leq F_i \\ P_i, Else \end{cases}$$

Optimal conservation of parrot-based algorithm is integrated with disaster management-based algorithm. Operative disaster management [9] in the time of floods and earthquakes encompasses an amalgamation of precautionary measures, preparation, retort, and rescue efforts. These stratagems aim to diminish the loss of human life, assets destruction, and disruption to societies. Human population should be safeguarded during the disaster period.

$$P = \{P_1, P_2, P_3, .., P_N\} \tag{103}$$

Strategies are signified by ordered list of fundamentals as,

$$P_i(t) = (A_i(t), F_i(t)) \tag{104}$$

$$A_i(t) = [A_{i1}, A_{i2}, A_{i3}, .., A_{iD}] \tag{105}$$

$$A_i(t) \rightarrow Sequence\ of\ actions$$

$$F_i(t) \rightarrow Fitness\ rate$$

$$p_d = log(t + M)/log(T) * 2$$

$$A_i(t+1) = A_G(t) + G_1(x) * (A_{o^i}(t) - A_o(t)) \tag{106}$$

$$p_d \rightarrow population\ of\ the\ society$$

$$A_G(t) \rightarrow stages\ of\ the\ plan$$

$$A_{o^i}(t) \rightarrow Implemented\ decisions$$

$$A_o(t) \rightarrow randomly\ selected\ population$$

$$G_1(x) \rightarrow factor\ for\ division$$

$$G_1(x) = \left(\delta/2^n * n! * \pi^{1/2}\right)^{\frac{1}{2}} \cdot H_n(\delta * x) \cdot e^{-.05*\delta^2*x^2}$$

Effervescent preeminent activities in the disaster management defined as,

$$DM(t) = \{DM_1, DM_2, .., DM_i, .., DM_{N'}\} \tag{107}$$

$$N' = N - p_d$$

$$DM_i \rightarrow fitness\ value\ of\ activity$$

Romanticism-based optimization is added with the procedure to boost the local search. Romanticism is a literary and creative effort is regarded as concentration on sensation, individuality, and the exaltation of nature, frequently divergent with the purpose grounded epitomes of the illumination. Romanticism also discovers attentiveness in beliefs and civilizations external to Europe, frequently venerating them as more trustworthy or nearer to the nature. Random population is created as

$$R = \begin{bmatrix} R_1 \\ \vdots \\ R_i \\ \vdots \\ R_N \end{bmatrix}_{N\times m} = \begin{bmatrix} r_{1,1} & \cdots & r_{1,m} \\ \vdots & \ddots & \vdots \\ r_{N,1} & \cdots & r_{N,m} \end{bmatrix} \tag{108}$$

$$r_{i,j} = lb_j + U\left(ub_j - lb_j\right)$$

$$F = \begin{bmatrix} F_1 \\ \vdots \\ F_i \\ \vdots \\ F_N \end{bmatrix}_{N\times 1} = \begin{bmatrix} F(R_1) \\ \vdots \\ F(R_i) \\ \vdots \\ F(R_N) \end{bmatrix}_{N\times 1} \tag{109}$$

Romanticism supported the exclusive understandings and perceptions of the individual, declining stiff community customs and anticipations. Prominence on the entity frequently led to narrative elements in their inscription, as playwrights drew stimulus from their own specific lives and familiarities.

$$R_{i,j}^{p1} = R_{i,j} + U \cdot \left(M_j - U_1 \cdot R_{i,j}\right) \tag{110}$$

$$P_{i,j}^{p1} \rightarrow new\,position\,of\ the\,author$$

$$M_j \rightarrow\ writing\ literature$$

$$R_i = \begin{cases} R_i^{p1}, F_i^{p1} \leq F_i \\ R_i, Else \end{cases}$$

Nature was also perceived as a retreat from the disingenuousness and restrictions of humanity, a domicile where entities could associate with their real selves.

$$W_i = \{R_k, F_k > F_i\} \tag{111}$$

$$W_i \rightarrow\ creation\,of\ literature$$

$$R_{i,j}^{p2} = R_{i,j} + U \cdot \left(R_{i,j} - U_2 \cdot R_{i,j}\right) \tag{112}$$

$$R_{i,j}^{p2} \rightarrow new\,position\,of\ the\,author$$

$$R_{i,j} \rightarrow ground\ reasons$$

$$R_i = \begin{cases} R_i^{p2}, F_i^{p2} \leq F_i \\ R_i, Else \end{cases}$$

Numerous Romantic authors also observed in the ancient, predominantly the primitive era, for stimulus, finding in it wisdom of furtive, escapade, and divine depth. Romantic authors discovered the paranormal, the secretive, and the glamorous, frequently integrating fundamentals of legends, fable, and folklore.

$$R_{i,j}^{p3} = R_{i,j} + W \cdot (1 - 2 \cdot W) \cdot \frac{ub_j - lb_j}{t} \tag{113}$$

$$R_{i,j}^{p3} \rightarrow new\ position\ of\ the\ author$$

$$R_i = \begin{cases} R_i^{p3}, F_i^{p3} \leq F_i \\ P_i, Else \end{cases}$$

8 Simulation Study

Optimal mathematical design of enhanced volunteer activities in a Monastery (OMVM), Human potentials-based optimization algorithm (HPOA), Poker strategy-based optimization algorithm (PSOA), optimal influence of Torah, Tanakh, and Talmud in Jewish life inspired algorithm (OTTJ), Sustainable conditions of Jewish population-based optimization algorithm (SCPO) and Optimal conservation of parrot-based algorithm (OCPO) are substantiated in 7 benchmark functions (f_1-Sphere/parabola, f_2-Ackley, f_3-Rosen rock, f_4-Generalized Griewank, f_5-Hartman 3, f_6-Six-hump camel-back, f_7-Goldstein-Price) [10]. Results are demonstrated in Table 1.

Optimal mathematical design of enhanced volunteer activities in a Monastery (OMVM), Human potentials-based optimization algorithm (HPOA), Poker strategy based optimization algorithm (PSOA), optimal influence of Torah, Tanakh, and Talmud in Jewish life inspired algorithm (OTTJ), Sustainable conditions of Jewish population-based optimization algorithm (SCPO) and Optimal conservation of parrot-based algorithm (OCPO) are applied to solve the power loss shrinking problem in electrical grid network [11]. OMVM, HPOA, PSOA, OTTJ, SCPO and OCPO are substantiated in the IEEE 30 bus system. Table 2 displays the valuation. Active power loss (LE (MW)), power deviation (RD (PU)) and stability of voltage (BI (PU)) assessment.

Table 1 OMVM, HPOA, PSOA, OTTJ, SCPO and OCPO valuation in 7 benchmark functions

F	HJD [10]	OMVM	HPOA	PSOA	OTTJ	SCPO	OCPO
1	0.8874	0.0019	0.0021	0.0022	0.0019	0.0018	0.0016
2	0.0918	0.0112	0.0116	0.0113	0.0115	0.0119	0.0120
3	22.9814	16.470	16.473	16.471	16.471	16.477	16.478
4	1.10 e-2	3.08 e-4	3.11 e-	3.12 e-4	3.09 e-4	3.13 e-4	3.10 e-4
5	−3.8629	−3.8629	−3.8629	−3.8629	−3.8629	−3.8629	−3.8629
6	−1.0318	−1.0311	−1.0315	−1.0313	−1.0314	−1.0317	−1.0319
7	3.0280	3.0112	3.0117	3.0114	3.0116	3.0119	3.0118

Table 2 OMVM, HPOA, PSOA, OTTJ, SCPO and OCPO assessment in IEEE 30 bus system

Technique	LE(MW)	RD(PU)	BI(PU)
PNH [12]	4.6304	0.1678	0.2524
PEY [13]	5.0763	0.6828	0.3297
GUO [14]	6.1300	0.4710	0.3521
MEA [15]	5.3048	0,4632	0.3614
OMVM	4.0015	0.0036	0.3944
HPOA	4.0019	0.0042	0.3938
PSOA	4.0022	0.0046	0.3934
OTTJ	4.0024	0.0049	0.3932
SCPO	4.0029	0.0051	0.3929
OCPO	4.0028	0.0053	0.3926

Computation time of OMVM, HPOA, PSOA, OTTJ, SCPO and OCPO are displayed in Table 3.

Table 3 OMVM, HPOA, PSOA, OTTJ, SCPO and OCPO computation time

Technique	IEEE30 BUS T(S)
OMVM	14.91
HPOA	15.04
PSOA	15.12
OTTJ	15.15
SCPO	15.23
OCPO	15.29

9 Conclusion

Volunteers in the monastery contribute to knowledge gaining activities especially to the learners belonging to deprived families and to certain rural area population who are unable to afford the cost of private education. Senior volunteers work as role models, providing direction and backing to the junior volunteers who have newly joined the service. Cognitive and emotional features, comprise judgement, hand appraisal, game theory, and mathematical abilities to comprehend varieties and possibilities. Important psychological characteristics are strength of mind, emotive control, concentration, and self-awareness to dodge tilt and make coherent judgments. Understanding the concept of game theory, possibility, bet sizing, and the equity in pot is vital for practical arithmetic and creating sound tactical choices. Strong problem-solving ability, analyzing multifarious circumstances, creating suppositions, and formulating the solutions are highly essential. Numerous Jewish sacraments and practices, such as Sabbath adherence and holidays, are engrained in the Torah. The Torah delivers a structure for creating moral decisions in several facets of life, from individual associations to commercial transactions. The Torah assists as the foundation for Jewish commandment (Halakha), which directs several characteristics of Jewish life, comprising nutritional laws, domestic law, and spiritual fulfillment. Jewish communities universally display fluctuating degrees of integration amongst themselves, with some groups preserving strong internal unity whereas others show more ingenuousness to exterior stimuli and intermarriage. Features like ancient perspective, topographical position, and spiritual compliance play a substantial role in determining these outlines. Certain societies like Bene Israel in India have efficaciously combined with their local people while conserving their Jewish distinctiveness. Pontiff allocates funds to organize medical camps for the parrots frequently to safeguard the parrots from various diseases. This attribute is scientifically defined in the exploitation segment of the algorithm.

Optimal mathematical design of enhanced volunteer activities in a Monastery (OMVM), Human potentials-based optimization algorithm (HPOA), Poker strategy-based optimization algorithm (PSOA), optimal influence of Torah, Tanakh, and Talmud in Jewish life inspired algorithm (OTTJ), Sustainable conditions of Jewish population-based optimization algorithm (SCPO) and Optimal conservation of parrot-based algorithm (OCPO) are substantiated in 7 benchmark functions and IEEE 30 bus systems.

AI Use Disclosure Statement The authors did not use any generative AI or AI-assisted technologies in the creation of this manuscript.

References

1. Gwyther K (2025) Utopian studies and the Hebrew Bible. Curr Biblic Res 23(2):107–124
2. Gummer H (2002) First record of willie wagtail (Rhipidura leucophrys) for New Zealand. Notornis 49(3):186–197
3. Tian Z, Lu Z, Huo D, Zhang Z, Sun Q, Wu H, Wang J, Kang Y, Xie Y, Tian Y, Yi K, Hou J (2025) Strategic analysis of avian conservation within key areas and key species - A case study of Baiyangdian. China. J Environ Manag 373(1):1–17
4. Cieśla-Korytowska M (2024) Romanticism towards history. Tematy i Konteksty 19(14):347–364
5. Avadhoota Datta Peetham (2025) Dattapeetham.org. Retrieved September 4, 2025, from https://www.dattapeetham.org/page/view/sri-ganapathy-sachchidananda-swamiji
6. Lin C-F, Lin Y-X, Chang S-H (2025) Medical image encryption using chaotic mechanisms: A study. Bioeng (Basel, Switz) 12(7):1–32
7. Burnes B (2004) Kurt Lewin and the Planned approach to change: A re-appraisal. J Manag Stud 41(6):977–1002
8. Wu Y, He Y, Yu L (2025) Adaptive crow search algorithm based on Levy flight. In: Proceedings of the 2025 6th international conference on computer information and big data applications, 1 (1), 779–782
9. Sobhaninia S, Meerow S, Keith L, Roy M, Trego S, Matos M (2025) A comparison of plan integration for flood and heat resilience: A case study of Baltimore, Maryland. Int J Disaster Risk Reduct: IJDRR 127(1):1–12
10. Yapıcı H, Çetinkaya N (2017) An improved particle swarm optimization algorithm using eagle strategy for power loss minimization. Math Probl Eng 2017(1):1–11
11. Kanagasabai L (2022) Novel empress SARANI optimization algorithm for active power loss reduction and voltage stability enhancement. Heliyon 10(22):1–64
12. Khan NH, Wang Y, Jamal R, Iqbal S, Ebeed M, Ghadi YY, Elbarbary ZMS (2024) Enhanced skill optimization algorithm: Solution to the stochastic reactive power dispatch framework with optimal inclusion of renewable resources using large-scale network. IET Renew Power Gener 18(S1):4565–4583
13. In S, Ang S, Choeung C, Ieng S, Cheng H, Huy V (2024) Optimal reactive power dis-patch using artificial gorilla troops optimizer considering voltage stability. Int J Electr Electron Res 12(3):1001–1009
14. Rao NT, Kumar KK, Kumar PP, Nuvvula RSS, Mutharasan A, Dhanamjayulu C, Shaik MR, Khan B (2024) Multiobjective optimal TCSC placement using multiobjective grey wolf optimizer for power losses reduction. Sci Rep 14(1):1–19
15. Hasanien HM, Alsaleh I, Tostado-Véliz M, Zhang M, Alateeq A, Jurado F, Alassaf A (2024) Hybrid particle swarm and sea horse optimization algorithm-based optimal reactive power dispatch of power systems comprising electric vehicles. Energy (Oxford, England) 286(1):1–20

Metric-Validated Scan-To-BIM Workflow for Reconstruction of As-Built Steel Structural Frames

Kamaliah Mohd Saha, Shek Poi Ngian, and Ng Chiew Teng

Abstract Digitalization in construction is transforming construction by converting actual site data into computable assets. The core of this transition is in the automated reconstruction of a Building Information Modeling (BIM) model from point clouds, converting raw scan data into traceable, analysis-ready as-built models. This study presents a rigorous scan-to-BIM workflow for steel structural frames that emphasizes automation and metric-based validation. High-density point cloud data were obtained using Faro Focus Laser Scanner 3D X130; registration and processing in Autodesk Recap Pro were maintained with quality metrics threshold (overlap > 60%, balance > 20%, and points < 6 mm > 99.7%) to ensure a chain of evidence; segmentation to isolate the elements, followed by semi-automated reconstruction in Autodesk Revit. A sample steel structural frame in a laboratory environment demonstrated that the workflow deliverables were appropriate for future investigations, such as for as-built structural performance, quality control, and asset maintenance. The contribution of this study is an automation-ready, reproducible workflow that connects data acquisition using a 3D laser scanner with accuracy assessment for a future analysis-ready BIM model. The workflow enhances the existing knowledge in scan-to-BIM research by providing a validated reference for academic and industry applications, hence supporting the digital transformation goals of Malaysia's Construction 4.0 Strategic Plan 2021–2025.

Keywords Laser scanner · Building information modeling (BIM) · Point cloud · As-built · Steel

K. M. Saha (✉) · S. P. Ngian · N. C. Teng
Faculty of Civil Engineering, UTM Skudai, Johor Bahru, Malaysia
e-mail: kamaliah@utm.my

S. P. Ngian
e-mail: shekpoingian@utm.my

N. C. Teng
e-mail: chiewteng.ng@utm.my

A. Azizi et al. (eds.), *Advanced Decision-Making Under Uncertainty*, Emerging Trends in Mechatronics, https://doi.org/10.1007/978-981-95-8696-7_2

1 Introduction

In an era marked by rapid technological evolution, automation and digitalization are progressively transforming the construction industry [1–5]. The implementation of 3D laser scanners to accurately capture as-built structures has become a pivotal factor in transforming traditional workflows. A significant outcome of this transition is the ability to automatically reconstruct Building Information Modeling (BIM) elements from point cloud data. This methodology is central to Construction Revolution (CR) 4.0, which aligns with Industry Revolution (IR) 4.0 paradigm, highlighting data-driven decision-making, digital workflows, and advanced integration across the architecture, engineering, and construction (AEC) lifecycle.

The automated reconstruction of BIM elements using point cloud data enables to reduce manual modeling, mitigates human error, and improves the accuracy of as-built models. This capability is crucial for existing inefficiencies in construction workflows where traditional methods [6–8] of data acquisition, such as rulers, measuring tapes, and total stations, are labor-intensive and human error-prone [6–8]. Leveraging advanced algorithms, particularly based on geometry recognition and computational optimization, researchers and practitioners seek to automate the segmentation, classification, and parametric modeling of the elements from point cloud data. This method not only facilitates model generation but also supports downstream applications, including progress monitoring, quality assurance, quality control, and as-built structural performance.

Furthermore, the automation of BIM modeling significantly contributes to promoting digitalization throughout the construction lifecycle. It facilitates interoperability, reduces the gap between design intent and as-built reality, and contributes to efficient project documentation. The integration facilitates decision-making [9–11] and construction efficiency [12].

In alignment with global BIM adoption trends, Malaysia's readiness reflects insights from established implementation in nations like Singapore and the United Kingdom, where national BIM mandates have markedly enhanced project coordination and asset management. The Construction Industry Development Board (CIDB) and Public Works Department (PWD) of Malaysia have established a cooperative framework to enhance interoperability and standardization via the Construction 4.0 Strategic Plan 2021–2025 [13] and PWD BIM Standards [14]. This framework has accelerated the integration of digital workflows in government initiatives, promoting consistent data environments, and facilitating automated systems. Consequently, Malaysia's BIM ecosystem is evolving towards global best practices, bridging the international methodology with specific construction needs through integrated policies, training, and research activities.

Despite the advancements, several challenges remain. This encompasses the management of extensive point cloud data, the assurance of semantic accuracy in element recognition, and achieving reliable automation. Additionally, the development of a robust workflow that can generalize across diverse environments remains a research priority, particularly developing the elements in correct categories and types with the use of generic elements, and the lack of studies employing a unified platform for the workflow.

This study seeks to develop an automated approach for modeling steel structural frame elements from point cloud data with Autodesk Revit [8, 15–18]. The use of Autodesk Recap Pro in the point cloud registration and processing brings some advantages, such as compatibility with Autodesk Revit to prevent data loss and ensure smooth information transfer. The proposed workflow leverages the seamless interoperability of the Autodesk ecosystem to efficiently convert raw scan data into intelligent BIM elements, facilitating accurate geometry reconstruction.

Within this transformation, scan-to-BIM serves as a critical intermediary layer that converts unstructured data into organized digital intelligence. The proposed workflow establishes a foundation for downstream applications, including deformation analysis and predictive maintenance.

2 Literature Review

Recent studies indicated that the integration of 3D laser scanners and BIM establishes a basis for creating accurate, analysis-ready as-built models, which are essential for project documentation, quality management, performance evaluation, and heritage management throughout a building's lifecycle. Digitalization facilitates the transformation of physical construction data into intelligent, computable assets, reducing reliance on traditional measurement methods.

Researchers such as [1, 2] have emphasized that the implementation of digital twins improved the traceability and sustainability of construction projects. Meanwhile, [3] illustrated how automation enhances real-time monitoring and decision-making. These studies collectively underscore the necessity of integrating digital workflows through the AEC sector to improve efficiency, accuracy, and productivity.

The application of 3D laser scanning in construction has increased significantly over the last decade. As noted by [12, 19], multi-platform and terrestrial scanning technologies have become essential for acquiring detailed geometric information of complex structures. The result of these point cloud datasets allows academics and practitioners to reconstruct physical structures within BIM environments with high accuracy. However, this process necessitates meticulous registration, processing, and data management to ensure geometric integrity.

Following registration and processing, the segmentation of point cloud data is essential for identifying elements for reconstruction. Segmentation enables the identification and extraction of individual elements, including beams, columns, etc., allowing for transition and strengthening matching into BIM environments [4, 18, 20]. Research conducted by [21, 22] revealed that automated segmentation significantly reduces human involvement and enhances semantic consistency in element classification.

The most critical phase in the scan-to-BIM workflow is the reconstruction of BIM elements from processed point clouds. This phase converts geometric information into intelligent parametric objects that represent the true dimensions, positions, and orientations of physical elements within a structure. Numerous studies, including [18, 23], have shown that the development of as-built BIM models using registered point clouds significantly improves the accuracy and efficiency of digital documentation.

A recent scan-to-BIM application in the Malaysian construction industry, indicated that the utilization of Autodesk Revit and Autodesk Recap Pro has become the standard practice due to their robust interoperability and efficient data interchange within the Autodesk ecosystem [24]. Autodesk Recap Pro serves as the primary tool for point cloud registration and processing, facilitating the alignment of multiple scans into a unified coordinate system. This tool additionally facilitates the evaluation of registration quality measures, which are essential for ensuring data fidelity before model reconstruction [25]. After the point cloud is registered and optimized, Autodesk Revit offers an efficient platform for reconstructing as-built BIM models utilizing parametric modeling features [26]. Its compatibility with Autodesk Recap Pro enables the seamless importing of point cloud data, hence simplifying the direct modeling of elements within the BIM environment. Furthermore, Autodesk Revit is able to handle parametric intelligence and analytical modeling, which enhances the accuracy of as-built representations. The integration of Autodesk Recap Pro and Autodesk Revit facilitates a cohesive digital workflow, reducing data loss, enhancing modeling accuracy, and promoting automation in as-built documentation and quality management.

However, these advancements still lack a replicable, comprehensive methodology that ensures both accuracy and usability in BIM software. Additionally, there is limited evidence of workflows that exploit the interoperability between Autodesk Recap Pro and Autodesk Revit to maintain data fidelity during the transition from raw point cloud to BIM model. Therefore, the research gap lies in the absence of a validated, interoperability-driven, and accuracy-measured scan-to-BIM framework that can be effectively implemented using commercially accessible tools. The research introduces a systematic workflow that effectively integrates metric-based validation for enhancing as-built modeling and future integration with digital twin systems.

3 Materials and Methods

This research was conducted in three (3) phases as illustrated in Fig. 1, demonstrating the systematic workflow from data acquisition to BIM model development. During the initial step, data were acquired with a 3D laser scanner, which recorded the geometric characteristics of the physical structure as dense point clouds. This phase is essential to ensure thorough spatial coverage and accuracy of the as-built data. The second phase involved the registration and processing of point cloud data to align multiple scans within a cohesive coordinate system, eliminate noise, and enhance geometric accuracy. This process ensured that the resulting dataset accurately represented the real-world structure. In the third phase, the processed point cloud was imported into another tool for the creation of the BIM model. Collectively, these

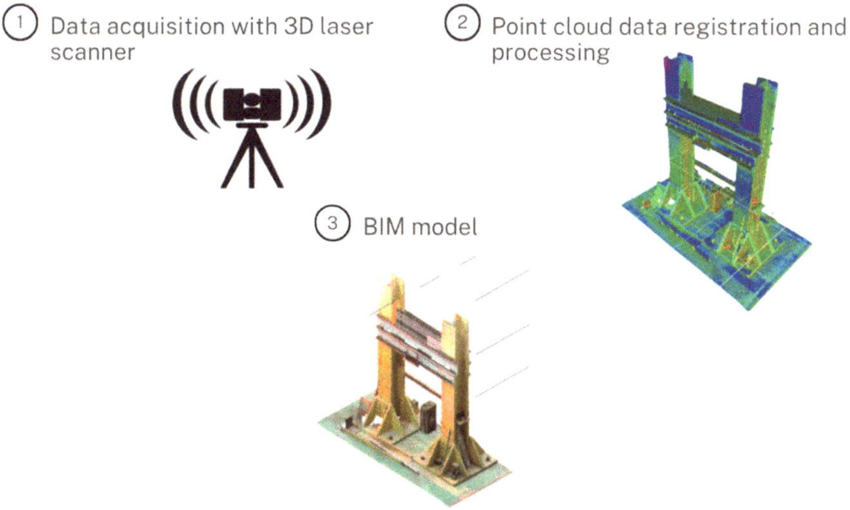

Fig. 1 Three (3) phases of the scan-to-BIM workflow

three (3) phases form an integrated scan-to-BIM workflow that facilitates accurate as-built modeling.

The chosen case study for this research was the sample steel structural frame at the Structure and Material Laboratory, Faculty of Civil Engineering, UTM Skudai, Johor, Malaysia, as illustrated in Fig. 2. The steel structural frame that was fabricated by Chee Kong Engineering & Construction Sdn. Bhd, Senai, Johor, Malaysia, consists of vertical steel columns, horizontal beams, and bolted connections. In this study, site limitations, including the restricted working area surrounding the frame, constrained the scanner's movement and necessitated adjustments to get comprehensive coverage, particularly obstructed by other nearby steel structural frames and apparatus. Secondly, the reflecting surfaces of the coated steel components resulted in minor issues, particularly at sharp edges and bolted regions, which necessitated post-processing. Furthermore, the indoor lighting presented challenges for the 3D laser scanner placement and alignment, requiring planning to ensure the complete capture of the elements.

In this study, all data analysis and BIM modeling were conducted on a mid-range computing workstation designed to accommodate the computing requirements of managing extensive 3D laser scanning datasets. The workstation had an MSI Cyborg 15 A13VEK, 13th Gen Intel® Core™ i7-13620H processor at 2.40 GHz, and was configured with a 32 GB of installed RAM. The system also operated with Windows 11 Home Single Language (Version 24H2, 64-bit), installed with 26,100.1457 OS build. The hardware specifications were chosen to optimize processing efficiency and accessibility, allowing the proposed workflow to be replicated by researchers or practitioners without requiring high-end computing equipment.

Fig. 2 Case study: sample steel structural frame at the structure and material laboratory, faculty of civil engineering, UTM Skudai, Johor, Malaysia

Although the hardware was not classified as high-end, it was sufficient for performing point cloud analysis, including importation, registration, and processing with Autodesk Recap Pro and BIM modeling with Autodesk Revit. Overall, the computing workstation offered a stable environment, demonstrating the capability for handling dense point cloud data and executing integrated BIM operations, hence confirming the feasibility of conducting extensive 3D laser scanning and BIM modeling processes on economical, mid-range computing systems.

3.1 Data Acquisition

For the first phase, the process commenced with data acquisition via the Faro Focus Laser Scanner 3D X130 as illustrated in Fig. 3 to capture the constructed environments. The selection of the 3D laser scanner was determined based on its proven accuracy, portability, capabilities, and user-friendly operation, which are essential for data collection as utilized by [27–36]. Detailed technical specifications for the Faro Focus Laser Scanner 3D X130 are tabulated in Table 1.

The scanning process was conducted in a controlled indoor laboratory facility. The Faro Focus Laser Scanner 3D X130 was mounted on a tripod and strategically positioned at multiple locations around the steel frame to provide comprehensive surface coverage and minimize occlusions. Three (3) criteria directed the planning of scan stations: (1) line-of-sight coverage, (2) overlapping fields of view among scan stations to ensure accurate point cloud data registration [38], and (3) accessibility within the spatial constraints of the laboratory. In this study, data were collected from

Fig. 3 Faro focus laser scanner 3D X130

fifteen (15) different scanning stations, as illustrated in Fig. 4, which were marked by a black dot. Reference targets were utilized throughout scanning to ensure accurate point cloud registration in accordance with the manufacturer's operational guidelines [39]. A minimum of three (3) targets was designated for each scanning station, to ensure an unobstructed line of sight from each scanner station and visible from more than one (1) scanner station, thereby enhancing the geometric redundancy essential for reliable registration. The scanner was set up with approximately 3 m to 5 m between stations, and required 2 h to complete. Each scan was set to a resolution of ¼ at $2^{\times}$ quality, to provide a balance between detail preservation and manageable file size, particularly in capturing minor features such as bolted connections and edge profiles. The raw scan data was then stored in its native format and visually validated on-site to confirm the completeness and accuracy of the capture.

In this process, data acquisition methods were based on time-of-flight (ToF) of laser pulses that reflected from objects [19, 40]. The operating principle of a 3D laser scanner utilizing ToF method. This method incorporated both a transmitter and a receiver. The transmitter delivered a laser pulse aimed at a target surface, known as the ground object. Upon encountering the object, the pulse was reflected and returned to the scanner and detected by the receiver. The scanner then quantified the duration between the pulse emission and its return. The range or distance to the object was determined using Eq. 1. This process allows the scanner to determine accurate distances to multiple points on the object's surface, capturing points in a coordinate system (x, y, z).

$$R = v.t/2 \tag{1}$$

Table 1 Technical specification for faro focus laser scanner 3D X130 [37]

Category	Specification
Power supply	19V (external), 14.4V (internal battery)
Power consumption	40W (normal), 80W (while battery charges)
Battery life	4.5 h
Operating temperature	5 °C to 40 °C
Weight	5.2 kg
Dimension	240 × 200 × 100 (mm)
Calibration	Annual
Range	0.6 m to 130 m indoor or outdoor with upright incidence to a 90% reflective surface
Measurement speed	122,000/244,000/488,000/976,000 points/sec
Ranging error	± 2 mm
Ranging noise @ 10 m	0.3 mm (90% reflective), 0.4 mm (10% reflective), compressed: 0.15 mm/0.2 mm
Ranging noise @ 25 m	0.3 mm (90% reflective), 0.5 mm (10% reflective), compressed: 0.15 mm/0.25 mm
Resolution	Up to 70 megapixels
Dynamic color feature	Automatic adaption of brightness
Step size (vertical/horizontal)	0.009° (40,960 3D-Pixel on 360°) / 0.009° (40,960 3D-Pixel on 360°)
Brightness control	Automatic
Field of view	Vertical: 300°, Horizontal: 360°
Data storage	SD, SDHC™, SDXC™; 32GB card included
Scanner control	Via touchscreen display and WLAN
Laser class	Class 1

where R is the slant distance or range, v is the speed of electromagnetic radiation, which is a known value, and t is the measured time interval.

3.2 *Point Cloud Data Registration and Processing*

The accurate registration and processing of point cloud data are essential for converting raw laser scanning outputs into dependable 3D spatial information applicable to BIM applications. The registration process ensures that multiple scan datasets obtained from different stations are geometrically aligned within a unified coordinate system, creating a comprehensive and spatially coherent representation of the scanned structure. Accurate registration and processing not only eliminates inconsistencies between the scan stations but also influences the dimensional accuracy and usability of the final as-built model. Following data acquisition, the point cloud data

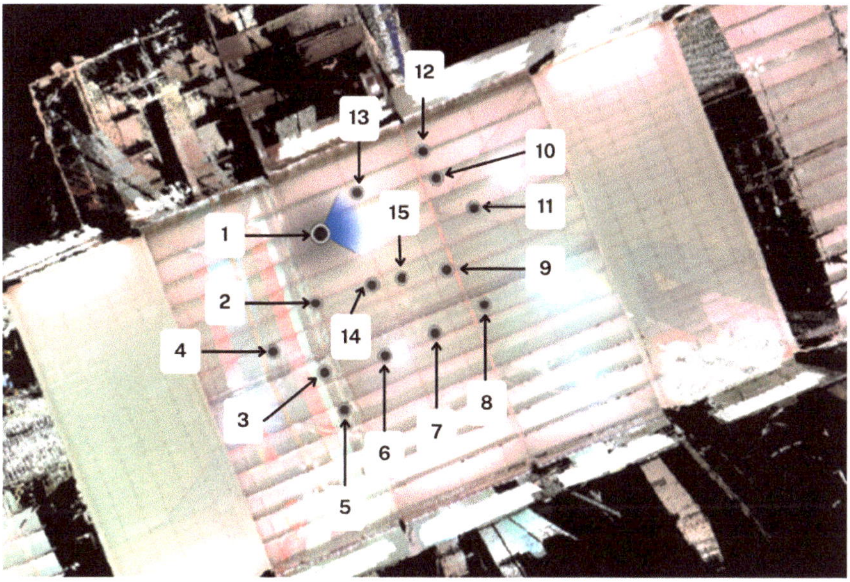

Fig. 4 Top view layout of scan stations for data acquisition

collected from various scan stations were subjected to a registration and processing workflow to ensure spatial consistency and geometric integrity. Authors in [41] highlighted that careful adjustment of processing parameters is essential to achieve high fidelity and usability of the resulting 3D model.

In this study, the registration process was carried out using Autodesk ReCap Pro, as utilized by [16, 18, 34, 42]. The process began with the importation of the scan data in.fls file format from fifteen (15) scan stations to align all the point cloud data within a unified coordinate system, as illustrated in Fig. 5. The process required approximately 30 min to complete. The registration method used was automatic registration. The accuracy of the registration was then assessed based on an alignment quality indicator in the registration quality report, which offered registration-quality metrics, including overlap, balance, and points < 6 mm [25], as described in Table 2. These metrics served as indicators of alignment accuracy and dataset reliability, ensuring that the resulting point cloud was geometrically consistent and appropriate for subsequent processing and modeling.

Once the scans were registered, the point cloud was exported in the standard format of.rcp [34], enabling seamless integration with Autodesk Revit. Upon completion, processing steps were conducted to improve the clarity and usability of the datasets. This encompassed eliminating stray points caused by reflective surfaces or ambient interferences and excludes irrelevant objects [43], such as scanning personnel, equipment, or any other fixtures. The cleaned dataset was then subjected to point cloud segmentation [20], which utilized the limit box method, as illustrated in Fig. 6, to isolate key structural components such as columns and beams, marked with different

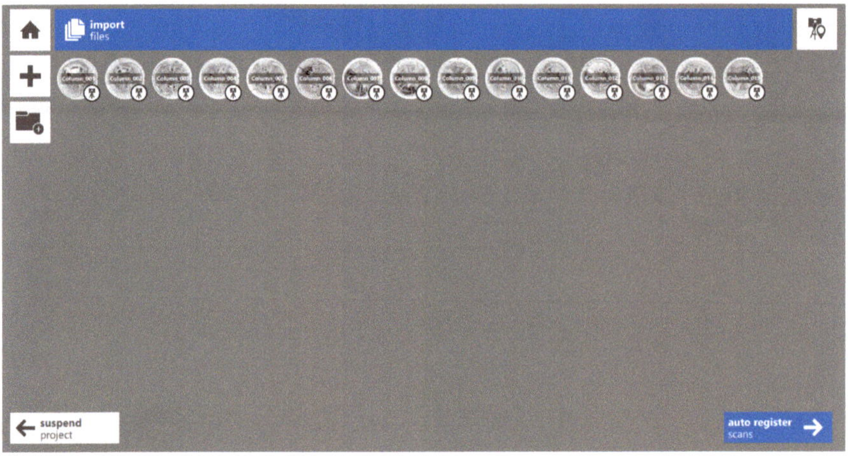

Fig. 5 Registration of point cloud data from fifteen (15) scan stations

Table 2 Registration quality report for autodesk recap pro [25]

Metric	Description	Recommended threshold
Overlap	Percentage of common features/area throughout the project	No fixed threshold documented, but [25] suggested a higher percentage of overlap ensures that features that are occluded behind objects in one scan are visible in one of the other scans
Balance	Percentage of common features from each scan in the overlapping region	>20%
Points <6 mm	Percentage of overlapping points within 6 mm of the corresponding feature in the project	>90%

color-coded regions. In this approach, a virtual bounding box was created around a selected region of the scanned environment to confine the working area and exclude extraneous surrounding data. This technique enhances the segmentation process by concentrating on specific structural elements while eliminating background noise and unnecessary points.

The segmentation was executed with region-based classification tools available in Autodesk Recap Pro. The process involved manually identifying and grouping points corresponding to particular elements. This step improved data clarity, facilitating more efficient processing and improved accuracy in following modeling within BIM or other digital environments.

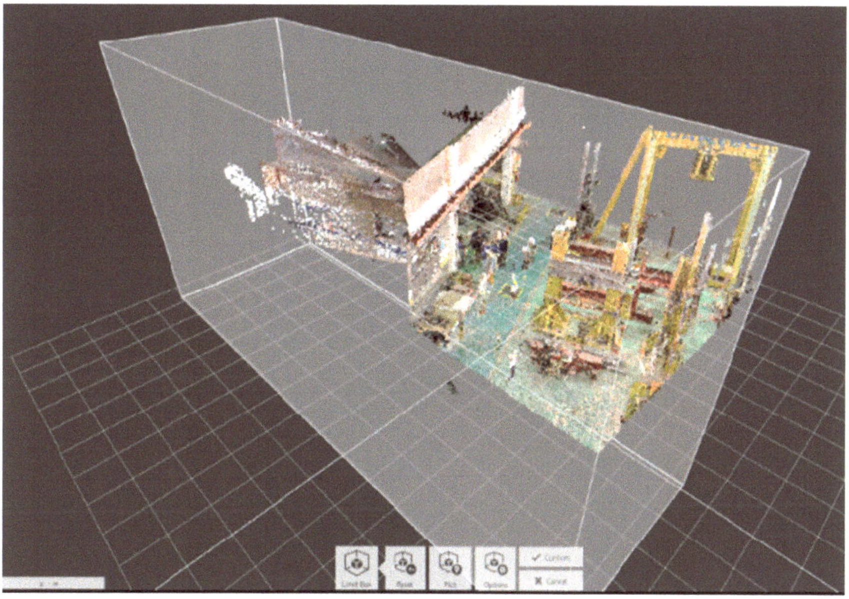

Fig. 6 Limit box method to isolate components for the segmentation process

3.3 BIM Model

The development of the BIM model represents the most pivotal stage of the workflow, where the registered point cloud data is converted into an intelligent, parametric digital model. This phase serves as the bridge between raw spatial data and actionable engineering information, facilitating accurate geometric representation, attribute assignment, and integration with structural analysis or facility management systems.

The final phase of the workflow involved the automated generation of a BIM model, which utilized the registered and segmented point cloud data. The registered point cloud was imported as an.rcp file into Autodesk Revit as illustrated in Fig. 7, to provide an analysis-ready as-built BIM. The workflow commenced with the input positioning of the point cloud data. There are three (3) options available in the Autodesk Revit: (a) Auto—By Shared Coordinate, (b) Auto—Origin to Internal Origin, and (c) Auto—Center to Center [44], which allow the user to choose how the point cloud is positioned in the project as described in Table 3. In this study, Auto-Center to Center has been chosen because the point cloud data was not reliable for survey control or a predetermined origin. Project units were also set to millimeters, and worksheets were created to manage visibility.

Subsequent to this, fine alignment was performed utilizing reference planes, length, width, and height to verify size and orientation. After the alignment process, level datums were derived, and section and elevation views were generated. In this study, heights were measured at multiple locations along each feature to confirm

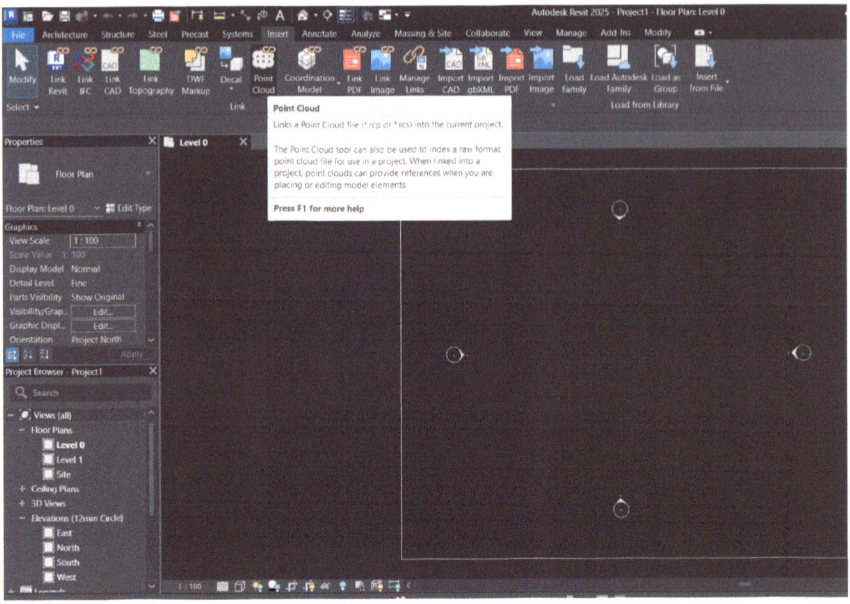

Fig. 7 Importation of point cloud data into autodesk revit

Table 3 Positioning type for importing point cloud in autodesk revit [44]

Positioning type	Description
Auto—by shared coordinate	Autodesk Revit assumes that the coordinates within the point cloud file are specified in the shared coordinate system utilized in the model. The point cloud origin will be positioned at the origin of the shared coordinates accessible via the survey base point. The point cloud will be aligned such that the north orientation in the cloud file corresponds to the True North if the model
Auto—origin to internal origin	Autodesk Revit positions the point cloud's origin at the Autodesk Revit project origin, which is represented as a project base point in the site plan
Auto—center to center	Autodesk Revit positions the centroid of the point cloud bounding box at the centroid of the model's bounding box. If the majority of the model is not visible, the center point may not be visible in the current view

parallelism and identify any tilt that occurred. The discrepancies were resolved by rotating the model. Modeling, then proceed with point cloud snapping for element placement in Autodesk Revit, and align it to geometric faces and establish level datums. Finally, columns and beams were modeled utilizing the Revit families.

4 Results

The implementation of the proposed workflow produced promising results, demonstrating both technological reliability and practical feasibility. In order to validate and assess the applicability of the proposed workflow, the sample steel structural frame at the Structure and Material Laboratory, Faculty of Civil Engineering, UTM Skudai, Johor, Malaysia, was selected as the case study. The case study was chosen not only for its structural significance, which encompasses columns and beams, but also for its accessibility. Furthermore, the utilization of standard industrial materials and sizes, closely mirrors typical construction practices, making the selected structure highly relevant for illustrating the potential of the proposed workflow.

4.1 Data Acquisition

The chosen case study was composed of two (2) primary columns located on the laboratory's floor and horizontally braced with beams for stability. Vertical element was fabricated from structural steel components of Universal Column (UC), horizontal elements of channel sections, and with bolted connections at the base. Dimensionally, the steel structural frame was approximately 3.60 m in height and 3.30 m in width at the base, as illustrated in Fig. 8.

The output consisted of a high-density point cloud in.fls format, accurately representing the 3D geometry of the structure, including detailed features of columns, beams, etc. The scanning resolution was set to optimize detail fidelity while maintaining data manageability, typically achieving a point cloud spacing of between 1 and 3 mm as illustrated in Fig. 9. This level of accuracy ensures that even fine structural features and surface anomalies of the scanned objects are reliably captured, establishing a reliable geometric basis for further modeling and analysis. The figure illustrates the density of data points distributed over the scanned surfaces, demonstrating the scanner's capacity to ensure consistency in spatial coverage.

4.2 Point Cloud Data Registration and Processing

Following data acquisition, the second phase focused on point cloud registration and processing utilizing Autodesk Recap Pro, where all individual scan stations were aligned and merged into a unified coordinate system. In this phase, a comprehensive registration quality report was generated, providing detailed metrics for each scan station, including overlap, balance, and points <6 mm. These values were essential for validating the geometric consistency between overlapping scans and ensuring reliable alignment. The report was summarized in Table 4, documenting the parameters across

Fig. 8 Data acquisition of physical structures

Fig. 9 Point cloud spacing ranging from 1 to 3 mm in.fls format

Table 4 Registration quality report in autodesk recap pro

Scan station	Overlap (%)	Balance (%)	Points <6 mm (%)
1	81.3	38.5	99.9
2	81.8	39.9	99.8
3	72.8	45.0	99.7
4	62.2	44.4	99.8
5	60.4	38.1	99.9
6	70.5	51.8	99.7
7	73.2	53.6	99.8
8	62.5	51.5	99.8
9	73.5	44.8	99.9
10	81.0	49.0	99.8
11	78.4	47.6	99.9
12	76.3	49.2	99.9
13	78.5	47.6	99.9
14	84.2	41.2	100.0
15	81.3	51.6	99.9

fifteen (15) scan stations, while the registered point cloud model after being subjected to processing is illustrated in Fig. 10.

The registration results for the 15 scan stations showed excellent compliance with the recommended thresholds. All scan stations achieved balance values ranging from 38.1 to 53.6%, exceeding the 20% threshold, while points < 6 mm consistently

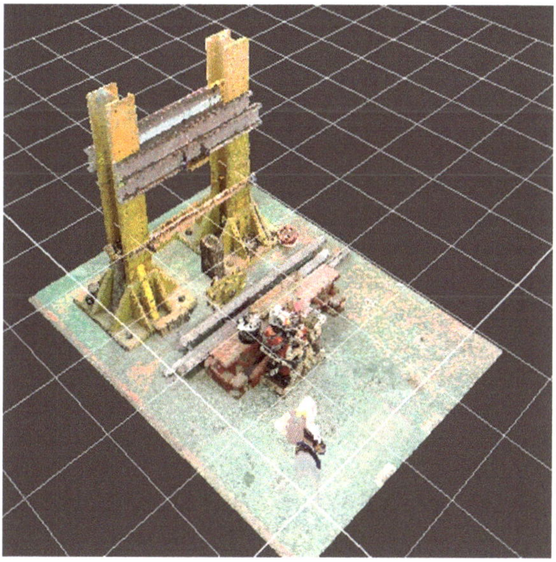

Fig. 10 Registered point cloud model in autodesk recap pro

exceeded 99.7%, significantly exceeding the 90% threshold, thereby confirming the excellent accuracy of the registration. The overlap percentage varied between 60.4 and 84.2%, with scan stations 4, 5, and 8 demonstrating comparatively lower overlap with ‘60%, indicating potential areas with reduced redundancy in occluded areas. However, the overall registration quality remained robust and well within acceptable limits for further processing and modeling. This level of registration accuracy established a reliable basis for as-built modeling in the scan-to-BIM workflow.

Figure 10 illustrates the registered point cloud model generated and visualized in Autodesk Recap Pro. The figure clearly illustrates a cohesive 3D model of the scanned steel frame, where individual scans obtained from different scan stations have been accurately aligned inside a singular coordinate system. This integration ensures geometric coherence and reduces positional inconsistencies in overlapping regions. The resultant model offers a comprehensive and high-fidelity digital representation of the physical structure. The clarity and comprehensiveness of the registered model signify effective scan alignment with sufficient overlap and balance among scanning stations.

The subsequent process involved segmentation to isolate each element, as illustrated in Fig. 11, performed in Autodesk Recap Pro, where individual elements of the point cloud model were isolated by different color coding for identification purposes. Each color represents a specific element, such as columns, beams, or etc., facilitating clearer differentiation. This process is crucial for reducing data complexity into manageable segments for further modeling in BIM environments. This significantly enhanced the clarity of the data and improved its suitability for the subsequent phase in Autodesk Revit. The final output was generated in.rcp format, ensuring interoperability within Autodesk platforms. The processed point cloud provided a high-resolution, geometrically consistent digital representation of the physical structure, serving as the essential input for BIM model creation.

4.3 BIM Model

The resulting BIM model, generated from the scan data, accurately represented the spatial configuration and dimensional characteristics of the physical structure. All elements were created as BIM elements in their families and positioned with their proper orientation. All structural elements were reconstructed in Autodesk Revit utilizing the geometry data collected from the registered point cloud.

As illustrated in Fig. 12, the columns were characterized as UC 356 × 406x393 (2 numbers), beams were channel C305 × 89x42 (4 numbers) and C102 × 51x10.42 (1 number), with stiffener plates. The columns were bolted to the concrete floor. In this study, all elements were modeled at the LOD 300 level of development, which consisted of their geometry, location, and type of elements [14]. All column and beam elements were automated model according to their Revit families, while the stiffener plate and base plate were modeled as custom types that were created based on their actual size, shape, and orientation.

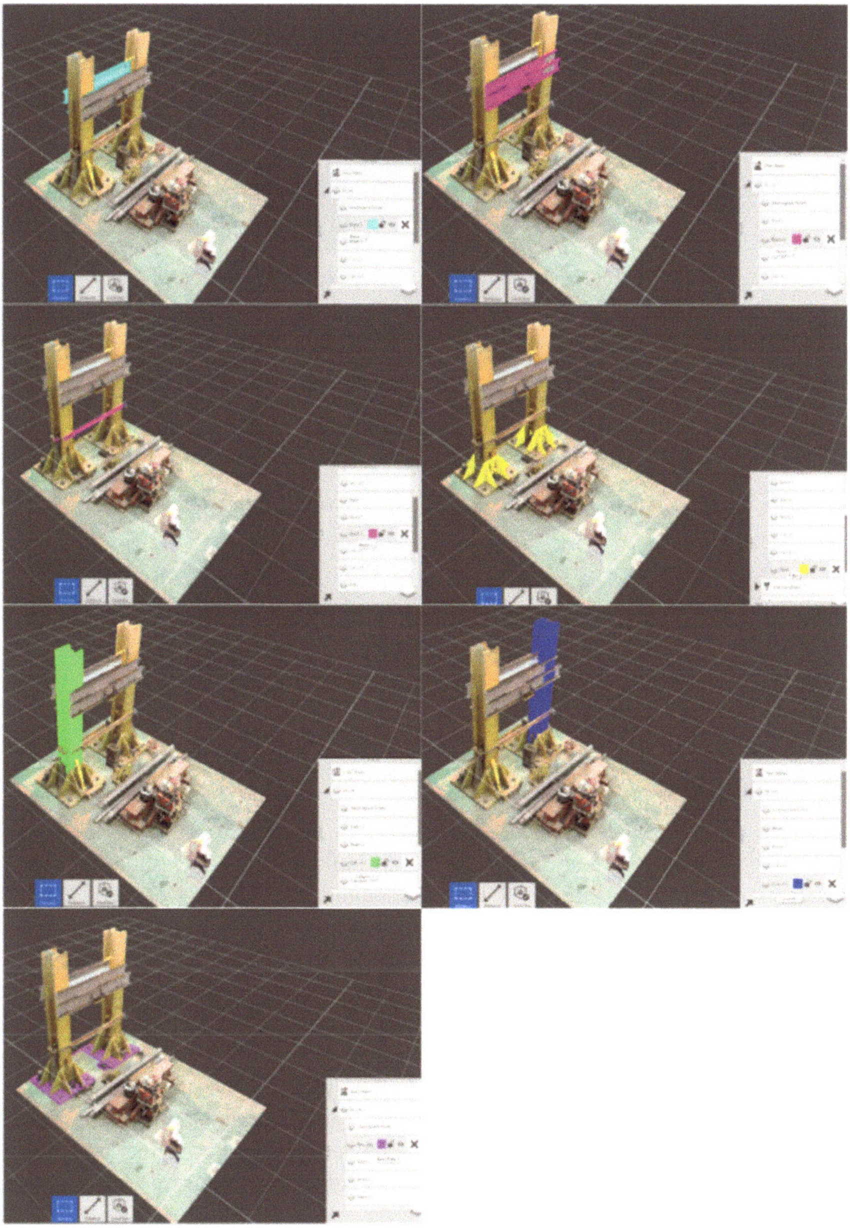

Fig. 11 Segmentation process to isolate each element, marked with color-coded in autodesk recap pro

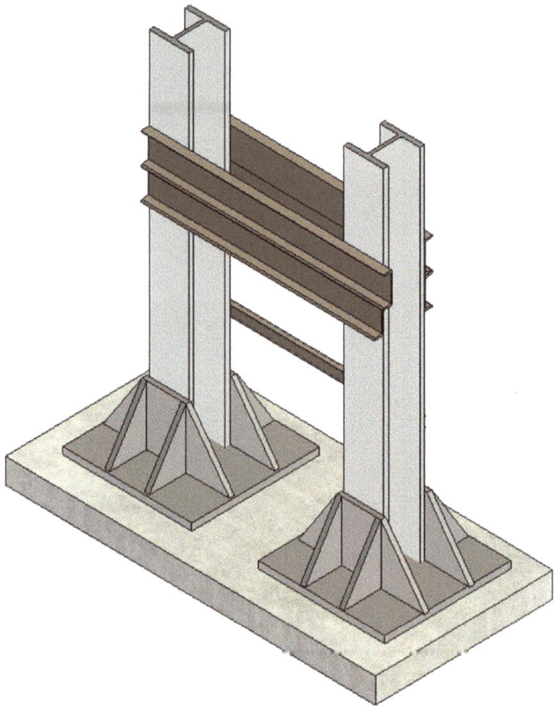

Fig. 12 As-built BIM of sample steel structural frame generated in Autodesk Revit from the registered point cloud

5 Discussions

This study aimed to produce an accurate 3D as-built model by integrating a 3D laser scanner with a BIM workflow. The findings demonstrate that the workflow achieved accurate registration of point cloud data from a 3D laser scanner by utilizing Autodesk Recap Pro. The quality of registration has proven to be a crucial factor influencing accuracy. Scan station configuration is a factor affecting both registration stability and subsequent modeling accuracy. Stations that observe the same objects from various angles and comparable distances produce more robust geometric restrictions, resulting in balanced and consistent data. In this study, most scan stations achieved well-connected and intersecting lines of sight. The few scan stations with a lower overlap of ~60% likely correspond with occluded or grazing-angle views, which reduce redundancy and may compromise paired constraints. While overall alignment remained robust, these data emphasize that the scan station is not just a count, but governs error propagation.

Segmentation in Autodesk Recap further enhanced fitness for purpose by isolating elements into color-coded regions. This process mitigates the impact of outlier influence and optimizes the streamlining to Autodesk Revit. The obtained segmentation clarity indicates that the processed point cloud is appropriate for semi-automated extraction of axis, surface, and section.

The as-built BIM accurately replicated the spatial arrangement and dimension attributes of the physical structure at LOD 300. By locating elements from the registered and segmented data, the respective true elements' orientation, the modeling avoided the common error of assuming idealized columns and beam axis. The resulting model serves as a traceable, geometry-accurate digital twin of the steel structural frame, prepared for the subsequent phase. In summary, the workflow exhibits methodological coherence and the outputs are thus not only visually consistent but also measurement-sound, offering a reliable basis for further acceptance checks and compliance assessment.

The resulting BIM model demonstrated a high degree of accuracy in replicating the geometric arrangement of the actual steel structure, hence validating the reliability of point cloud data in defining spatial relationships and element alignment. The integration of scan data into Autodesk Revit enabled the generation of a digital replica that accurately reflected real-world conditions. This signifies that the point cloud dataset served as an accurate and analysis-ready as-built BIM model.

Moreover, each element was allocated its respective material and section properties according to its Revit family type, which facilitates accurate structural representation at LOD 300. The systematic modeling process enhances the geometric integrity and provides semantic consistency between the physical and digital, which is essential for subsequent analysis and decision-making phases. This digital workflow seamlessly bridges the gap between physical site conditions with virtual project environments, minimizing manual intervention and enhancing traceability throughout the construction lifecycle.

This study's findings underscore several best practices essential for reliable point cloud modeling. The proposed workflow serves as an educational resource for instructing on 3D laser scanners, BIM integration, and digital twin modeling, in addition to its technical execution. The scalability of the proposed workflow has significant implications for industrial implementation. While the current study focused on a laboratory-scale construction, the methodological principles can be generalized to more complex structures, including warehouses, bridges, and industrial facilities. Hence, this workflow offers both scholarly and professional significance by connecting theoretical digitalization to practical application.

6 Conclusions

In conclusion, the three (3) phase scan-to-BIM process generated a high-quality, interoperable as-built dataset and model of a steel structural frame. Phase 1 obtained dense measurements throughout the entire frame envelope. Phase 2 accomplished robust registration in Autodesk Recap Pro, within the threshold specified. Phase 3 produced a LOD 300 Revit model that accurately instantiates parametric families for all elements, preserving true orientation and connectivity.

Practically, the workflow established: (a) a reliable data acquisition to registration process that delivers accurate alignment consistency suitable for element reconstruction, (b) a segmentation methodology that enhances model fidelity, and (c) a systematic BIM authoring technique that integrates custom and generic elements for repeatable elements.

These findings also encourage interdisciplinary integration, wherein BIM-based models can be integrated with other technologies such as Internet of Things (IoT), real-time monitoring sensors, or artificial intelligence (AI). These interfaces can automate quality management, predict actual structural performance, and improve preventative maintenance via digital twins. By integrating geometric accuracy with real-time data streams, future work may advance this workflow towards intelligent facility management systems that continuously evaluate asset condition and safety compliance.

Overall, this research enhances the digital transformation of the construction industry by establishing a systematic, data-driven workflow that connects field-acquired spatial data to intelligent BIM environments. The workflow's reproducibility, compatibility with mid-range hardware, and compliance with quality thresholds position it as a practical model for public and industrial applications. In addition, the workflow provides valuable guidance for practitioners and researchers seeking to establish standardized benchmarks for accuracy, interoperability, and performance assessment in future implementations.

Future work should extend validation beyond laboratory settings to encompass active building sites that present complex structures and curved elements. A systematic scan station planning investigation is necessary to manipulate overlap and range, enabling the reporting of confidence intervals of registration quality metrics. In modeling, robust learning-based semantic segmentation should be employed to mitigate the limit box method bias. Finally, provide a compliance system that codifies project regulations and integrates the measured as-built model with structural analysis, enabling the quantification and prioritization of performance impacts from any observable deviations.

Several challenges will influence implementation and should be managed, including occlusion and access constraints that can depress overlap. In the registration process with Autodesk Recap Pro, utilizing limit box segmentation, residual misfits may occur at the edges or boundaries, and excessive outlier rejection poses the risk of over-pruning real geometry.

Acknowledgements The authors would like to express sincere gratitude to the Faculty of Civil Engineering, UTM, for granting permission to use laboratory samples and facilities for the work. Special appreciation is extended to research collaborators and colleagues for their valuable insights and feedback that greatly enhanced the quality of this study. This work was supported by the Research Grant Hi-Tech (F4+), Q.J130000.4622.00Q65.

AI Use Disclosure Statement The authors used generative AI technologies only for proofreading. This includes minor corrections to grammar, spelling, and syntax. No AI was used for ideation, data analysis, or writing original content.

References

1. Trubina N et al (2024) Digital technologies and material passports for circularity in buildings: an in-depth analysis of current practices and emerging trends, LNCE. Springer Nature Switzerland, 489. https://doi.org/10.1007/978-3-031-57800-7_64
2. Long W, Bao Z, Chen K, Thomas Ng S, Yahaya Wuni I (2023) Developing an integrative framework for digital twin applications in the building construction industry: A systematic literature review. Adv Eng Informatics, 59(2023), p 102346. https://doi.org/10.1016/j.aei.2023.102346
3. Srivastava A et al (2022) Imperative role of technology intervention and implementation for automation in the construction industry. Adv Civ Eng. 2022. https://doi.org/10.1155/2022/6716987
4. Rao AS et al (2022) Real-time monitoring of construction sites: Sensors, methods, and applications. Autom Constr, 136(2021), p. 104099. https://doi.org/10.1016/j.autcon.2021.104099
5. Musarat MA, Khan AM, Alaloul WS, Blas N, Ayub S (2024) Automated monitoring innovations for efficient and safe construction practices. Results Eng. 22(2023), p 102057. https://doi.org/10.1016/j.rineng.2024.102057
6. Ma Z, Liu Y, Li J (2023) Review on automated quality inspection of precast concrete components. Autom Constr. 150, p 104828. https://doi.org/10.1016/j.autcon.2023.104828
7. Li H, Zhang C, Song S, Demirkesen S, Chang R (2020) Improving tolerance control on modular construction project with 3d laser scanning and bim: A case study of removable floodwall project. Appl Sci 10(23):1–21. https://doi.org/10.3390/app10238680
8. Tran H, Nguyen TN, Christopher P, Bui DK, Khoshelham K, Ngo TD (2021) A digital twin approach for geometric quality assessment of as-built prefabricated façades. J Build Eng. 41. https://doi.org/10.1016/J.JOBE.2021.102377
9. Hakimi O, Liu H, Abudayyeh O (2024) Digital twin-enabled smart facility management: A bibliometric review. Front Eng Manag. 11(1):32–49. https://doi.org/10.1007/s42524-023-0254-4
10. Almatared M, Liu H, Abudayyeh O, Hakim O, Sulaiman M (2024) Digital-twin-based fire safety management framework for smart buildings. Buildings, 14(1). https://doi.org/10.3390/buildings14010004
11. Ismaeel WSE (2022) Challenges for applying artificial intelligence in the field of urban design and planning; case of Egypt. 2022 Eng Technol Sustain Archit Inter Des Environ ETSAIDE 2022. https://doi.org/10.1109/ETSAIDE53569.2022.9906351
12. Kukko A, Kaartinen H (2020) Multiplatform mobile laser scanning. In laser scanning: an emerging technology in structural engineering, B. Riveiro and R. Lindenbergh, Eds., CRC Press
13. CIDB, construction 4.0 strategic plan (2021–2025) next revolution of the Malaysian construction industry. CIDB, 2020
14. PWD, BIM: Piawaian JKR, JKR 29300–0074–21, 2nd ed. Unit Building Information Modeling (BIM), Bahagian Pengurusan Projek Kompleks, Cawangan Perancangan Aset Bersepadu, Public Work Deparment, 2021
15. Garcia-Gago J, Sánchez-Aparicio LJ, Soilán M, González-Aguilera D (2022) HBIM for supporting the diagnosis of historical buildings: case study of the Master Gate of San Francisco in Portugal. Autom Constr, 141. https://doi.org/10.1016/J.AUTCON.2022.104453
16. Moyano J, Gil-Arizón I, Nieto-Julián JE, Marín-García D (2021) Analysis and management of structural deformations through parametric models and HBIM workflow in architectural heritage. J Build Eng. 45(March):2022. https://doi.org/10.1016/j.jobe.2021.103274
17. Choi M, Kim S, Kim S (2024) Semi-automated visualization method for visual inspection of buildings on BIM using 3D point cloud. J Build Eng, 81(2023), p 108017. https://doi.org/10.1016/j.jobe.2023.108017
18. Rocha G, Mateus L (2024) Using dynamo for automatic reconstruction of BIM elements from point clouds. Appl Sci, 14(10)

19. Petrie G, Toth CK (2018) Introduction to laser ranging, profiling and scanning. In: Topographic Laser Ranging and Scanning, 2nd ed., Shan J, Toth CK, Eds., Second edition. Boca Raton : Taylor & Francis, CRC Press, 2018.: CRC Press. pp 1–28. https://doi.org/10.1201/9781315154381
20. Einizinab S et al (2023) Enabling technologies for remote and virtual inspection of building work. Autom Constr. 156. p 105096. https://doi.org/10.1016/j.autcon.2023.105096
21. Liu J, Li D, Feng L, Liu P, Wu W (2019) Towards automatic segmentation and recognition of multiple precast concrete elements in outdoor laser scan data. Remote Sens 11(11):1–27. https://doi.org/10.3390/rs11111383
22. Mohammadi M, Rashidi M, Mousavi V, Karami A, Yu Y, Samali B (2021) Quality evaluation of digital twins generated based on uav photogrammetry and tls: Bridge case study. Remote Sens 13(17):1–22. https://doi.org/10.3390/rs13173499
23. Mohammadi M, Rashidi M, Yu Y, Samali B (2023) Integration of TLS-derived bridge information modeling (BrIM) with a decision support system (DSS) for digital twinning and asset management of bridge infrastructures. Comput Ind. 147. https://doi.org/10.1016/J.COMPIND.2023.103881
24. CIDB, Malaysia building information modelling (BIM) Report 2021, 2022
25. Autodesk Inc., Autodesk ReCap | Registration Quality Report | Autodesk. Accessed: Jul. 30, 2025. Available: https://help.autodesk.com/view/RECAP/ENU/?guid=scan_register_quality
26. Autodesk, Autodesk Revit: BIM software to design and make anything. Accessed: May 10, 2025. Available: https://www.autodesk.com/asean/products/revit/overview
27. Jung J et al (2014) Productive modeling for development of as-built BIM of existing indoor structures. Autom Constr 42:68–77. https://doi.org/10.1016/j.autcon.2014.02.021
28. Bosché F, Guenet E (2014) Automating surface flatness control using terrestrial laser scanning and building information models. Autom Constr 44:212–226. https://doi.org/10.1016/j.autcon.2014.03.028
29. Wang Q, Cheng JCP, Sohn H (2015) Automated quality inspection of precast concrete elements with irregular shapes using terrestrial laser scanner and BIM technology. 32nd Int Symp Autom Robot Constr Min Connect to Futur Proc. https://doi.org/10.22260/isarc2015/0035
30. Kim MK, Wang Q, Park JW, Cheng JCP, Sohn H, Chang CC (2016) Automated dimensional quality assurance of full-scale precast concrete elements using laser scanning and BIM. Autom Constr 72:102–114. https://doi.org/10.1016/j.autcon.2016.08.035
31. Wang Q, Kim MK, Cheng JCP, Sohn H (2016) Automated quality assessment of precast concrete elements with geometry irregularities using terrestrial laser scanning. Autom Constr 68:170–182. https://doi.org/10.1016/j.autcon.2016.03.014
32. Wang Q, Cheng JCP, Sohn H (2016) Automatic reconstruction of as-built BIM from laser scanned data of precast concrete elements for dimensional quality assessment. ISARC 2016—33rd Int Symp Autom Robot Constr, no Isarc, pp 114–122. https://doi.org/10.22260/isarc2016/0015
33. Yang L, Cheng JCP, Wang Q (2020) Semi-automated generation of parametric BIM for steel structures based on terrestrial laser scanning data. Autom Constr. 112(2019), p. 103037. https://doi.org/10.1016/j.autcon.2019.103037
34. Aftab I, Kapitány K, Lovas T (2023) Automating scan to BIM operations using grasshopper. Period Polytech Civ Eng. 67(4):1187–1192. https://doi.org/10.3311/PPci.22390
35. Liu J, Cui N, Cheng G, Li D, Ma X, Liao Y (2023) Towards the automated virtual trial assembly of large and complex steel members using terrestrial laser scanning and BIM. Eng Struct. 291, p 116448. https://doi.org/10.1016/j.engstruct.2023.116448
36. Cheng G, Liu J, Cui N, Hu H, Xu C, Tang J (2023) Virtual trial assembly of large steel members with bolted connections based on point cloud data. Autom Constr. 151, p 104866, 2023, https://doi.org/10.1016/j.autcon.2023.104866
37. FARO Technologies Inc., Specifications FARO ® Laser Scanner Focus 3D X 130, The Powerful X Series Laser Scanner. 2016
38. Sanhudo L et al (2019) A framework for in-situ geometric data acquisition using laser scanning for BIM modelling. J Build Eng. 28(August):2020. https://doi.org/10.1016/j.jobe.2019.101073

39. FARO Technologies Inc., Laser Scanner Best Practices—FARO® Knowledge Base. Accessed: Sep. 29, 2025. Available: https://knowledge.faro.com/Hardware/Focus/Focus/Laser_Scanner_Best_Practices
40. Gallego Ripoll N (2012) Laser scanner: eSafety & ITS applications. in *Laser Scanner Technology*, J. A. Munoz Rodriguez, Ed., InTech. pp 51–70
41. Liu Z, Liu Z, Sun Z (2024) Massive point cloud processing for efficient construction quality inspection and control. Sensors, 24(21). https://doi.org/10.3390/s24216806
42. Yin C, Wang B, Gan VJL, Wang M, Cheng JCP (2021) Automated semantic segmentation of industrial point clouds using ResPointNet++. Autom Constr. 130, p 103874. https://doi.org/10.1016/j.autcon.2021.103874
43. Kuschnerus M, Schröder D, Lindenbergh R (2021) Environmental influences on the stability of a permanently installed laser scanner. Int Arch Photogramm Remote Sens Spat Inf Sci—ISPRS Arch. 43(B2–2021), pp 745–752. https://doi.org/10.5194/isprs-archives-XLIII-B2-2021-745-2021
44. Autodesk, Help | Insert a Point Cloud File. Accessed: Aug. 13, 2025. Available: https://help.autodesk.com/view/RVT/2024/ENU/?guid=GUID-B89AD692-C705-458F-A638-EE7DD83D694C

Application of the Computational Intelligence Models to Modify the Q Classification System Based on the Rock Mass Waves Velocity

Mohammad Rezaei and Pouya Koureh Davoodi

Abstract Accurate evaluation of rock mass quality plays a vital role in the design, stability analysis, and construction of underground excavations such as tunnels and caverns. The Q classification system is one of the most widely used empirical approaches in rock engineering. However, determining the Q value traditionally demands extensive field investigations and laboratory testing, which are often time-consuming, expensive, and subjective. This research aims to develop a fast, reliable, and non-destructive method to indirectly estimate the Q index by establishing its correlation with the compressional (V_p) and shear (V_s) wave velocities of rocks. To achieve this, three computational intelligence models including simulated annealing (SA), particle swarm optimization (PSO), and a hybrid PSO–SA algorithm were developed and trained using actual data obtained from the Beheshtabad water transfer tunnel in southwestern Iran, a site characterized by diverse geological formations. The predictive capability of each model was assessed using several statistical criteria, including R^2, RMSE, MAE, MAPE, A20, IOA, and IOS, alongside supplementary score and REC curve analyses. The modeling was performed in two scenarios: one using V_p and the other using V_s as the sole input variable. In both cases, the hybrid PSO–SA algorithm exhibited superior performance. When V_s was used as input, the model achieved an R^2 of 0.9629 and an RMSE of 0.6342, outperforming the V_p-based model ($R^2 = 0.9504$, RMSE $= 0.7232$). The results suggest that V_s serves as a more reliable indicator for estimating rock mass quality under the studied conditions. Overall, the proposed hybrid metaheuristic model offers a robust, cost-effective, and non-invasive framework for predicting the Q value, significantly reducing the dependence on traditional field-based measurements.

M. Rezaei (✉)
Department of Mining Engineering, Faculty of Engineering, University of Kurdistan, Sanandaj, Iran
e-mail: m.rezaei@uok.ac.ir

P. K. Davoodi
Department of Mining Engineering, Faculty of Engineering, Urmia University, Urmia, Iran
e-mail: p.kourehdavoodi@urmia.ac.ir

A. Azizi et al. (eds.), *Advanced Decision-Making Under Uncertainty*, Emerging Trends in Mechatronics, https://doi.org/10.1007/978-981-95-8696-7_3

Keywords Rock mass quality · Q-system · Seismic wave velocity · Metaheuristic algorithms · Hybrid PSO-SA model

1 Introduction

A comprehensive understanding of rock mass quality is one of the fundamental requirements in the design, construction, and stability evaluation of underground structures such as tunnels, mining openings, and subsurface storage facilities. Rock mass classification systems have long been recognized as practical tools for simplifying complex geological data into interpretable parameters that assist engineers in decision-making and support design. Among these systems, the Q-classification, introduced by Barton et al. [1], has become one of the most widely used empirical methods in rock engineering projects worldwide. The Q-system assesses the quality of a rock mass using six key parameters: rock quality designation (RQD), joint set number (J_n), joint roughness number (J_r), joint alteration number (J_a), joint water reduction factor (J_w), and stress reduction factor (SRF). These parameters are combined into a single dimensionless number (Q) that reflects the overall condition of the rock mass and its stability potential [1].

This classification system has been successfully utilized in a wide range of tunneling and underground construction projects, supporting both preliminary assessments and detailed design phases across various geological environments. Despite its proven practicality, determining the Q-value through conventional procedures demands extensive field surveys, geological mapping, core logging, and laboratory testing. These processes are inherently time-consuming, costly, and highly dependent on the subjective judgment of field engineers. Moreover, the natural heterogeneity of rock masses and the uncertainties associated with in-situ measurements often introduce significant variability into the final Q-value determination. These challenges have encouraged researchers to develop indirect, data-driven, and predictive methods capable of estimating the Q-value from parameters that are easier, faster, and more objective to obtain [1].

Among different physical and geophysical approaches, seismic wave velocity measurements i.e., compressional (V_p) and shear (V_s) wave velocities have attracted considerable attention as potential indicators for evaluating rock mass quality. These parameters are inherently sensitive to the intrinsic properties of rocks, including lithological composition, fracture intensity, porosity, moisture content, and in-situ stress conditions. Generally, rocks with higher V_p and V_s values correspond to dense, well-cemented, and mechanically strong formations, whereas lower velocities signify weaker, highly fractured, or weathered zones. Seismic techniques provide additional advantages, being non-destructive, rapid, and suitable for both surface and subsurface applications [2–5]. Hence, developing empirical or intelligent correlations between Q-values and seismic velocities offers an opportunity to reduce fieldwork intensity while maintaining a reliable and reproducible rock mass classification framework.

However, the relationship between seismic velocities and the Q index is nonlinear, multi-dimensional, and site-dependent, rendering conventional regression-based approaches inadequate for accurate prediction. To address this issue, recent years have witnessed a rapid expansion of computational intelligence (CI) techniques in rock mechanics and tunneling research. CI encompasses a diverse set of algorithms inspired by natural and biological systems, designed to process complex, uncertain, and nonlinear datasets. Examples include artificial neural networks, fuzzy logic, evolutionary computation, genetic algorithms, and hybrid metaheuristic optimization methods. Such techniques have been effectively applied in geotechnical fields to model phenomena such as rock strength estimation, slope stability, excavation support design, and permeability prediction [5–14]. Among these, two of the most recognized metaheuristic algorithms are simulated annealing (SA) and particle swarm optimization (PSO). The PSO algorithm, inspired by the collective behavior of birds flocking or fish schooling, optimizes potential solutions by sharing information among individuals in the population. It is widely valued for its simplicity, rapid convergence, and computational efficiency. In contrast, the SA model draws inspiration from the metallurgical process of the annealing, where a material is gradually cooled to reach a stable, low-energy configuration. Its probabilistic search strategy enables it to escape local minima, making it well-suited for global optimization tasks, including those encountered in rock and tunnel engineering [5, 15].

Although PSO and SA algorithms each exhibit distinct advantages, recent studies suggest that hybridization of the two algorithms can significantly enhance the optimization performance. By merging the exploratory strength of the PSO model with the refinement capability of SA, the hybrid PSO–SA framework achieves improved accuracy, convergence stability, and robustness in complex and uncertain environments. This synergy allows the hybrid method to balance global exploration and local exploitation, producing more consistent and precise results. Accordingly, the present research introduces a novel framework for indirect estimation of the Q-value by utilizing seismic parameters (V_p and V_s) within three computational intelligence models including PSO, SA, and their integrated hybrid PSO–SA variant. The performance and precision of these models were evaluated using real-world data collected from the Beheshtabad water transfer tunnel in southwestern Iran. This tunnel provides an excellent case study, as it traverses multiple geological formations with diverse structural characteristics and discontinuity patterns, making it ideal for testing the applicability of intelligent modeling and metaheuristic techniques [5, 15].

The proposed modeling framework comprises three main stages, summarized as follows:

1. Data compilation and preparation: Extraction and organization of Q, V_p, and V_s values from the Beheshtabad tunnel database, ensuring representative coverage of varying geological and mechanical conditions.
2. Model development: Construction of PSO, SA, and hybrid PSO–SA algorithms to explore the nonlinear correlations between seismic velocity parameters and Q-values. The models were trained and validated using real datasets obtained from field measurements.

3. Model evaluation and validation: The predictive accuracy of each model was quantitatively assessed using multiple statistical criteria, including the coefficient of determination (R^2), root mean square error (RMSE), mean absolute error (MAE), and mean absolute percentage error (MAPE). Additional indicators such as the index of agreement (IOA), index of simulation (IOS), and accuracy within 20% (A20) were employed to assess model reliability. Furthermore, scoring systems and Regression Error Characteristic (REC) curves were applied to provide a comprehensive evaluation of predictive performance.

This study outputs contribute to the advancement of rock mechanics and tunneling engineering issues by:

- Proposing a rapid, cost-effective, and non-invasive approach for estimating Q-values from seismic parameters, thereby reducing dependency on subjective field assessments.
- Comparing the performance and accuracy of three advanced computational intelligence models in predicting the V_p–V_s–Q relationship under varying geological conditions.
- Demonstrating the effectiveness and precision of hybrid metaheuristic algorithms in complex geotechnical modeling scenarios, advocating their broader use in the design and construction of underground structures.

2 Background and Literature Review

2.1 The Q-system: Definition and Applications

Originally proposed by Barton et al. [1], the Q-system has become one of the most widely adopted empirical approaches for rock mass classification in tunneling and underground excavation projects. This method evaluates rock mass quality (Q) using a semi-empirical formulation that integrates six fundamental parameters: rock quality designation (RQD), joint set number (J_n), joint roughness number (J_r), joint alteration number (J_a), joint water reduction factor (J_w), and stress reduction factor (SRF). The relationship among these parameters is expressed through the following general equation to calculate Q [1]:

$$Q = \frac{RQD}{J_n} \times \frac{J_r}{J_a} \times \frac{J_w}{SRF} \tag{1}$$

Each parameter in this formulation captures essential aspects of the geological, structural, and environmental characteristics influencing the rock mass. The resulting Q-value provides a quantitative measure of rock mass stability and is extensively used in the design of tunnel support systems, including the selection, spacing, and thickness of rock bolts, shotcrete layers, and reinforcement elements [15].

One of the notable advantages of the Q-system compared with other empirical classification techniques such as the Rock Mass Rating (RMR) and the Geological Strength Index (GSI) is its comprehensive incorporation of in situ stress conditions, groundwater influence, and joint properties, which play a decisive role in controlling rock behavior under varying environmental conditions. Nevertheless, the practical application of the Q-system is not without challenges. The accurate evaluation of parameters like SRF and J_w demands specialized expertise, detailed field investigations, and continuous data collection, all of which are time-consuming and costly. Furthermore, incomplete or unreliable information regarding joint orientation, groundwater condition, or in situ stress can substantially reduce the accuracy of the computed Q-values in the previous studies [16].

Given the above-mentioned limitations, there has been a growing research emphasis on the development of the data-driven and intelligent predictive models that can estimate Q-values indirectly using other influencing parameters that are easier to measure or available through the non-invasive methods.

2.2 *Application of Artificial Intelligence in Predicting the Q Index*

To overcome the inherent constraints of traditional empirical rock mass classification methods, recent studies have increasingly adopted artificial intelligence (AI) and machine learning (ML) algorithms to predict key rock mass parameters, including the Q index. These approaches are particularly appealing due to their ability to capture complex, nonlinear interdependencies among geotechnical and geomechanical factors, while offering strong generalization capabilities across diverse geological settings. For instance, Fatemi Aghda et al. [17] examined correlations between P-wave velocity, electrical resistivity, and rock mass quality indicators such as the Q index, and reported strong relationships suggesting that Q can be reliably estimated through these indirect parameters [17]. Similarly, Kianpour et al. [18] applied intelligent modeling to predict Q-values in limestone formations, demonstrating improved accuracy in classification, particularly in areas with limited accessibility [18]. In another study, Taban et al. [19] employed Gene Expression Programming (GEP) to develop simplified models for estimating Q-values based on a reduced set of significant variables, thereby minimizing the dependence on extensive field measurements during early design phases [19]. Although early research efforts primarily relied on traditional regression analysis and qualitative assessments, recent developments have introduced hybrid AI–metaheuristic approaches that combine the strengths of machine learning models with global optimization algorithms. For example, Santos et al. [20] and Ho et al. [21] developed robust predictive frameworks using Support Vector Machines (SVM) in conjunction with exploratory optimization algorithms, achieving superior accuracy compared with empirical equations, particularly in complex geological environments.

Despite these advancements, the integration of petrophysical and dynamic rock properties such as compressional (V_p) and shear (V_s) wave velocities into the AI-based predictive models remains relatively limited. This gap underscores the need for research that investigates the direct correlation between Q-values and seismic parameters, which are measurable through rapid, non-destructive geophysical methods and can provide valuable quantitative insights into rock mass conditions.

2.3 The Research Gap and Study Motivation

Although the Q-system has long been recognized as an effective and reliable tool for rock mass classification and tunnel design, its practical use is often constrained by the cost, time, and difficulty associated with collecting the required field data. A review of the existing literature reveals that very few comprehensive models directly relate the Q index to dynamic rock properties, particularly seismic parameters such as P-wave (V_p) and S-wave (V_s) velocities. These parameters, obtainable from seismic surveys and borehole geophysical logging, have yet to be fully explored as predictive inputs for Q estimation.

The present study seeks to address this research gap by applying computational intelligence techniques specifically, simulated annealing (SA), particle swarm optimization (PSO), and their hybrid PSO–SA model, to model the complex relationship between Q and rock mass wave velocities. This framework harnesses the predictive power of metaheuristic algorithms to provide a faster, more accessible, and objective method for estimating Q-values, particularly during early design stages or in areas where detailed geological information is scarce.

Table 1 summarizes several existing empirical and intelligent models that describe quantitative relationships among the three major rock mass classification systems including Q, RMR, and GSI. These comparative studies highlight the ongoing efforts to harmonize and convert between the systems, thereby improving consistency in design and engineering analysis.

However, as the correlations shown in Table 1 indicate, while numerous studies have investigated the interrelationships among Q, RMR, and GSI, far fewer have examined the direct connection between the Q index and the physical or dynamic characteristics of rock masses. This lack of research represents a critical limitation, considering that geophysical and petrophysical parameters are often much easier, faster, and more economical to acquire.

To bridge this gap, the present study aims to develop robust, efficient, and practical predictive models for the Q index using parameters that are cost-effective and readily measurable in most geotechnical investigations. By employing advanced metaheuristic optimization algorithms, namely PSO, SA, and their hybrid PSO–SA integration, this research proposes a powerful alternative framework for Q estimation that minimizes dependence on subjective field observations while maintaining scientific reliability and engineering relevance.

Table 1 Existing relationships between Q-index and other rock mass classification indices

Equation	R^2	Refs.
$Q = 0.08e^{0.049GSI}$	–	[22]
RMR = 9LnQ + 44	0.77	
RMR = 5.9LnQ + 43	0.81	[23]
RMR = 5LnQ + 60.8	–	[24]
RMR = 10.5LnQ + 41.8	0.66	[25]
RMR = 7.5lnQ + 42	–	[26]
RMR = 5.3lnQ + 50.81	–	[27]
RMR = 6.3InQ + 41.6	0.78	[28]
RMR = 12.11LnQ + 50.81	–	[29]
RMR = 9.19lnQ + 43.89	–	[30]
RMR = 9lnQ + 49	–	[31]
RMR = 7lnQ + 44	–	[32]
RMR = 15lnQ + 50	–	[33]
RMR = 7LnQ + 36	–	[34]
RMR = 5.97lnQ + 49.5	–	[35]
RMR = 3.7lnQ + 53.1	0.86	[36]
RMR + 6.4lnQ + 49.6	0.72	[37]
RMR = 2.8lnQ + 45.19	0.67	[38]
RMR = 6.3LnQ + 43	–	[39]
RMR = 15.5LnQ + 39.7	0.84	[40]
RMR = 6.6535LnQ + 35.533	0.49	[41]
RMR = 5.4LnQ + 55.2	0.55	
RMR = 5.7Ln + 43.65	0.714	
RMR = 5.37lnQ + 40.48	0.53	[42]
RMR = 8.15LnQ + 44.88	0.859	[43]
RMR = 9.19lnQ + 43.89	–	[44]
RMR = 8.09LnQ + 43.08	0.835	[45]
RMR = 12.11LnQ + 50.81	–	
RMR = 5.9LnQ + 43	0.81	
RMR = 4.52LnQ + 43.635	0.74	[46]
RMR = 5LnQ + 60.8	–	
RMR = 6.55Ln + 59.53	–	[47]
RMR = 10.5LnQ + 41.8	0.66	
RMR = 5.614LnQ + 49.395	0.82	[48]
RMR = 7.5lnQ + 42	–	
RMR = 8.8832LnQ + 43.26	0.90	[49]
RMR = 5.3lnQ + 50.81	–	
RMR = 12.334LnQ + 26.018	0.77	[50]

(continued)

Table 1 (continued)

Equation	R^2	Refs.
RMR = 6.3InQ + 41.6	0.78	
RMR = 6.63LnQ + 32.76	0.88	[51]
RMR = 9LnQ + 44	0.77	

3 Study Area Description

This section presents an overview of the geographical setting and geological characteristics of the study area, which forms the basis of the research. It also describes the methods used to evaluate the Q index and the compressional (V_p) and shear (V_s) wave velocities. Furthermore, it summarizes the statistical properties and range of the collected datasets, which provide the fundamental framework for the subsequent analyses and modeling processes undertaken in this study.

3.1 Location of the Case Study

The Beheshtabad water transfer tunnel represents a major component of Iran's national inter-basin water transfer project, designed to convey water from the Zagros Mountains in the west to the central plateau, where arid conditions and water scarcity are severe. The project's primary goal is to help meet the growing domestic, agricultural, and industrial water demands in central Iran by diverting part of the Beheshtabad river's flow to the Zayandeh-Rood basin. Geographically, the tunnel traverses the Chaharmahal and Bakhtiari and Isfahan provinces. The entrance portal lies at the downstream of the Beheshtabad River (32°02′ N latitude, 50°38′ E longitude), while the outlet is situated approximately 65 km northeast, emerging near the Cham Asman dam on the upper reaches of the Zayandeh-Rood river (32°22′ N, 51°12′ E) [5, 14, 52, 53].

Along its alignment, the tunnel cuts through a variety of lithological units, encompassing igneous, metamorphic, and sedimentary rocks. This geological diversity, coupled with the frequent presence of fault zones, fractures, and tectonic deformations, results in significant variability in the physical and mechanical properties of the encountered rock masses. Such heterogeneity poses considerable challenges for tunnel design and stability analysis. Consequently, a comprehensive geomechanical assessment, particularly involving P-wave (V_p) and S-wave (V_s) velocity measurements and Q-system classification, is essential for understanding the mechanical response of rock masses under excavation-induced stresses [5, 14, 52, 53].

The combination of structural complexity and lithological diversity along the Beheshtabad tunnel alignment necessitates careful consideration in both stability evaluation and support system design. Data collected from field surveys, core logging, and seismic testing have therefore played a central role in identifying the zones of weakness and determining suitable reinforcement strategies to ensure safe excavation in these complex geological conditions.

3.2 Geological and Structural Overview of the Beheshtabad Tunnel Path

From a geological standpoint, the Beheshtabad tunnel begins in the Folded Zagros Zone and terminates in the Sanandaj–Sirjan structural zone, both of which are part of Iran's tectonically active mountain belts. Extensive field observations and exploratory drilling data reveal that the tunnel intersects multiple rock formations, including limestones, marls, sandstones, metamorphic schists, and intrusive igneous rocks. These formations vary greatly in strength, weathering degree, and fracture density, all of which influence tunneling performance and excavation stability. The regional structure is dominated by several major fault systems and fold axes, which have created zones of weakened rock, high fracture density, and varying groundwater inflow. The tunnel, approximately 64 km in total length, is divided into eight distinct segments, each roughly 8 km long, to facilitate geotechnical evaluation and design. Each segment exhibits unique geological and structural features that have been identified through geological mapping, resistivity profiling, and magnetic surveys [5, 14, 52, 53].

The magnetic survey conducted in this studied area further validated the existence of multiple faults, beddings, and horsts. The Beheshtabad tunnel route is a particular geotechnical interest due to its complex structural configurations. Within the Jahangbin Mountain range (km 16–24 of the tunnel route), extensive fracture zones exceeding 500 m in length were identified through electrical resistivity measurements. These findings were further validated through multiple resistivity profiles obtained using different electrode arrays and supplemented by geomagnetic mapping. The results revealed several faults with clearly defined boundaries, corresponding closely to previously mapped fault traces. The interpreted resistivity data indicated the values generally below 100 Ω-meters in the southern portion of the investigated zone, suggesting the presence of highly fractured and possibly saturated rock masses. These lower resistivity values gradually transition into high-resistivity limestone formations belonging to the Jahangbin Range, situated beneath the Qarl Limestone unit [52, 53].

Magnetic surveys conducted along the same section revealed pronounced anomalies in total magnetic intensity, identifying several fault lines and horst–graben structures. Notably, an area of elevated magnetic intensity was detected between km 17 and 18, where surface anomalies indicate the presence of major fracture systems. This zone has been emphasized in previous geophysical studies due to its structural significance. Gradient measurements across the region also confirmed additional faults and horsts, marked as "G" on the corresponding geological map. Further electrical soundings performed along the southern part of the route, between km 40 and 48, confirmed similar findings. The pseudo-resistivity section derived from these measurements revealed several fault zones and significant resistivity contrasts, including one particularly prominent fault that correlates with structural patterns observed in the magnetic survey data. The combination of these results underscores the complex structural framework of the Beheshtabad tunnel path, which must be carefully considered in both excavation planning and support system design [52, 53].

3.3 Seismic Testing and Q-index Evaluation Methodology

A comprehensive field investigation program was conducted to evaluate the geomechanical and geophysical characteristics of the rock mass along the Beheshtabad Water Transmission Tunnel alignment. This section details the procedures used to determine the Q-classification index, as well as the methodology applied to measure compressional (V_p) and shear (V_s) wave velocities using the downhole seismic testing technique. The collected data provided the foundation for developing empirical correlations between rock mass quality and seismic wave propagation properties.

The Q-classification system serves as a semi-empirical indicator of rock mass quality, integrating multiple geological, structural, and environmental factors to assess the overall stability of the rock mass. In this study, the Q-index was calculated using the previously introduced Eq. (1), which incorporates six key parameters: joint orientation, joint spacing, surface condition, alteration degree, groundwater condition, and stress reduction factor. All parameters were derived from geological logging of boreholes, visual inspection of discontinuities, and structural mapping performed along the tunnel alignment.

A total of 150 independent data points were compiled to determine Q-values. These were obtained from core recovery records and joint characterization data collected at multiple drilling sites distributed evenly along the tunnel corridor. This dataset provides a representative depiction of the lithological and structural variability across the entire alignment and serves as a reliable basis for subsequent analyses linking the Q-index with seismic parameters.

To determine seismic velocities, the downhole seismic method was employed. In this approach, a seismic source is positioned at the surface while a multi-component geophone is installed within a vertical borehole at successive depths. When seismic waves are generated at the surface, their arrival times at the receiver are recorded at each depth interval. The wave velocity for both compressional (V_p) and shear (V_s) components is then computed by dividing the vertical distance between sensor levels by the corresponding travel time difference. This method provides accurate and reliable velocity profiles, reflecting the in situ elastic behavior of the rock mass under natural stress and moisture conditions.

A total of 150 measurements were obtained for both V_p and V_s parameters at different borehole locations throughout the tunnel zone, encompassing a broad spectrum of geological formations characteristic of the Beheshtabad region. The recorded data revealed substantial variation in wave velocities, reflecting the influence of lithological diversity, fracture density, degree of alteration, and moisture content, factors that are also key components of the Q-classification system.

Altogether, 150 corresponding data triplets (Q, V_p, and V_s) formed the core dataset used for modeling. Based on these measurements, mathematical models were constructed using metaheuristic optimization algorithms to explore the quantitative relationships between Q–V_p and Q–V_s pairs. The models were designed to capture the nonlinear response of rock masses according to their seismic characteristics. The predictive relationships obtained can provide a practical means of the estimating rock mass quality using the solely geophysical inputs, thereby offering a fast, cost-effective, and objective approach for the preliminary assessment and design of underground structures. This approach reduces the uncertainty associated with traditional field-based evaluations and improves the reliability of geotechnical predictions.

3.4 *Statistical Characteristics of the Utilized Data*

A descriptive statistical analysis was performed on the dataset comprising the Q index and seismic parameters (V_p and V_s) to obtain an initial understanding of the behavior, range, and variability of the main variables employed in this research. Four statistical measures, minimum value, maximum value, arithmetic mean, and standard deviation were calculated for each parameter. These metrics provide insights into the dispersion and central tendency of the data and form the basis for subsequent modeling and correlation analyses.

The Q-index values, representing the overall quality of the rock mass, ranged from 0.01 to 12.5, with an average of 3.06 and a standard deviation of 2.87. This wide range and relatively high standard deviation highlight the heterogeneous nature of the rock mass along the Beheshtabad tunnel route, which is consistent with the complex geological setting of the region.

The compressional wave velocity (V_p), measured through downhole seismic testing, varied between 1738 and 4655 m/s, with an average velocity of 3225 m/s and a standard deviation of 829.9 m/s. Such variations in V_p can be attributed to differences in lithology, degree of fracturing, and moisture content, which collectively influence the elastic behavior and density of subsurface materials.

Likewise, the shear wave velocity (V_s) ranged from 1212 to 2195 m/s, with an average of 1710 m/s and a standard deviation of 280 m/s. Compared with V_p, the V_s data exhibit a narrower range, suggesting a relatively consistent shear response of the rock mass under low-strain dynamic conditions.

A bar chart in Fig. 1 visually represents the distribution of the statistical indicators for Q, V_p, and V_s. The figure juxtaposes the minimum, maximum, mean, and standard deviation values for each parameter, offering a clear comparative overview of their variability. The visual comparison illustrates the extent of variation and the degree of data dispersion among the three variables, thereby providing valuable insight into their roles in rock mass characterization and seismic response assessment.

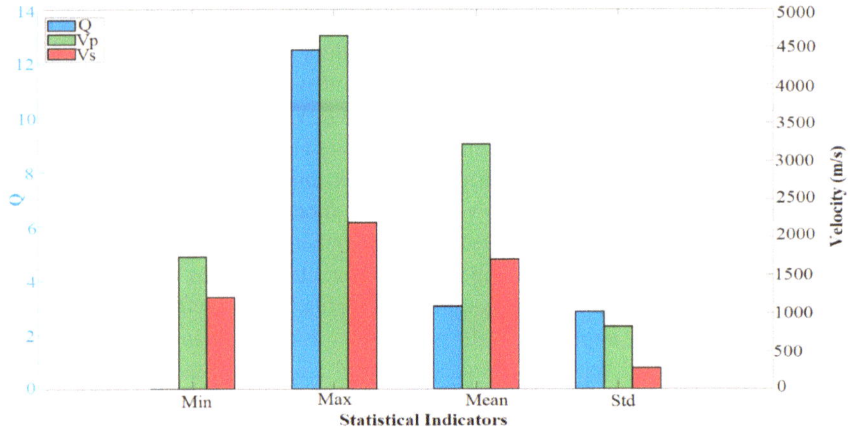

Fig. 1 Statistical comparison of Q, V_p, and V_s

Overall, the statistical characteristics confirm the presence of significant geomechanical and geophysical variability within the study area. This variability reinforces the need for employing advanced modeling techniques, such as metaheuristic algorithms, to accurately capture and interpret the complex, nonlinear relationships between rock mass quality and seismic parameters. These results form a solid foundation for the subsequent correlation modeling and validation processes conducted in the following sections of this study.

4 Methodology

This section presents the methodological framework adopted to develop and evaluate predictive models for estimating rock mass quality (Q) based on geophysical parameters. The proposed approach employs three metaheuristic algorithms to establish robust mathematical relationships between the input variables (V_p and V_s) and the output variable (Q). The primary objective is to formulate reliable and adaptable models capable of accurately describing the nonlinear correlations between seismic wave velocities and the rock mass classification index.

A detailed explanation of each algorithm used in the modeling process is provided, including their conceptual basis, search mechanisms, and operational procedures. Subsequently, the key hyperparameters of each algorithm are defined and optimized to ensure superior performance. The processed dataset is then introduced into each algorithm to determine the most suitable mathematical expressions linking the input and output variables. In the final phase, the predictive performance of the developed models is rigorously assessed and compared using several statistical performance indicators, including the coefficient of determination (R^2), mean absolute percentage error (MAPE), root mean square error (RMSE), mean absolute relative error (MARE), and mean square error (MSE). These indices collectively evaluate both accuracy and reliability of the models. The corresponding formulas for these metrics are given in Eqs. (2)–(6).

$$R^2 = 1 - \frac{\sum_{i=1}^{n} (y_i - \hat{y}_i)^2}{\sum_{i=1}^{n} (y_i - \overline{y}_i)^2} \tag{2}$$

$$MAPE = \frac{100\%}{n} - \sum_{i=1}^{n} \left| \frac{y_i - \hat{y}_i}{y_i} \right| \tag{3}$$

$$RMSE = \sqrt{\frac{1}{n} \sum_{i=1}^{n} (y_i - \hat{y}_i)^2} \tag{4}$$

$$MARE = \frac{1}{n} \sum_{i=1}^{n} \frac{|y_i - \hat{y}_i|}{|y_i|} \tag{5}$$

$$MSE = \frac{1}{n} \sum_{i=1}^{n} (y_i - \hat{y}_i)^2 \tag{6}$$

In the formulas provided, y_i corresponds to the ith observed value, $\hat{y}_i$ to the ith predicted value, n indicates the total number of observations, $\bar{y}$ is the mean of the observed data [5, 14].

4.1 Description of the Applied Algorithms

In this study, three metaheuristic optimization algorithms were utilized i.e., particle swarm optimization (PSO), simulated annealing (SA), and a hybrid PSO–SA algorithm to derive interpretable mathematical relationships between the input variables (V_p and V_s) and the output variable (Q). These algorithms were selected due to their proven ability to handle nonlinear optimization problems and their established success in solving geotechnical and geophysical modeling challenges where conventional regression methods often fail to capture complex interdependencies.

The PSO algorithm draws inspiration from the collective motion observed in flocks of birds or schools of fish. In this optimization process, each potential solution is represented as a particle that moves within a multi-dimensional search space. The particles continuously adjust their positions by considering both their own best-known positions (personal experience) and the best positions found by neighboring particles (social experience). Through this cooperative mechanism, the swarm converges toward an optimal or near-optimal solution.

In the context of this study, each particle represents a mathematical function that models the relationship between V_p or V_s and the Q index. The objective function of the PSO algorithm is to minimize the discrepancy between the observed and predicted Q values. Over successive iterations, the algorithm updates particle velocities and positions according to the global and local best solutions, thereby refining the predictive equation toward higher accuracy. The operational structure of the PSO algorithm, including initialization, fitness evaluation, velocity update, and position adjustment, is illustrated in Fig. 2. This flow demonstrates how the swarm dynamically explores and exploits the search space to identify the most accurate and stable correlation between seismic velocities and rock mass quality parameters.

The concept of simulated annealing (SA) is derived from the physical process of thermal annealing, in which a heated material is slowly cooled to achieve a state of minimal internal energy. Analogously, in computational optimization, SA begins with an initial candidate solution and iteratively explores the surrounding solution space through random perturbations. At each iteration, the algorithm may accept not only better solutions but occasionally worse solutions as well based on a probabilistic acceptance function that decreases over time, known as the cooling schedule. This probabilistic acceptance mechanism prevents the search process from becoming trapped in local optima, thereby improving the likelihood of locating the global optimum. As the temperature parameter decreases gradually, the probability of accepting inferior solutions diminishes, allowing the system to converge toward the most stable and optimal solution.

In the present study, the SA algorithm was employed to refine and enhance candidate mathematical equations obtained during the modeling process. By iteratively adjusting parameters and accepting controlled variations, SA effectively improved the accuracy and generalization capability of the predictive models that estimate Q-values based on seismic wave velocity data. The operational logic and workflow of the SA algorithm, comprising initialization, random perturbation, acceptance testing, and cooling are schematically presented in Fig. 3.

A hybrid PSO–SA algorithm was developed in this research to integrate the global exploration strength of particle swarm optimization (PSO) with the local refinement precision of simulated annealing (SA). This hybridization leverages the complementary advantages of both algorithms: PSO efficiently explores the broader search space to identify promising regions, while SA performs fine-tuning in these regions to enhance model precision and minimize residual error. In the first phase of the hybrid process, PSO conducts an extensive global search, dynamically guiding particles toward optimal regions within the multidimensional solution space. Once a set of promising candidate solutions is identified, the second phase activates the SA mechanism, which iteratively refines these solutions through probabilistic local adjustments. This sequential coupling improves both the convergence rate and the stability of the resulting models, making the hybrid framework particularly effective in nonlinear and high-dimensional optimization problems common in geotechnical modeling. The overall computational flow of the PSO–SA hybrid algorithm, including initialization, global exploration, local refinement, and convergence, is illustrated in Fig. 4. This dual-stage strategy enables the model to achieve

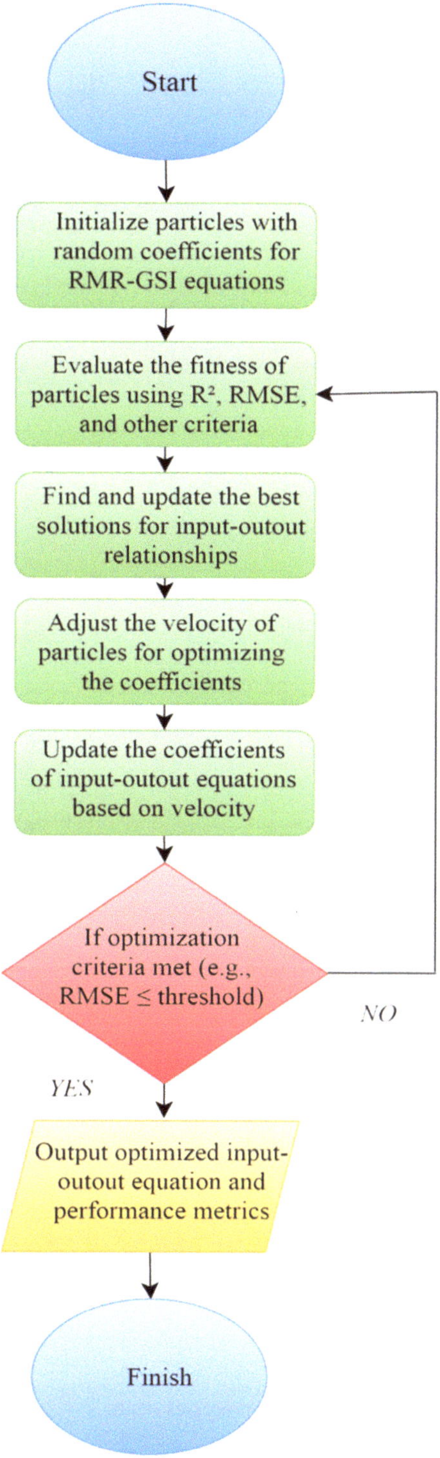

Fig. 2 PSO algorithm flowchart

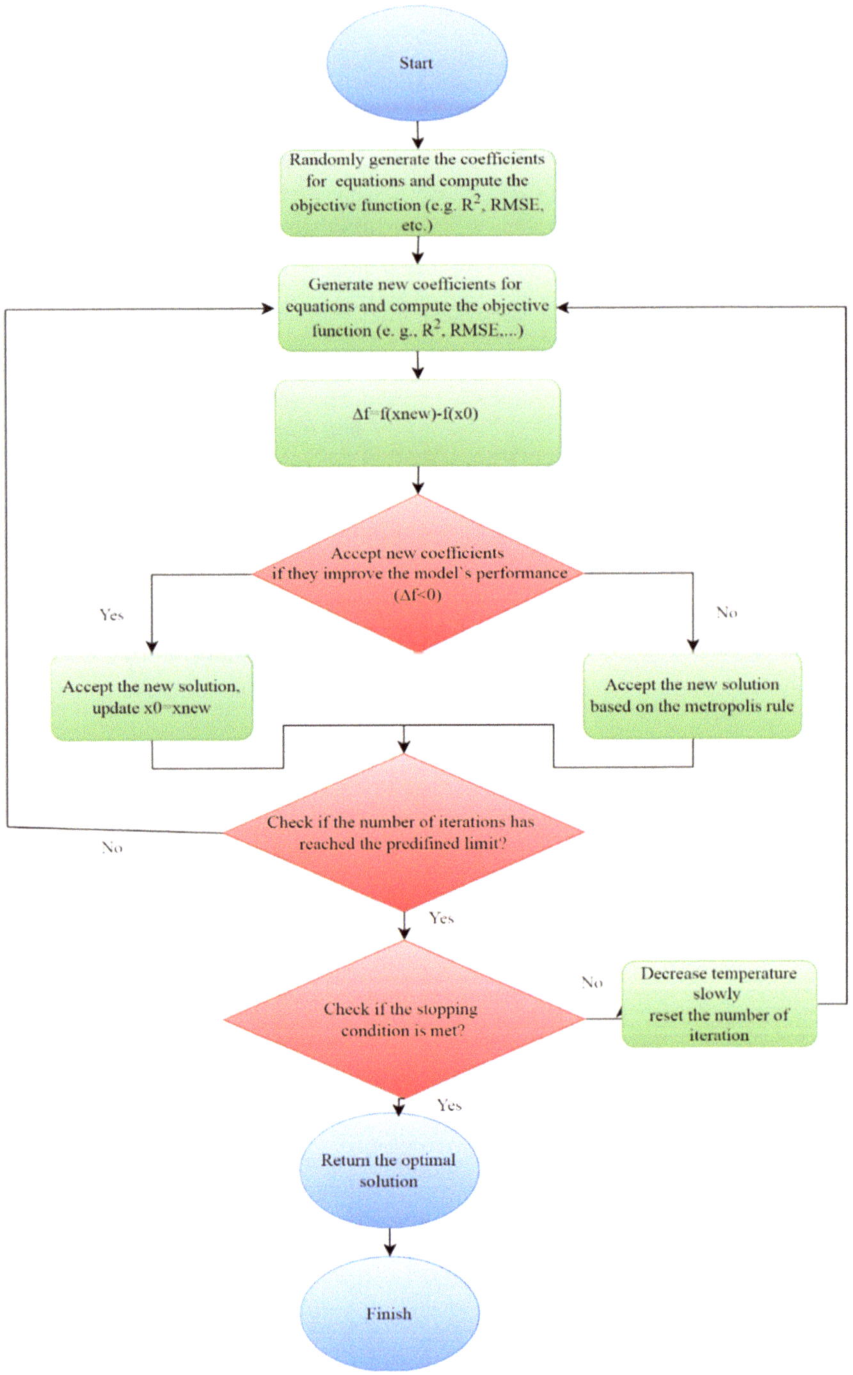

Fig. 3 SA algorithm processes stages

an optimal balance between exploration and exploitation, ensuring high reliability and adaptability across diverse geological conditions.

The predictive models developed through the three metaheuristic approaches i.e., PSO, SA, and PSO–SA yielded explicit mathematical expressions that describe the relationship between the seismic velocity inputs (V_p and V_s) and the Q index. These models were evaluated and compared using five statistical performance metrics: the correlation coefficient (R), Root Mean Square Error (RMSE), Mean Absolute Relative Error (MARE), Mean Absolute Percentage Error (MAPE), and Mean Square Error (MSE). Each metric provides unique insights into model performance:

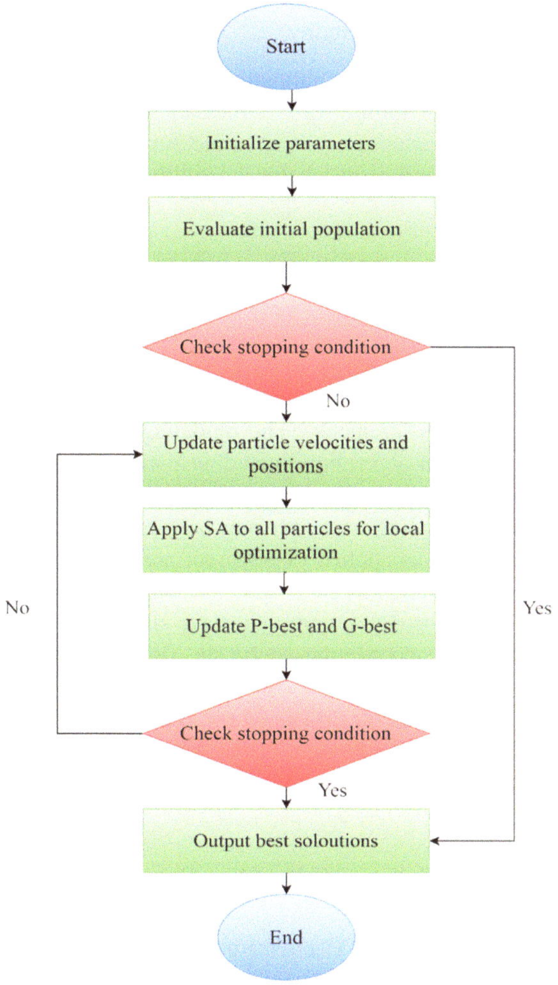

Fig. 4 Hybrid PSO-SA algorithm flowchart

- The correlation coefficient (R^2) quantifies how well the predicted Q-values align with the observed values, reflecting the proportion of variance explained by the model.
- The RMSE measures the overall magnitude of prediction errors in the same units as the response variable, providing an intuitive sense of model deviation, though it is somewhat sensitive to outliers.
- The MARE and MAPE indicators measure relative and percentage errors, respectively, allowing for meaningful model comparisons across datasets of different scales.
- The MSE, which emphasizes squared deviations, helps identify and penalize large prediction errors, making it particularly useful for detecting outliers and evaluating overall model stability.

By employing a combination of these quantitative validation metrics, the study ensured that the derived mathematical relationships between Q and the seismic parameters were not only accurate but also robust and generalizable under varying geological and geomechanical conditions.

4.2 Modeling of Q

This section focuses on the development of mathematical models designed to estimate the rock mass quality index (Q) based on compressional (V_p) and shear (V_s) wave velocity measurements. The modeling framework employed three metaheuristic optimization algorithms including particle swarm optimization (PSO), simulated annealing (SA), and their hybrid combination (PSO–SA) to capture and represent the complex nonlinear relationships between the seismic velocities and Q-values. Each algorithm was applied to construct predictive equations linking the input parameters (V_p and V_s) to the output variable (Q). Model performance was subsequently evaluated through a comprehensive statistical analysis using five well-established metrics: the correlation coefficient (R), Root Mean Square Error (RMSE), Mean Absolute Relative Error (MARE), Mean Absolute Percentage Error (MAPE), and Mean Square Error (MSE). These evaluation criteria enabled a rigorous comparison of the three algorithms, allowing the identification of the model that most accurately predicts Q-values from seismic wave velocity data.

4.2.1 Modeling of Q Based on V_p

In this stage, compressional wave velocity (V_p) served as the sole input variable for predicting the rock mass quality index (Q) using the three metaheuristic algorithms including PSO, SA, and PSO–SA. Each algorithm generated an explicit mathematical expression describing the relationship between V_p and Q, and the corresponding results are graphically illustrated in Figs. 5, 6 and 7, where Fig. 5 presents the PSO model output, Fig. 6 shows the SA-based model, and Fig. 7 depicts the integrated PSO–SA model results. The extracted mathematical Q-V_p relations from the PSO, SA, and PSO-SA models are presented in Eqs. (7) to (9), respectively.

$$Q = 0.0000001733V_p^2 + 0.0031618311V_p - 5.7068 \quad (7)$$

$$Q = 0.0000001096V_p^2 + 0.0035990985V_p - 6.3755 \quad (8)$$

$$Q = 0.0000000538V_p^2 + 0.0039472465V_p - 6.8590 \quad (9)$$

To assess and compare the predictive capability of each model and their extracted equations, the five aforementioned statistical indices i.e., R^2, RMSE, MARE, MAPE, and MSE were computed. Together, these metrics provide a comprehensive understanding of both the accuracy and the generalization performance of the developed models.

The PSO model achieved an R^2 value of 0.9142, signifying a strong correlation between V_p and Q. The RMSE and MSE were recorded as 0.9603 and 0.9222, respectively, representing the average magnitude of squared prediction errors. The MARE and MAPE values were 0.1421 and 24.82%, respectively, indicating moderate relative and percentage deviations between the observed and predicted Q-values. Overall, these results confirm that the PSO algorithm effectively captured the general relationship between V_p and Q, though there remains some room for improvement in reducing prediction error and enhancing model precision.

The model developed using the simulated annealing (SA) algorithm exhibited superior performance compared with the PSO model across all evaluation indices.

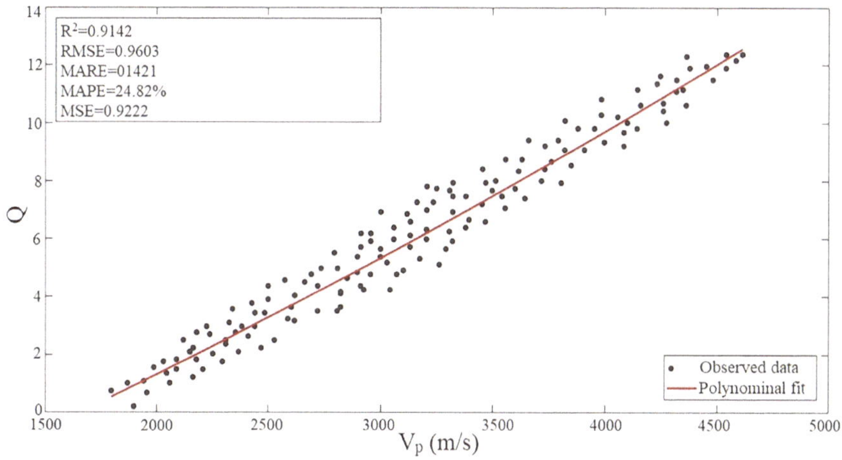

Fig. 5 Correlation between the P-wave velocity (V_P) and the Q-value based on the PSO metaheuristics algorithm

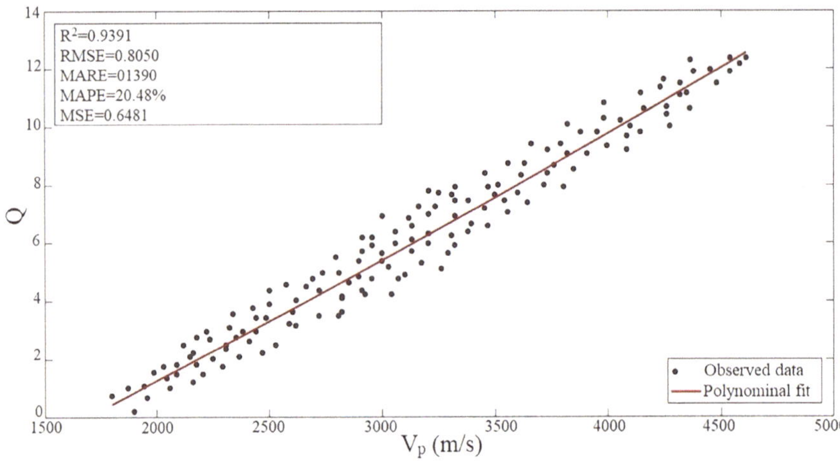

Fig. 6 Correlation between the P-wave velocity (V_p) and the Q-value based on the SA metaheuristics algorithm

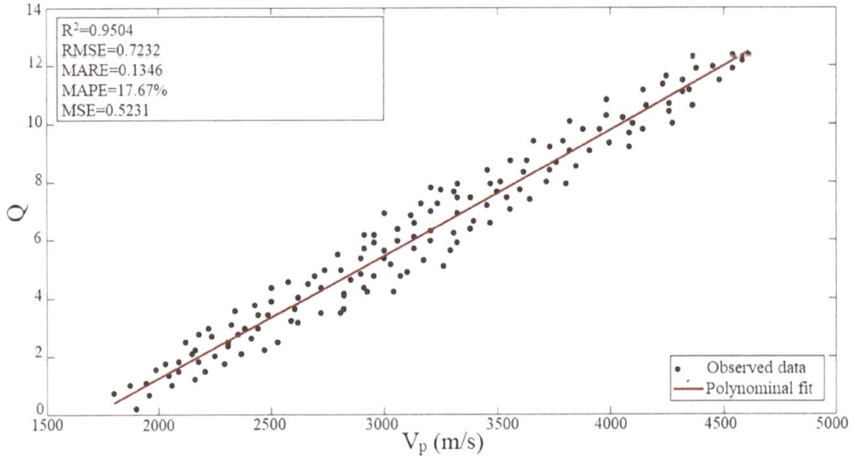

Fig. 7 Correlation between the P-wave velocity (V_p) and the Q-value based on the hybrid PSO-SA metaheuristics algorithm

The R^2 value increased to 0.9391, reflecting stronger explanatory power and a more accurate fit to the observed data. Correspondingly, RMSE and MSE values decreased to 0.8050 and 0.6481, demonstrating improved prediction accuracy. Furthermore, MARE and MAPE were reduced to 0.1390 and 20.48%, respectively, signifying a notable reduction in both relative and percentage error levels. These findings highlight the capability of SA to achieve higher precision and stronger predictive consistency than PSO when modeling Q as a function of V_p.

The hybrid PSO–SA algorithm, which combines the global exploration capacity of PSO with the local refinement capability of SA, produced the best overall performance among the three approaches. The R^2 value reached 0.9504, the highest recorded in this comparison, indicating an exceptionally strong agreement between predicted and observed Q-values. The RMSE and MSE decreased further to 0.7232 and 0.5231, respectively, while MARE and MAPE dropped to 0.1346 and 17.67%. These metrics collectively demonstrate that the hybrid PSO–SA framework achieved the most accurate and stable predictions, outperforming both individual algorithms.

The results clearly emphasize the advantage of hybridization in metaheuristic modeling. By merging the exploratory power of PSO with the refinement ability of SA, the hybrid model effectively overcame the limitations of each standalone algorithm, delivering enhanced robustness, faster convergence, and higher predictive accuracy. In summary, while all three models demonstrated satisfactory predictive performance, the hybrid PSO–SA algorithm consistently outperformed the individual PSO and SA methods across all evaluation indices, validating its capability to capture the nonlinear and complex relationship between V_p and Q with high reliability and generalization strength.

4.2.2 Modeling of Q Based on V_s

In this section, the shear wave velocity (V_s) was employed as the input variable for developing predictive models of the rock mass quality index (Q). Similar to the analysis performed using compressional wave velocity (V_p), three separate metaheuristic algorithms i.e., particle swarm optimization (PSO), simulated annealing (SA), and the hybrid PSO–SA approach were independently implemented to establish quantitative relationships between V_s and Q. Based on this modeling, three mathematical Q-V_s relations were developed from PSO, SA, and PSO-SA models, as given in Eqs. (10) to (12), respectively. The developed models and their resulting equations were evaluated using the same five statistical performance indicators comprising R^2, RMSE, MARE, MAPE, and MSE to maintain consistency and comparability across all analyses (Figs. 8, 9 and 10).

$$Q = -0.0000050063V_s^2 + 0.03306767152V_s - 30.8496 \quad (10)$$

$$Q = -0.0000043197V_s^2 + 0.0286349451V_s - 29.3508 \quad (11)$$

$$Q = 0.0000016820V_s^2 + 0.0089185228V_s - 13.4046 \quad (12)$$

The PSO model developed using V_s as the input variable achieved an R^2 value of 0.8755, signifying a reasonably strong but weaker correlation compared to the V_p-based PSO model ($R^2 = 0.9142$). The RMSE and MSE values were 1.1571 and 1.3389, respectively, both notably higher than those obtained in the V_p-based case (0.9603 and 0.9222). Moreover, the MARE value reached 0.2959, while the

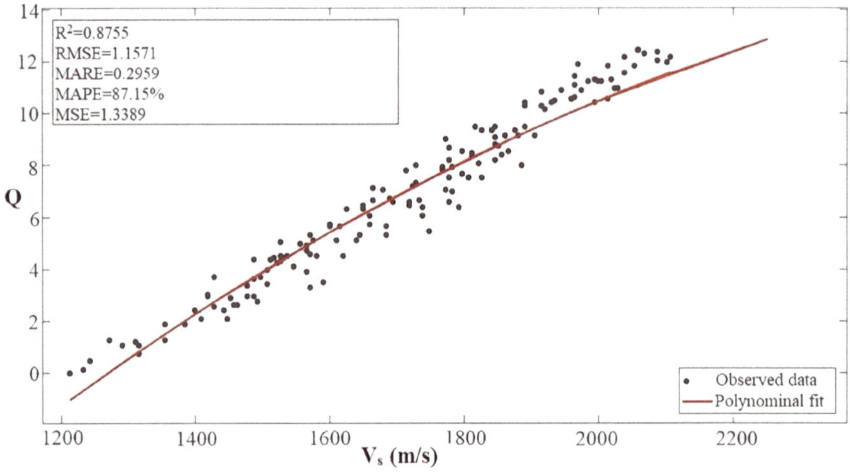

Fig. 8 Relationship between the S-wave velocity (V_S) and Q based on the PSO metaheuristics algorithm

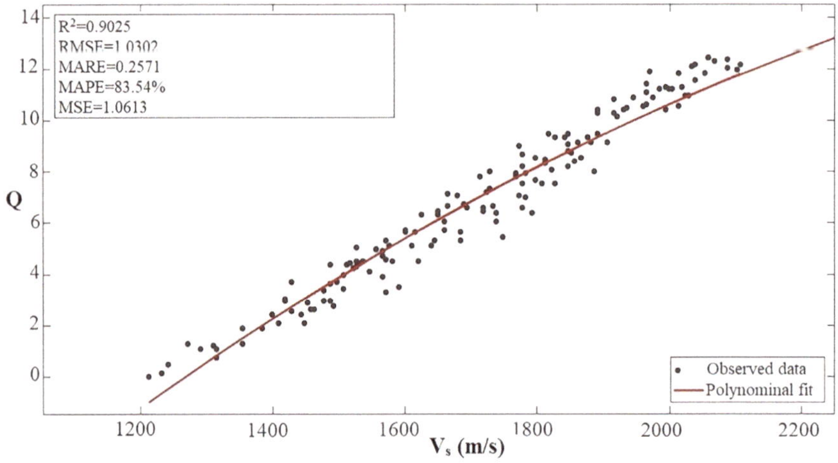

Fig. 9 Relationship between the S-wave velocity (V_S) and Q based on the SA metaheuristics algorithm

MAPE increased substantially to 87.15%, indicating a relatively poor fit and higher predictive uncertainty. These results suggest that the PSO algorithm, while capable of identifying general trends, struggled to accurately represent the relationship between V_s and Q. The diminished performance highlights the sensitivity of PSO to input data characteristics and indicates that V_s alone may be a less stable predictor of Q when modeled through this particular optimization technique.

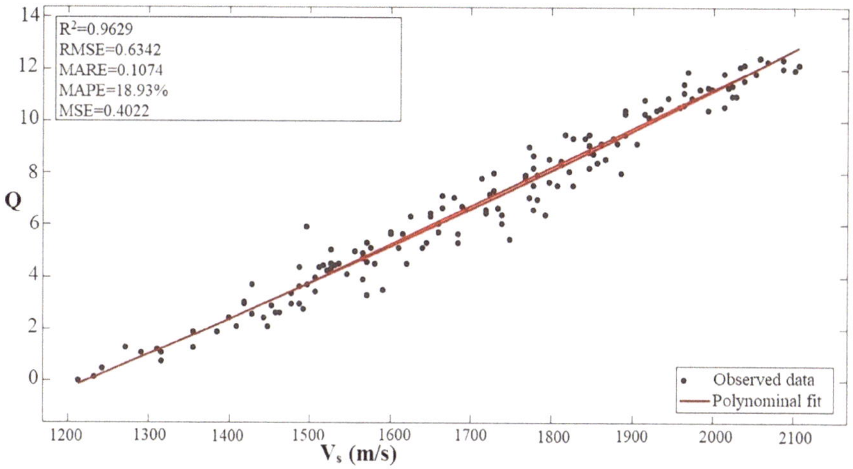

Fig. 10 Relationship between the S-wave velocity (V_S) and Q based on the hybrid PSO-SA metaheuristics algorithm

Consistent with the findings from the V_P-based modeling, the simulated annealing (SA) approach exhibited improved predictive accuracy over the PSO algorithm. The R^2 value increased to 0.9025, demonstrating a stronger correlation between V_s and Q. Likewise, RMSE and MSE values decreased to 1.0302 and 1.0613, respectively, while the MARE and MAPE values dropped to 0.2571 and 83.54%. Despite these improvements, the MAPE value remained relatively high, indicating continued challenges in achieving precise predictions using V_s as the sole input. These findings confirm that while the SA algorithm enhances performance compared to PSO, it still exhibits limitations in accurately mapping the nonlinear V_S–Q relationship.

The hybrid PSO–SA algorithm once again demonstrated superior performance, substantially outperforming both individual algorithms. The model achieved an R^2 value of 0.9629, the highest among all V_s-based models and even exceeding the V_P-based hybrid model ($R^2 = 0.9504$). The RMSE and MSE were further reduced to 0.6342 and 0.4022, respectively, indicating markedly improved precision. Similarly, MARE and MAPE values decreased significantly to 0.1074 and 18.93%, reflecting a substantial reduction in both relative and percentage prediction errors. These results confirm that the hybrid framework successfully integrates the exploratory capability of PSO with the local refinement strength of SA, producing highly accurate and consistent predictions of Q-values from V_s data.

When comparing the V_S-based results to those obtained from the V_P-based models in Sect. 4.2.1, a clear trend emerges. The PSO and SA models demonstrated reduced performance with V_s input, as evidenced by higher MAPE and MSE values, implying that V_p serves as a more stable and reliable predictor of Q in standalone optimization frameworks. However, the hybrid PSO–SA model not only maintained its superior performance with V_s datasets but actually surpassed the V_P-based hybrid model across several statistical indicators. This finding underscores the robustness and adaptability of the PSO–SA hybrid approach, which effectively compensates for the inherent variability in the V_s dataset.

The hybrid model's ability to extract meaningful patterns from both compressional and shear wave velocities illustrates its versatility and confirms its value as a comprehensive predictive tool in the complex field of rock mass quality estimation. The results emphasize that while V_s alone may present modeling challenges for individual algorithms, the hybrid metaheuristic framework is capable of uncovering the underlying nonlinear relationships with high accuracy, minimal error, and enhanced generalization potential.

5 Evaluation and Validation of the Derived Models

In this section, the mathematical formulations generated by the applied metaheuristic algorithms are thoroughly evaluated and validated using a variety of analytical techniques and statistical indicators. The main objective is to determine the reliability, accuracy, and general applicability of each model in predicting the rock mass quality index (Q) based on the given wave velocity data.

The first step in this evaluation involves comparing the models through Relative Error Curve (REC) analysis. This method provides a clear visual insight into how each model performs under different error thresholds. By plotting the cumulative percentage of predictions that fall within progressively increasing error limits, the REC curve facilitates a direct assessment of model robustness. A curve that rises more steeply and reaches higher values indicates superior performance, as it reflects a larger proportion of predictions falling within smaller error margins.

Following the REC analysis, a scoring assessment was carried out to quantitatively compare the three models using five core statistical parameters: R^2, RMSE, MARE, MAPE, and MSE. Each model was assigned a score from one to three for each metric. For R^2, which reflects the goodness of fit, the model with the highest value received a score of three, the second-best two, and the lowest one. Conversely, for the four error-based metrics (RMSE, MARE, MAPE, and MSE), the model achieving the lowest error was awarded a score of three, the next lowest received two, and the highest error was given one point.

The scores across all five metrics were then summed to produce a composite score for each model. This total score served as the basis for ranking the algorithms, with the highest-scoring model identified as the most accurate and reliable for predicting the Q index. This structured scoring system ensured a fair and consistent evaluation, providing a clear overview of the relative performance of each modeling approach.

Ultimately, the models are evaluated with three sophisticated statistical metrics: the agreement index IOA, the objective skill score IOS, and the A20 index. The IOA measures the degree of agreement between observed and anticipated values, with numbers nearing one signifying almost complete alignment. The IOS offers a comprehensive skill score considering both precision and dependability, aiding in differentiating models that might excel in typical scenarios yet underperform in extreme situations. The A20 index evaluates the share of forecasts with an absolute relative error that stays within twenty percent of the actual value, acting as a useful standard for acceptable prediction quality in engineering contexts.

By combining these methods, a thorough and multidimensional evaluation of the developed models is achieved. The results from this analysis provide a solid basis for determining which modeling approach delivers the most reliable, accurate, and broadly applicable predictive equations for estimating the Q index from geophysical input data.

5.1 Relative Error Curve (REC) Analysis

In this section, the predictive capability of the developed models was assessed through Relative Error Characteristic (REC) analysis, a method that provides both graphical and quantitative insights into model performance. Unlike single-value statistical measures such as RMSE or MAPE, which capture only the average error magnitude, the REC curve reflects the distribution of prediction errors across different thresholds. This makes it a powerful and complementary tool for evaluating how consistently each model performs throughout the entire dataset. The REC curve illustrates the cumulative percentage of predictions that fall within various levels of relative error, thereby offering a holistic view of model accuracy and stability. A steeper and more elevated REC curve signifies a model that achieves a higher proportion of accurate predictions within tighter error margins, thus indicating superior predictive reliability.

Figure 11 presents the REC comparison for the three metaheuristic algorithms comprising particle swarm optimization (PSO), simulated annealing (SA), and the hybrid PSO–SA when the compressional wave velocity (V_p) was used as the input variable. The Area Under the Curve (AUC), a quantitative measure of the model's overall reliability, was computed for each case. The PSO model achieved an AUC value of 2.776, while the SA model slightly surpassed it with an AUC of 2.889. The hybrid PSO–SA model, however, demonstrated the most robust performance, attaining an AUC of 3.210. This higher AUC value reflects the hybrid algorithm's greater capacity to yield accurate Q-index predictions, as a larger fraction of its estimations fell within smaller relative error limits.

Similarly, Fig. 12 displays the REC curves for the models when shear wave velocity (V_s) was used as the input parameter. In this case, the AUC values were determined as 2.4928 for PSO, 2.7870 for SA, and a notably higher 3.283 for the hybrid PSO–SA model. The superior performance of the hybrid approach in this scenario underscores its effectiveness in capturing the more intricate and nonlinear relationships that typically characterize the V_s–Q interaction. The enhanced AUC achieved by the PSO–SA algorithm indicates that its combination of global search ability (from PSO) and local refinement capacity (from SA) allows for a more comprehensive exploration of the solution space, leading to improved prediction precision and robustness.

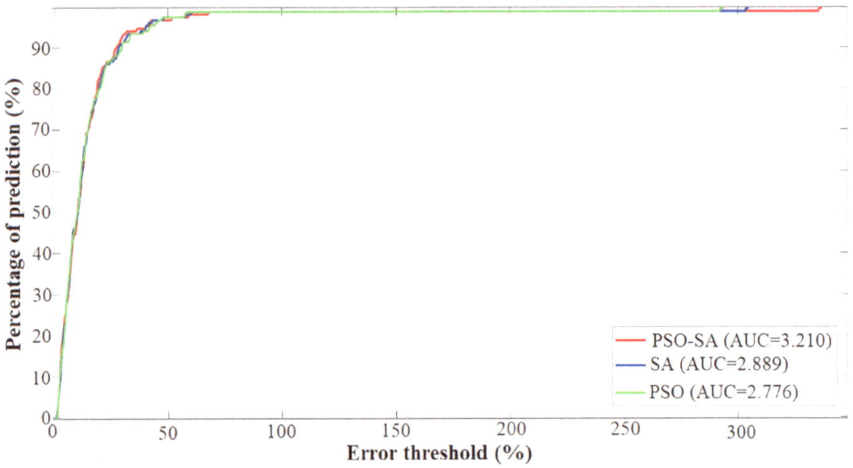

Fig. 11 REC curves of PSO, SA, and PSO-SA models based on V_p input

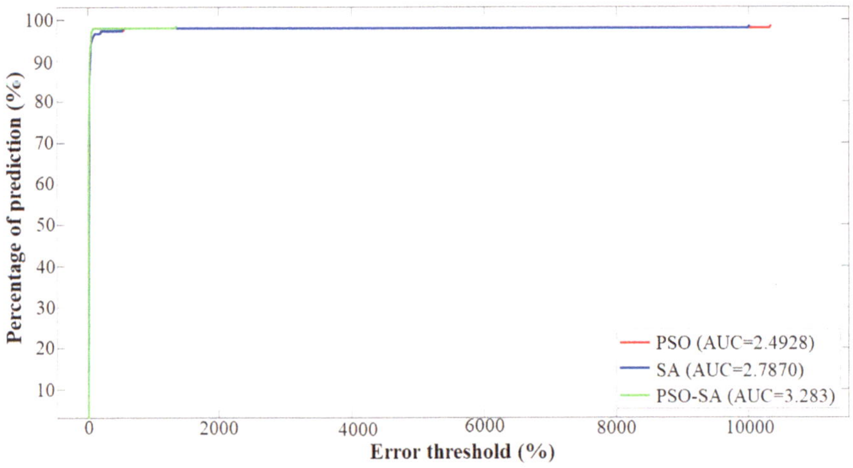

Fig. 12 REC curves of PSO, SA, and PSO-SA models based on V_s input

When comparing the models across both input scenarios, a clear trend emerged: the standalone PSO and SA models generally performed better when V_p served as the predictor variable. However, the hybrid PSO–SA model displayed a remarkable increase in AUC when V_s was employed, demonstrating its ability to adapt effectively to more complex seismic behavior. This result highlights a key advantage of hybrid optimization approaches, while simpler algorithms may perform adequately under straightforward conditions, hybrid systems exhibit greater adaptability, accuracy, and consistency when faced with the nonlinearities and heterogeneities inherent in geomechanical and seismic datasets.

5.2 Score Analysis

In this subsection, a structured score-based evaluation was performed to systematically compare the performance of three implemented metaheuristic algorithms i.e., PSO, SA, and their hybrid PSO–SA in predicting the rock mass quality index (Q) using seismic wave velocities (V_p and V_s) as input variables. This comparison employed five widely used statistical metrics: R2 (coefficient of determination), RMSE (root mean square error), MARE (mean absolute relative error), MAPE (mean absolute percentage error), and MSE (mean squared error). For each metric, a score ranging from 1 to 3 was assigned to the models independently, allowing a fair and quantitative comparison.

For the goodness-of-fit indicator R^2, the model with the highest value received a score of 3, the second-best earned 2, and the model with the lowest R^2 received 1. Conversely, for the error-based metrics (RMSE, MARE, MAPE, and MSE), the model with the lowest error was awarded a score of 3, the next-best received 2, and the model with the highest error received 1 [5, 14].

When modeling Q using compressional wave velocity (V_p), the statistical metrics for each algorithm were as follows:

- PSO: $R^2 = 0.9142$, RMSE = 0.9603, MARE = 0.1421, MAPE = 24.82%, and MSE = 0.9222
- SA: $R^2 = 0.9391$, RMSE = 0.8050, MARE = 0.1390, MAPE = 0.48%, and MSE = 0.6481
- PSO-SA: $R^2 = 0.9504$, RMSE = 0.7232, MARE = 0.1346, MAPE = 17.67%, and MSE = 0.5231.

Following the scoring approach, the PSO–SA hybrid model outperformed the other two, achieving the highest score of 3 in all five metrics and a perfect total score of 15. The SA algorithm performed moderately, scoring 2 in each metric for a total of 10 points. The PSO model showed the weakest performance, receiving a score of 1 across all criteria, totaling 5 points.

Figure 13 illustrates this comparative evaluation, showing a bar chart of the scores assigned to each algorithm across the five evaluation metrics when V_p was used as the input.

When modeling Q based on shear wave velocity (V_s), the statistical results for each algorithm were:

- PSO: $R^2 = 0.8755$, RMSE = 1.1571, MARE = 0.2959, MAPE = 87.15%, and MSE = 1.3389
- SA: $R^2 = 0.9025$, RMSE = 1.0302, MARE = 0.2571, MAPE = 83.54%, and MSE = 1.0613
- PSO-SA: $R^2 = 0.9629$, RMSE = 0.6342, MARE = 0.1074, MAPE = 18.93%, and MSE = 0.4022.

Again, the hybrid PSO–SA model consistently delivered superior results across all five metrics, achieving the maximum score of 15. The SA algorithm maintained

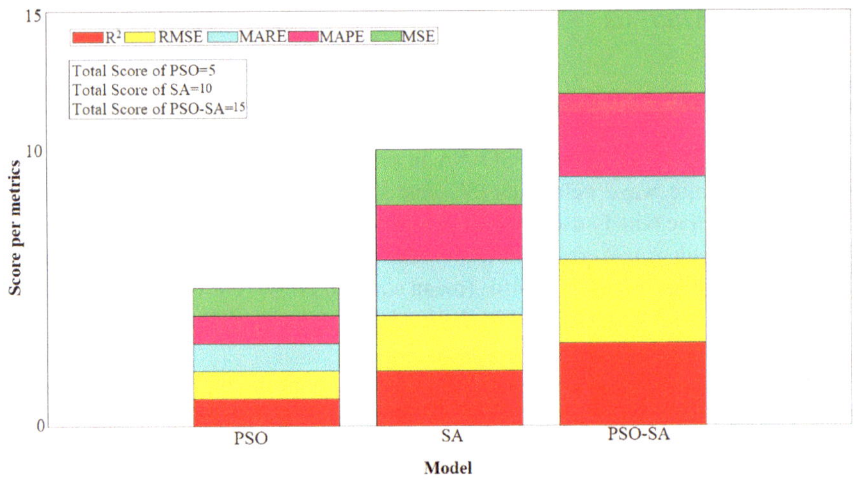

Fig. 13 Comparative performance ratings of the SA, PSO, and hybrid PSO-SA algorithms for Q prediction using V_p

moderate performance with a total of 10 points, while the PSO model, performing the weakest, earned 5 points.

Figure 14, presents the visual comparison of the score distributions for the three algorithms under V_s-based modeling, clearly highlighting the predictive advantage and generalization capability of the PSO–SA hybrid algorithm.

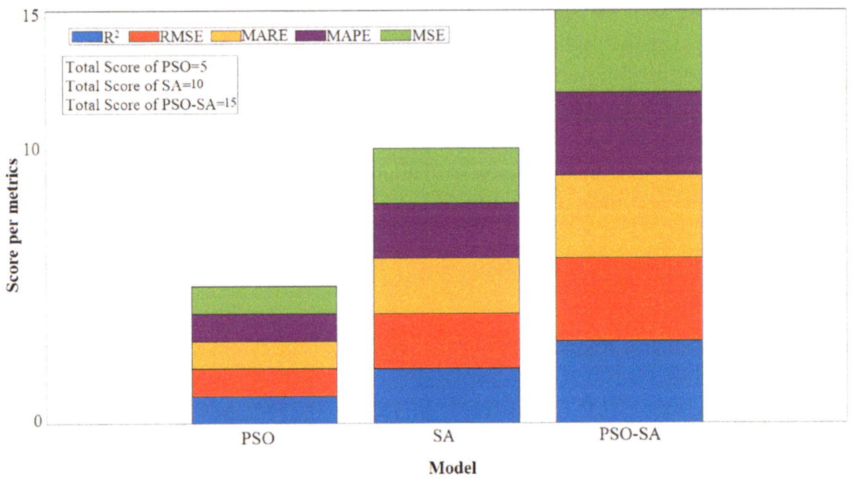

Fig. 14 Comparative performance ratings of the SA, PSO, and hybrid PSO-SA algorithms for Q prediction using V_s

Overall, this score analysis provides a concise yet comprehensive summary of the models performance (both error and accuracy mode) by integrating the multiple statistical measures into a unified ranking framework. It further emphasizes the clear superiority of the PSO–SA hybrid approach under both V_p- and V_s-based conditions, demonstrating that the combination of global and local search strategies in PSO–SA enhances prediction accuracy and robustness in estimating the rock mass quality index (Q).

5.3 Evaluation of Model Accuracy Using A20, IOA, and IOS Indices

In this section, the accuracy and reliability of the developed models were evaluated using three statistical indices: A20, IOA, and IOS. These indices were selected for their specific characteristics and their widespread application in predictive modeling, particularly in geotechnical and geophysical studies.

The A20 (Accuracy within ± 20%) index measures the percentage of predicted values that fall within ±20% of the actual values. This metric is particularly important in engineering applications, where maintaining low relative errors is critical. By quantifying the proportion of predictions within an acceptable tolerance range, A20 provides a straightforward and intuitive measure of model performance.

The IOA (Index of Agreement) is a normalized metric that assesses the level of agreement between predicted and observed values. It ranges from 0 (no agreement) to 1 (perfect agreement) and is widely employed in environmental, geological, and geophysical modeling due to its sensitivity to deviations from the mean.

The IOS (Index of Simulation) functions similarly to IOA but uses the mean of the predicted values in its calculation. This adjustment often leads to more accurate assessments when the data exhibit considerable variability. Using both IOA and IOS together provides a more comprehensive evaluation of model performance in capturing the actual trends of the data. The A20, IOA, and IOS indices were computed according to Eqs. (13) to (15), respectively. In Eqs. (13) to (15), Q_i, P_i, and N denote the actual value, estimated value, and total data count, respectively. Also, $\overline{Q}_i$ and $\overline{P}_i$ represent the average values of actual and estimated outputs, respectively [5, 14].

$$a20 = \frac{1}{N}\sum_{i=1}^{N}\delta_i,\ Where\ \delta_i = 1\ if:\left|\frac{P_i - Q_i}{Q_i}\right| \le 0.2,\ Otherwise:\delta_i = 0 \tag{13}$$

$$IOA = 1 - \frac{\sum_{i=1}^{N}(P_i - Q_i)^2}{\sum_{i=1}^{N}(|P_i - \overline{Q}_i|) + |Q_i - \overline{P}_i|)^2} \tag{14}$$

Table 2 Computed values of A20, IOA, and IOS indices for PSO, SA, and PSO-SA models using compressional wave velocity (V_p) as input

Index			Model
IOS	IOA	A20	
0.9881	0.9881	0.8467	PSO
0.9896	0.9896	0.8400	SA
0.9966	0.9966	0.8800	PSO-SA

$$IOA = 1 - \frac{\sum_{i=1}^{N} (P_i - Q_i)^2}{\sum_{i=1}^{N} (|P_i - \overline{P}_i|) + |Q_i - \overline{Q}_i|)^2} \tag{15}$$

Table 2 summarizes the calculated values of these three metrics for the PSO, SA, and hybrid PSO–SA models based on V_p (compressional wave velocity). According to the table, the SA model achieves an A20 value of 0.8467, indicating that approximately 84.67% of its predictions fall within ±20% of the actual values. The PSO model shows a slightly lower A20 of 0.8400, while the hybrid PSO–SA model surpasses both, reaching an A20 of 0.8800.

Regarding IOA and IOS, the differences between the models are more pronounced. The SA model scores 0.9881 for both IOA and IOS, whereas the PSO model performs slightly better with values of 0.9896. The hybrid PSO–SA model demonstrates the highest accuracy, achieving 0.9966 for both indices, highlighting its strong predictive capability when V_p is used as the input.

Table 3 presents the same evaluation for models based on V_s (shear wave velocity) as input. A similar trend is observed: the SA model achieves an A20 of 0.8320, the PSO model slightly lower at 0.8180, and the hybrid PSO–SA model again performs best with an A20 of 0.8610.

For IOA, the SA, PSO, and PSO–SA models yield 0.9839, 0.9862, and 0.9938, respectively, confirming the superior performance of the hybrid model even with V_s-based input. The IOS values follow the same pattern, with scores of 0.9845, 0.9857, and 0.9932, respectively.

The results summarized in Tables 2 and 3 clearly indicate that the hybrid PSO–SA model consistently outperforms the individual SA and PSO algorithms under both V_p- and V_s-based scenarios. This finding underscores the effectiveness of hybrid

Table 3 Computed values of A20, IOA, and IOS indices for PSO, SA, and PSO-SA models using shear wave velocity (V_s) as input

Index			Model
IOS	IOA	A20	
0.9845	0.9839	0.8320	PSO
0.9857	0.9862	0.8180	SA
0.9932	0.9938	0.8610	PSO-SA

metaheuristic approaches in enhancing the predictive accuracy of rock mass quality indicators, such as Q, in engineering geological applications.

5.4 Constraints and Suggestions

Despite the promising results achieved through the application of computational intelligence models for the indirect estimation of the Q-value, several limitations of this study should be considered when interpreting and generalizing the findings.

First, the dataset used in this research was exclusively obtained from the Beheshtabad water conveyance tunnel. Although this site exhibits considerable geological variability, it may not fully represent all geological conditions encountered in other underground projects. Consequently, the generalizability of the developed models to different geological environments could be limited.

Second, the modeling process relied solely on seismic wave velocity variables (i.e., V_p and V_s) as input factors. While these seismic indicators are crucial for characterizing rock mass behavior, other parameters such as uniaxial compressive strength, in-situ stress conditions, porosity index, and the mechanical properties of discontinuities can also significantly affect Q-value predictions. Restricting the model inputs to only V_p and V_s may therefore oversimplify the inherently complex behavior of rock masses.

Third, although metaheuristic algorithms such as PSO, SA, and the hybrid PSO–SA are highly effective in optimization tasks, they are sensitive to the selection of initial configuration parameters (e.g., population size, cooling rate, number of iterations). Inappropriate parameter settings can lead to slow convergence or entrapment in local minima, which may compromise model performance.

Based on these limitations, the following recommendations are proposed for future research:

- Expanding the dataset by incorporating samples from underground projects in diverse geological settings to improve the generalizability of the models.
- Integrating seismic inputs with additional mechanical and structural rock parameters to enhance the accuracy of Q-value predictions.
- Exploring more advanced hybrid algorithms, such as combinations of PSO with evolutionary strategies or machine learning models, to further improve model performance and robustness.
- Conducting uncertainty analysis and cross-validation procedures to rigorously assess the reliability of model predictions under varying geological and operational conditions.

These suggestions aim to guide future studies toward more robust and widely applicable predictive models for rock mass quality assessment, ultimately supporting safer and more efficient underground engineering practices.

6 Conclusion

This study presented an indirect approach for estimating the rock mass Q-value by utilizing compressional (V_p) and shear (V_s) wave velocities through meta-heuristic algorithms. Three computational models comprising PSO, SA, and the hybrid PSO–SA algorithms were developed and trained using real data obtained from the Beheshtabad water conveyance tunnel. Statistical evaluation using robust indicators such as R^2, RMSE, IOA, MAE, A20, MAPE, and IOS consistently demonstrated that the hybrid PSO–SA model outperformed the individual algorithms under both V_p- and V_s-based scenarios.

The modeling process was carried out in two separate phases: one using only V_p as input and the other using only V_s. In both scenarios, the hybrid PSO–SA model exhibited significantly higher predictive accuracy. Comparative analysis revealed that the PSO–SA model with V_s input achieved an R^2 of 0.9629 and an RMSE of 0.6342, whereas the same model with V_p input yielded an R^2 of 0.9504 and an RMSE of 0.7232. These results indicate that V_s had a more pronounced influence than V_p on Q-value prediction in the context of this study. Complementary reliability metrics, including A20, IOA, and IOS, alongside score analysis and REC curve evaluations, further confirmed the superior performance of the hybrid model under both input conditions.

A key strength of this research lies in the hybrid PSO–SA model, which combines the global search capability of PSO with the local refinement mechanism of SA. This integration achieved a well-balanced trade-off between accuracy and stability, resulting in reduced errors and improved convergence compared to the standalone algorithms. Moreover, the model's design, based on the real-world data and two easily measurable seismic parameters, enhances its practical applicability to similar underground engineering projects.

Overall, the findings of this study highlight the potential of computational intelligence models, particularly hybrid metaheuristic algorithms, as powerful tools for estimating geomechanical properties of rock masses. These models offer rapid, accurate, and non-destructive predictions, thereby reducing reliance on time-consuming and subjective field surveys. With further development, including the integration of additional geophysical and mechanical inputs, such intelligent frameworks can play a pivotal role in advancing the automation, reliability, and precision of rock engineering design.

AI Use Disclosure Statement The authors declare that no artificial intelligence tools were used for generating the scientific content, data analysis, or interpretation in this chapter. Minor language and grammar proofreading were performed using Chat GPT (Open AI, 2025 version) to enhance readability and linguistic clarity only. All research concepts, analyses, discussions, and conclusions presented herein are entirely the authors' original intellectual work.

References

1. Barton NR, Lien R, Lunde J (1974) Engineering classification of rock masses for the design of tunnel support. Rock Mech 6(4):189–239
2. Rezaei M, Koureh Davoodi P, Najmoddini I (2019) Studying the correlation of rock properties with P-wave velocity index in dry and saturated conditions. J Appl Geophys 169:49–57
3. Rezaei M, Koureh Davoodi P (2021) Determining the relationship between shear wave velocity and physicomechanical properties of rocks. Int J Min Geo-Eng 55(1):65–72
4. Rezaei M, Hosseini SP, Jahed Armaghani D, Khandelwal M (2025) Improving the P-wave velocity determination by considering the effects of joint properties in artificial rock samples. J Min Environ 16(3):1009–1025
5. Koureh Davoodi P, Hajizadeh F, Rezaei M (2025) Evaluating the RMR correlation with the rock mass wave velocity using the meta-heuristics algorithms. Sci Rep 15(1):17716
6. Rezaei M, Ahmadi SR, Nguyen H, Jahed Armaghani D (2025) Improved determination of the S-wave velocity of rocks in dry and saturated conditions: application of machine-learning algorithms. Transp Geotech 49:101371
7. Ding X, Hasanipanah M, Rezaei M (2025) Assessment of mechanical properties of rock using deep learning approaches. Measurement 250:117180
8. khoshjavan S, mazlumi M, Rezai B, Rezai M (2010) Estimation of hardgrove grindability index (HGI) based on the coal chemical properties using artifical neural networks. Orient J Chem 26(4):1271–1280
9. Majdi A, Rezaei M (2013) Application of artificial neural networks for predicting the height of destressed zone above the mined panel in longwall coal mining. In: 47th U.S. Rock mechanics/geomechanics symposium, 23–26 June. San Francisco, California, USA, pp 1665–1673
10. Rezaei M (2020) Feasibility of novel techniques to predict the elastic modulus of rocks based on the laboratory data. Int J Geotech Eng 14(1):25–34
11. Asadizadeh M, Rezaei M (2021) Surveying the mechanical response of non-persistent jointed slabs subjected to compressive axial loading utilising GEP approach. Int J Geotech Eng 15(10):1312–1324
12. Rezaei M, Rajabi M (2021) Assessment of plastic zones surrounding the power station cavern using numerical, fuzzy and statistical models. Eng Comput 37:1499–1518
13. Wang Y, Rezaei M, Abdullah RA, Hasanipanah M (2021) Developing two hybrid algorithms for predicting the elastic modulus of intact rocks. Sustainability 15(5):4230
14. Koureh Davoodi P, Hajizadeh F, Rezaei M (2025) Application of the metaheuristic algorithms to quantify the GSI based on the RMR classification. Sci Rep 15:28950
15. Barton N, Grimstad E (1994) The Q-system following twenty years of. application in NMT support selection. Felsbau 12(6):428–436
16. Barton N (2002) Some new Q-value correlations to assist in site characterization and tunnel design. Int J Rock Mech Min Sci 39:185–216
17. Fatemi Aghda SM, Kianpour M, Talkhablou M (2019) Efficiency of seismic wave velocity and electrical resistivity in estimation of limestone rock mass quality indices (Q, Qsrm) (Case study: Asmari formation, SW Iran). J Min Environ 10(2):529–541
18. Kianpour M, Fatemi Aghda SM, Talkhablou M (2019) Prediction model of limestone rock mass quality, using seismic wave velocity (Case study: Sarvak formation in Bakhtiari dam site). Iran J Earth Sci 11(4):277–289
19. Taban MH, Hajiazizi M, Ghobadian R (2021) Prediction of Q-value by multi-variable regression and novel genetic algorithm based on the most influential parameters. Civ Eng Infrastruct J 54(2):267–280
20. Hu J, Zhou T, Ma S, Yang D, Guo M, Huang P (2022) Rock mass classification prediction model using heuristic algorithms and support vector machines: a case study of Chambishi copper mine. Sci Rep 12:928
21. Santos AEM, Lana MS, Pereira TM (2022) Evaluation of machine learning methods for rock mass classification. Neural Comput Applic 34:4633–4642

22. Wijaya BJ, Warmada IW, Indrawan GB (2023) Correlation between GSI, RMR, and Q systems of rock masses at the construction site of spillway tunnel at Meninting Dam. I AIP Conf Proc 2629:030007
23. Rutledge JC, Preston RL (1978) Experience with engineering classifications of rock. In: Proceedings of the international tunnelling symposium. Tokyo, pp A3.1–A3.7
24. Cameron-Clarke IS, Budavari S (1981) Correlation of rock mass classification parameters obtained from borecore and In-situ observations. Eng Geol 17(1–2):19–53
25. Abad J, Morales A, Nunez-Lagos R, Pacheco AF (1983) A dynamical mechanism for the two-neutrino double beta decay. Nuovo Cim A 75(2):173–187
26. Baczynski NRP (1980) Applications of various rock mass classifications to unsupported openings at Mount Isa, Queensland: a case study. Third Australia-New Zealand conference on Geomechanics, Wellington, pp 2–137
27. Udd JE, Wang H (1985) A comparison of some approaches to the classification of rock masses for geotechnical purposes. In: The 26th U.S. symposium on rock mechanics (USRMS), Rapid City, South Dakota, USA
28. Kaiser PK, MacKay C, Gale AD (1986) Evaluation of rock classifications at B. C. Rail tumbler ridge tunnels. Rock Mech Rock Eng 19:205–234
29. Choquet P, Hadjigeorgiou J (1993) The design of support for underground excavations. Compr Rock Eng 4(313–348):1993
30. Abad J, Celada B, Chacon E, Gutierrez V, Hidalgo E (1983) Application of geomechanical classification to predict the convergence of coal mine galleries and to design their supports. In: The 5th ISRM congress, ISRM-5CONGRESS-1983-162, Melbourne, Australia
31. Al-Harthi AA, Al-Amri RM, Shehata WM (1999) The porosity and engineering properties of vesicular basalt in Saudi Arabia. Eng Geol 54(3–4):313–320
32. El-Naqa A (1994) Estimation of transmissivity from specific capacity data in fractured carbonate rock aquifer, central Jordan. Environ Geol 23(1):73–80
33. Barton N (1995) The influence of joint properties in modelling jointed rock masses. In: The 8th ISRM congress, ISRM-8CONGRESS-1995–194. Tokyo, Japan
34. Tuğrul A (1998) The application of rock mass classification systems to underground excavation in weak limestone, Ataturk Dam. Turkey Eng Geol 50(3–4):337–345
35. Sunwoo C, Hwang S-H (2020) Correlation of rock mass classification methods in Korean rock mass. In: Frontiers of rock mechanics and sustainable development in the 21st Century, 1st edn, CRC Press
36. Sari D, Pasamehmetoglu AG (2004) Proposed support design, Kaletepe tunnel. Turkey Eng Geol 72(3–4):201–216
37. Kumar N, Samadhiya NK, Anbalagan R (2004) Application of rock mass classification systems for tunneling in Himalaya, India. Int J Rock Mech Min Sci 41(Supplement 1):852–857
38. Coşar S (2004) Application of rock mass classification systems for future support design of the Dim Tunnel near Alanya. Middle East Technical University, Ankara, Turkey, p 217
39. Ranasooriya J (2009) The reliability of rock mass classification systems as underground excavation support design tools. PhD thesis, Curtin University, China
40. Abdullatif O (2010) Geomechanical properties and rock mass quality of the carbonate Rus Formation, Dammam dome, Saudi Arabia. Arab J Sci Eng 35(2)
41. Castro Caicedo ÁJ, Pérez Pérez DM (2013) Correlations between the RMR and Q geomechanical classifications at" La Línea" exploratory tunnel, Colombian Central Andes. Boletín Ciencias la Tierra 34:42–50
42. Mahdevari S, Ayati F, Hashemi M (2010) Estimation of the required thrust for tunnel excavation by using 3D numerical modeling-a case study of the Beheshtabad tunnel. In: The ISRM international symposium-6th Asian rock mechanics symposium, ISRM-ARMS6-2010-110. New Delhi, India
43. Laderian A, Abaspoor MA (2012) The correlation between RMR and Q systems in parts of Iran. Tunn Undergr Sp Technol 27(1):149–158
44. Perez Ruiz DA (2016) Modern optimization with R. J Stat Softw 70. Book Review 3

45. Rafiee R (2013) Tunnels stability analysis using binary and multinomial logistic regression (LR). J Geol Min Res 5(4):97–107
46. Sayeed I, Khanna R (2015) Empirical correlation between RMR and Q systems of rock mass classification derived from Lesser Himalayan and Central crystalline rocks. In: International conference on engineering geology in New Millennium, IIT-Delhi, New Delhi, India, p 29th
47. Senra K (2019) Correlations between geomechanical properties to amphibolites and schists from South of Minas Gerais state. In: The 14th ISRM congress, ISRM-14CONGRESS-2019-196. Foz do Iguaçu, Brazil
48. Soufi A, Bahi L, Ouadif L, Kissai JE (2018) Correlation between Rock mass rating, Q-system and rock mass index based on field data. MATEC Web Conf 149(2):02030
49. Rezaei M, Latifi S (2018) Development of a new empirical relation between RMR and Q rock mass classification systems. Iran J Eng Geol 12(2):100–103
50. Sadeghi S, Sharifi Teshnizi E, Ghoreishi B (2020) Correlations between various rock mass classification/characterization systems for the Zagros tunnel-W Iran. J Mt Sci 17:1790–1806
51. Hassanpour J, Khoshkar AS, Farasani MG, Hashemnejad A (2022) Investigating the relationships between rock mass classification systems based on data from mechanized tunneling projects in Iran. Bull Eng Geol Environ 81:147
52. Rezaei M, Habibi H (2023) Designing of the Beheshtabad water transmission tunnel based on the hybrid empirical method. Struct Eng Mech 86(5):621–633
53. Zayandab Engineering Consulting Company (2010) Geotechnical report of beheshtabad tunnel. (In Persian)

Analyzing Floor Level Variation Based on Indoor Environmental Quality and Human Perception in an Educational Institute Using Machine Learning

G. M. Rahad, Md. Shoriful Islam Shuvo, Jobayer Hossain, Mohammad Nyme Uddin, and Md. Raisul Islam Riyad

Abstract Buildings are becoming taller, increasing complexity in indoor environmental conditions and occupant perceptions. This study classifies floor level in a multi-storey educational building in Dhaka, Bangladesh, using environmental measurements and occupant perception data. A total of 1,087 observations were collected over four months during the summer via physical surveys and air-quality monitors. Three machine learning classifiers such as Decision Tree, Random Forest, and XGBoost were trained and evaluated using confusion matrices; SHAP (Shapley Additive exPlanations) was used for feature ranking and model interpretability. All models achieved accuracies above 96%. The most influential features for floor-level classification were number of fans, number of lights, humidity, temperature, and CO_2 concentration. These findings demonstrate the capability of interpretable ML to distinguish vertical variations in indoor environments and identify key drivers of inter-floor differences. The results offer actionable insights to inform the design of adaptive, occupant-centric, and energy-efficient multi-storey educational buildings.

G. M. Rahad · J. Hossain · Md. R. I. Riyad
Department of Civil Engineering, International University of Business Agriculture and Technology, Dhaka, Bangladesh
e-mail: 22206044@iubat.edu

J. Hossain
e-mail: 22206064@iubat.edu

Md. R. I. Riyad
e-mail: 22206042@iubat.edu

G. M. Rahad · Md. S. I. Shuvo · J. Hossain · M. N. Uddin · Md. R. I. Riyad
Building Energy & Environmental Management (BEEM) Enhanced by AI, Dhaka, Bangladesh
e-mail: shuvo35-866@diu.edu.bd

Md. S. I. Shuvo
Department of Software Engineering, Daffodil International University (DIU), Dhaka, Bangladesh

M. N. Uddin (✉)
Department of Architectural Engineering, Imam Mohammad Ibn Saud Islamic University (IMSIU), Riyadh 11432, Saudi Arabia
e-mail: mnuddin@imamu.edu.sa

A. Azizi et al. (eds.), *Advanced Decision-Making Under Uncertainty*, Emerging Trends in Mechatronics, https://doi.org/10.1007/978-981-95-8696-7_4

Keywords Floor level · Machine learning · Indoor air quality · Human perception

1 Introduction

There is a significant global trend toward increasing building heights over recent decades, as proven not only in commercial and residential buildings but also in educational buildings. Vertical development is promoted by several interrelated conditions, such as lack of land in megacities with high population growth rates, advances in structural engineering and construction technology, and the desire for architectural identity and institutional glory. Also, the unit cost of horizontal travel exceeds the unit cost of vertical travel; a tall building will economize [1]. Having multistoried educational facilities is in alignment with the view that it can effectively centralize academic activities like classrooms, labs, offices, and recreational spaces in one vertical entity, which in turn results in smoother operations. Occupants generally spend approximately 90% of their time within the indoor environment [2]. And it is where a high Indoor Environmental Quality (IEQ) is requisite in everything from our operational efficiency to well-being [3, 4]. To understand the overall comfort level of occupants [3] the important comfort parameters, thermal comfort plays a significant role [5, 6]. Moreover, this concept is extremely important in low-income hot and humid cities such as Dhaka, Bangladesh. These environments also include naturally ventilated educational buildings (NVEBs), as mechanical heating, ventilation, and air conditioning (HVAC) systems are very expensive to install and maintain. In addition, NVEBs are often more environmentally friendly and energy-efficient, as they rely heavily on natural air flows (prone to changes at floor heights and weather). This can result incapable comfort levels within the building. Educational institutions are one of the few building types that are occupied by a mix of students, teachers, and staff who use the buildings for long periods. Temperature and air quality act as direct influencers of attention length, cognitive function, and well-being in general; hence, thermal comfort is quite paramount.

In recent times, applications of machine learning (ML) have remarkably achieved success in various disciplines, including the development of effective prediction models for thermal and visual comfort (TVC). Machine learning techniques are often referred to as black box models. This has become increasingly recognized in architecture and building design. Their capacity to handle complex challenges efficiently with greater accuracy has made them widely adopted by researchers and practitioners in the field [7, 8]. In this sense, evolving tools in machine learning are particularly favorable for predicting floor level, providing the ability to estimate indoor conditions and improving how buildings work in ways that outdated methods can't. Decision-makers can use this machinery to plan and run NVEBs so that they can build educational high-rise buildings that are not only space-efficient and visually exceptional but also always support the comfort, health, and cognitive success of residents on all floors.

Based on existing research on the built environment, the following gaps have been identified:

- More focus on TVC in NVEBs, while limited analysis on floor level classification based on environmental parameters and human perception.
- Very less number of research on the built environment in developing, low-income, hot, and humid regions like Bangladesh.

This study establishes the application of interpretable machine learning models, specifically for Random Forest (RF), Decision Tree (DT), and XGBoost, to predict floor level in a naturally ventilated classroom of an educational building located in Dhaka, Bangladesh. The study is conducted using a real-world served dataset of instances collected over four months in the summer season. Traditional researches were conducted focusing on comfort analysis. By shifting the predictive focus from comfort classification to floor level stratification, the research identifies key feature-specific drivers of vertical variability in indoor environments. The use of SHAP (SHapley Additive exPlanations) enhances model interpretability by revealing how features such as temperature, humidity, CO_2 concentration, and adaptive responses influence the likelihood of a given floor level. All the models achieved a superior prediction accuracy, which is described in the result discussion part. These insights not only challenge the limitations of conventional flat-floor assumptions in building performance analysis but also support the superiority of ML models and low-cost, adaptive design strategies that account for floor level variations in naturally ventilated educational buildings.

The remaining parts of this paper are structured as follows: Sect. 2 provides a comprehensive literature review of floor level and vertical stratification related information. Section 3 outlines the research methodology, including data collection, cleaning, model selection, cross-validation, and accuracy assessment. Section 4 presents the results and discussion of the study. Finally, Sect. 5 concludes the paper.

2 Literature Review

The vertical stratification of indoor environmental quality (IEQ) in multistory educational buildings, especially in hot, humid, and high-density urban settings such as Dhaka, combines two fast-evolving research threads: (1) measurement and health/productivity impacts of indoor air quality (IAQ) and thermal comfort in educational institutions, and (2) application of modern, interpretable machine learning (ML) tools to classify, predict, and explain IEQ patterns. This review synthesizes recent findings in both domains and highlights how interpretable ML models, such as Decision Tree, Random Forest, and XGBoost, along with evaluation tools like the confusion matrix and interpretability techniques such as SHAP, can support the detection of floor level differences and guide low-cost sensor deployments in both naturally ventilated educational settings.

Classrooms are a critical microenvironment for children's health and cognitive performance: many studies link poor ventilation and elevated CO_2 (and associated pollutants) to lower test scores, increased absenteeism, and health complaints [9, 10]. Intervention studies, such as the FRESH ventilation trial, have demonstrated that mechanical ventilation can substantially lower classroom CO_2 concentrations and improve IAQ metrics, with potential downstream benefits for health and performance [11]. Similarly, experimental laboratory work has shown that modest elevation of CO_2 (1000–2500 ppm) can damage decision-making and cognitive function as well, suggesting a dual role for the gas ventilation indicator/direct stressor for complex tasks [12]. Recent research has demonstrated that floor elevation plays a critical role in shaping indoor environmental conditions [13]. Similarly, a vertical temperature increases of 0.1–0.2 °C per floor in high-rise buildings was found, further supporting the importance of vertical modeling [14]. These studies collectively indicate that assuming uniform indoor conditions across floors can result in misleading evaluations of building performance and occupant comfort. Recent critical reviews on IAQ in schools have compiled evidence that good ventilation strategies (mechanical or well-managed natural ventilation) play a central role in safeguarding children's respiratory health and learning outcomes [15].

Thermal comfort is similarly impactful. In hot and humid climates (e.g., South Asia), thermal stress can meaningfully reduce students' attention and productivity; adaptive comfort approaches have been used to establish more realistic comfort bands for naturally ventilated classrooms, but variability by building height and floor can create systematic differences that conventional "flat-floor" assumptions miss [16, 17]. Studies in dense urban and hot-climate schools point to seasonality, occupant behavior (window opening, fans), and building layout as strong drivers of thermal comfort variance [18]. It is reported a 35% deviation between PMV predictions and actual comfort votes in Indian classrooms, further exposing the limitations of traditional models in dynamic settings [19].

In educational buildings with mixed ventilation strategies, floor level influences occupant exposure and perceived comfort: upper floors may receive more daylight and heat gain, lower floors may suffer from pollutant ingress from street level, and ventilation effectiveness can vary by floor depending on window availability and local airflows [20, 21]. The practical implication is that single-zone or single-floor analyses can mask important vertical heterogeneity; designers and facility managers need methods that detect and explain floor-specific IEQ patterns so they can target low-cost interventions (fans, opening patterns, local mechanical inputs).

Recent years have seen an explosion in low-cost sensor (LCS) deployments in schools and citizen-science programs (e.g., SAMHE, national school campaigns) to map classroom CO_2, $PM_{2.5}$, temperature, humidity, and VOCs at scale. Low-cost sensing is attractive for widespread monitoring in resource-constrained settings (e.g., Dhaka), but raw LCS outputs need calibration and quality control. Multiple studies show that ML-based calibration (including Random Forests and ensemble methods) substantially improves LCS accuracy and drift correction versus simple linear corrections, enabling reliable day-to-day exposure estimation in classrooms. Large projects also demonstrate the value of coupling contextual data (occupant numbers, windows/fans, schedules) with sensor streams to interpret IAQ patterns and inform interventions [22, 23].

ML methods, Decision Trees (DT), Random Forests (RF), and gradient-boosting algorithms such as XGBoost have been successfully applied to classification and regression tasks for indoor comfort, pollutant estimation, and occupancy detection. Each algorithm has tradeoffs: DTs are transparent but can overfit; RFs reduce variance and handle nonlinearities well; XGBoost provides strong predictive performance through boosting, often outperforming single-tree models on complex datasets [24–27]. However, high accuracy alone is not sufficient for built environment applications; stakeholders (designers, facility managers) require interpretable outputs to translate model findings into actionable design or operational changes. In a recent study [28], a machine learning algorithm was developed to create a thermal comfort model for occupant comfort prediction. The model utilized the ASHRAE RP-884 dataset to classify key features and predict the thermal sensation level. This has driven the recent uptake of explainable-AI tools such as SHAP (SHapley Additive exPlanations), which quantify feature contributions at both the global and instance level and provide human-readable explanations of model outputs [29].

SHAP has been used in building and IAQ domains to (a) rank features that drive comfort or pollutant predictions and (b) explain why particular rooms or times are classified a certain way (e.g., a floor predicted "hot" due to high solar gains and few fans) [30]. Combining confusion matrix reporting with SHAP feature attribution provides both a quantitative measure of classifier reliability and a transparent explanation of what drives errors and correct predictions, a valuable combination when recommending targeted interventions (add fans on X floor, change lighting policy, adjust window schedules).

Several field studies have already combined low-cost sensor networks and ML models to predict comfort or pollutant status in educational institutions and naturally ventilated buildings. For example, studies applying RF and XGBoost have achieved high classification accuracy for thermal and visual comfort categories, while SHAP analyses reveal that simple layout and behavioral variables (number of fans, lights, proximity to window) often rank among the top predictors, a result consistent with field evidence that layout and occupant behavior mediate vertical IEQ differences [16, 21, 29]. But in spite of these benefits, a lot of educational high-rise designs still rely on traditional beliefs of consistent indoor conditions across floors, an approach that ignores the fact that elevation can have an extensive impact on environmental factors like temperature, humidity, CO_2 levels, lighting power, and wind velocity [13].

Nevertheless, few traditional design valuation indexes and post-occupancy evaluation studies delve into the specific changes in IEQ from floor to floor, leading to gaps in knowledge and practical advice. To close this gap, dependable data-driven methods are needed to create precise, floor-specific environmental predictions by fusing reliable measurements with subjective occupant feedback. This combination of structural design, climate response, and ventilation strategy shows the economic limitations and sustainability challenges in educational buildings in these typical climates [31, 32]. In buildings like this, it is normal to get changes in environmental parameters like temperature, humidity, CO_2 concentration, and lighting level with the movement from one floor to another. These differences happen for a few reasons: upper floors often get more sunlight, air tends to flow differently at different heights, and warm air naturally rises while cooler air stays lower, a process known as the stack effect. Over a day, these small shifts can add up, shaping how comfortable the spaces feel and how well the indoor environment supports the people using them [33].

3 Methodology

This study adopts a systematic, multi-phase methodological framework aimed at developing and evaluating predictive models for floor level classification in different classrooms in a naturally ventilated educational building (NVEB) located in Dhaka, Bangladesh. The framework integrates subjective occupant comfort responses with objective environmental measurements and layout-related data to assess whether vertical stratification of indoor environmental quality (IEQ) can serve as a reliable predictor of floor elevation. The overall research design is presented through a flowchart in Fig. 1.

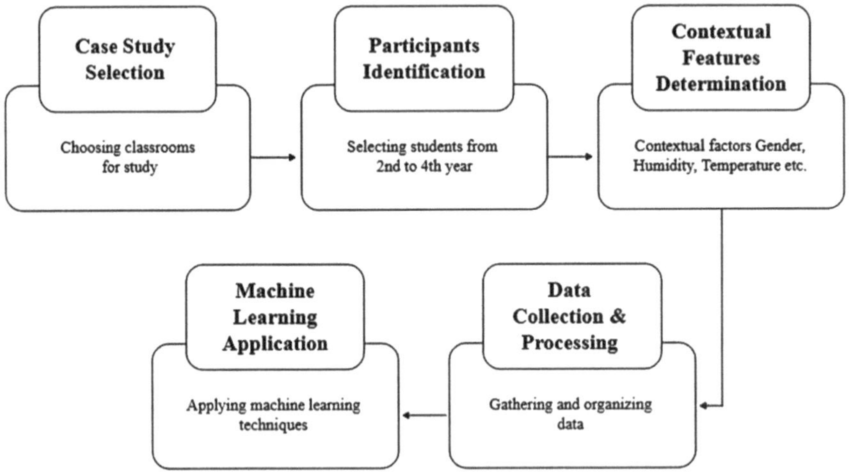

Fig. 1 Methodology flow diagram

3.1 Field Survey

Case study building and location. The case study building is a mid-rise building with a central academic facility located beside the Turag River in Dhaka, Bangladesh. The building represents a typical urban academic style often seen in hot and humid tropical climates. It is a 12-story reinforced concrete structure built to hold many occupants in a compact space, which leads to high density, and is approximately 15 years old. The building has very few mechanical ventilation systems, as it is a green campus. Instead, it mostly relies on natural ventilation methods for maintaining its indoor air quality and thermal comfort of occupants. Also, the building has the advantage of daylight due to well-placed windows and openings in the classrooms, lobby, and other spaces. This helps to cut down the need for artificial lighting during the day. Dhaka experiences a hot and humid tropical climate. The city has a diverse monsoonal season, with summer temperatures averaging around 31 °C (May–August) and winter temperatures averaging around 23 °C (December–February) [29]. The building has very few mechanical ventilation systems, as it is a green campus. Instead, it mostly relies on natural ventilation methods for maintaining its indoor air quality and thermal comfort of occupants. Also, the building has the advantage of daylight due to well-placed windows and openings in the classrooms, lobby, and other spaces. This helps to cut down the need for artificial lighting during the day. This combination of structural design, climate response, and ventilation strategy shows the economic limitations and sustainability challenges in educational buildings in these typical climates.

Participant recruitment and sampling. Second- to fourth-year students were chosen as the participants, aged between 18 and 27 years. The freshmen are not considered participants in data collection because they are not very accustomed to the environment and are always in a cheerful mood. This relatively narrow age range was chosen to limit variability in physiological thermal responses, while capturing differences related to floor-specific exposure and academic schedules. In addition, data were collected from the fourth to the tenth floors, although the building has 12 floors. The lower four floors, that means from the ground floor to the third floor, were not considered in order to avoid the effect of tree shading. Similarly, the top two floors were not included because they have a centrally controlled HVAC system. The participants were selected in random order, and voluntary recruitment after the informed consent procedure. Classrooms on different floors were selected to include students, and a stratified sampling strategy was applied to have balanced representation from lower, middle, and upper floors. This has the net effect of incorporating spatial and perceptual diversity into the dataset [34].

Survey design and measured parameters. Predictor variables were selected based on a literature review covering vertical IEQ variation and occupant comfort modeling. Twelve input features, encompassing environmental factors such as temperature (°C), humidity (%), lighting level (Lux), and CO_2 concentration (PPM); occupant demographics including gender; adaptive behaviors like thermal comfort votes, and seating orientation; as well as layout-related data including number of windows, number of fans, and number of lights was considered during data collection for this study. Ultimately chosen features were grouped into occupant, environmental, and layout categories (Table 1).

Data collection protocol. Data were collected over four consecutive summer months (May–August), during which natural ventilation was the dominant cooling strategy and solar gains were highest. The calibrated smart sensors used for environmental measurements were set at 0.75 m above floor level to represent the breathing zone of seated occupants. The data was collected from the participants during 11:00 a.m.–3:00 p.m. at peak occupancy periods. Device specifications used for data collection are shown in Table 2.

Table 1 Key parameters on layout, occupant, and environmental

Category	Parameters	Unit	Description
Layout	Occupant-to-window distance	ft	Distance between occupant and window
	Number of fans	Numerical	Total number of fans in the classroom
	Number of lights	Numerical	Total number of lights in the classroom
	Number of windows	Numerical	Total number of windows in the classroom
	Floor level	Numerical	Number of floors in the case study building
Occupant	Thermal comfort	7-point scale	Thermal sensation votes (−3(cold) to +3(hot))
	Room orientation	Cardinal directions	Seating orientation of occupant on chair (i.e., North–South, East–West, West–East, North–East)
	Gender	–	Gender of the students
Environmental	Temperature	°C	Maximum and minimum indoor temperature during data collection
	Humidity	Numerical (%)	Maximum and minimum indoor humidity during the data collection
	CO_2	PPM	Maximum and minimum indoor CO_2 concentrations during the data collection
	Lighting level	Lux	Amount of lighting in lux at the sitting location

Table 2 Instrumentation specifications included

Devices	Parameters
Air quality monitoring device (INKBIRDPLUS AK3)	CO_2: (400–5000) ppm; accuracy: ±50 ppm; resolution: 1 ppm; temperature: (0–50) °C; accuracy: ±1 °C; humidity: (0–99)%; accuracy: ±2%
Lux meter (FLUS MT-912)	Range: 0–200,000 lx; accuracy: ±4%

Time-synchronized comfort perception and measurement data were collected through a structured questionnaire of Google Form immediately after environmental measurements, comprised subjective comfort data as well as the information on their environmental metrics.

3.2 Data Preparation

The raw dataset ($n = 1{,}087$) underwent quality assurance procedures to ensure suitability for ML applications. Outlier removal using the interquartile range (IQR) method for continuous variables. Missing value imputation: above computed median value for continuous features, and mode for categorical. Categorical Entry normalization so you can detect if the name was written in lower case or upper case, and avoid disputes for that.

3.3 Model Development

Model selection rationale. Three supervised ML classifier algorithms were chosen based on their performance in similar IEQ and comfort-prediction contexts:

Random Forest (RF): A random forest algorithm uses ensemble learning, which consists of multiple decision trees working simultaneously. Each tree is constructed from a sample of the original dataset. Random Forest uses averaging to combine all of the trees, thus stabilizing the predictions and preventing the overfitting risk caused by a single decision tree. To conclude, this model has effectively a good result over other algorithms especially when the datasets are very complex having both numerical and categorical features. As a result of its random selection, it has more noise situations, so this model is good for classification and regression.

Decision Tree (DT): This algorithm is one of the simplest and most powerful models because it splits data into branches of the tree. As the name "decision tree" suggests, one can simply follow a decision path that begins from the root and ends at leaf nodes. Even though the decision trees tend to perform well when working with numerical or categorical data, if they are not properly taken care of, they may overfit. This makes them less stable compared to ensemble methods.

XGBoost: This is an example of a boosting algorithm. As the name suggests, it buys multiple sequential trees, and each new tree is tailored to correct the errors made by its predecessor. It is tailored for speed and overall efficiency and works wonders with structured datasets. It is recognized for its accuracy and ability to handle overfitting, mainly by regularizing, shrinking or leaving out some of the data. Consequently, it is among the most used methods for performing both classification and regression.

Model training and evaluation. Model test and train: Each model was trained using the training subset and evaluated against the testing subset. Following preprocessing, the dataset was divided into 80% for training and 20% for testing subsets using stratified sampling to maintain proportional representation of each floor level. The dependent or target variable (floor level) was encoded as a categorical label for classification. Primarily using confusion matrices and overall accuracy, the performance was assessed.

Cross-validation: GridsearchCV was used for hyperparameter optimization. To select the best hyperparameters for the model by balancing between accuracy and generalization, thus preventing overfitting. In the Random Forest model, hyperparameter tuning concentrated on the number of trees, the maximum depth of each tree, and the minimum samples required at the leaf nodes. In the case of the Decision Tree, its major adjusted features included the maximum depth of the tree and the splitting criterion, choosing between Gini impurity and entropy. Several other key features were adjusted for XGBoost, with the most notable being the learning rate, tree depth, number of estimators, and the subsampling ratio, all of which drastically improved the model's accuracy, efficiently processed intricate patterns in the dataset, and overall, enhanced the model.

3.4 Feature Importance Analysis

SHAP (SHapley Additive exPlanations) is a method that ameliorates the interpretability of models by attributing contribution scores to each feature. This score is a metric indicating how much impact each feature has on the prediction made by our model. SHAP was such a big deal, it could now get more than just a "black box" output and see through which variables we got the final outcome of classification. The results suggested that the features of fan count, light count, humidity, CO_2 level, and temperature were crucial to predict the floor level. These results match existing research on vertical differences in indoor environments, where variations in ventilation, lighting, and thermal conditions have been shown to change noticeably across building floors. This consistency with previous studies enhances the reliability of the findings and highlights the practical importance of environmental features in predicting floor levels [27].

3.5 Model Performance Test

The performance of machine learning models is estimated by considering several evaluation matrices. Finding accuracy is the most common approach among them. It is the ratio of correctly classified samples to the total number of samples. A more accurate machine learning model is identified by its higher accuracy value.

$$\text{Accuracy} = (\text{TP} + \text{TN})/(\text{TP} + \text{FP} + \text{TN} + \text{FN}) \tag{1}$$

$$\text{Precision} = \text{TP}/(\text{TP} + \text{FP}) \tag{2}$$

$$\text{Recall} = \text{TP}/(\text{TP} + \text{FN}) \tag{3}$$

$$\text{F1} = 2 * (\text{Precision} * \text{Recall})/(\text{Precision} + \text{Recall}) \tag{4}$$

Here, TP means the true positive (i.e., the number of correctly classified positive samples); FP means false positive, TN stands for true negative, and FN represents false negative.

3.6 Methodological Significance

The proposed method shows how interpretable machine learning can analyze vertical differences in indoor environmental quality (IEQ) in naturally ventilated educational buildings (NVEBs). By combining feedback from occupants with objective measurements from sensors, the framework offers more than just predictions; it provides practical insights for building management. The process results in a plan based on evidence, good for that specific floor conditions one after the other. For instance, repositioning fans for improved airflow, changing lighting layout to minimize excessive lights and darkness in the environment, or adjusting ventilation strategies to maintain climate conditions and CO_2 levels based on height. This method highlights the importance of combining perception with data analysis. It shows the real-world potential for creating healthier and more adaptable learning environments.

4 Result Discussion

This section presents the analytical findings from the application of three interpretable machine learning models, Random Forest (RF), Decision Tree (DT), and XGBoost, for predicting floor level in an NVEB. The combined evaluation approach ensures that the models are assessed not only in terms of predictive accuracy but also in their ability to provide actionable, interpretable insights for building performance optimization.

4.1 SHAP (SHapley Additive ExPlanations)

Decision Tree (DT). The SHAP diagram of the DT classifier, Fig. 2a, sheds light on the effective features concerning floor level predictions of the naturally ventilated educational building. The three most important parameters once again were the number of fans (mean SHAP values 0.85), humidity (%) (mean SHAP values 0.60), and temperature (°C) (mean SHAP values 0.45), clearly exhibiting their dominating influence on floor level predictions. To consider: as humidity and temperature are basic environmental indoor air quality indicators, the number of fans is a layout-related variable that impacts airflow perception and thermal experience. The number of lights and room orientation also had a moderate impact (mean SHAP ~0.30), emphasizing the utility of interior layout properties.

However, unlike RF, the DT model focused more power on a smaller number of features. This is as to be expected, given that DTs produce rule-based partings and put the higher ROI features first in terms of contributions to class separation. Intuitively, this implies that the splits early on in the DT structure are probably just based on the number of fans and humidity, in which case, strong performance will still be achieved, though fewer features have dictated the decision path.

The number of lights and room orientation also resulted in moderate SHAP values (~0.30). Room orientation, while of lesser importance for RF, does seem to have a greater bearing in DT, perhaps because the model decisions follow a single nest path, meaning environmental variables that change predictably with building orientation and solar exposure are emphasized.

Random Forest (RF). Figure 2b shows the SHAP diagram for the RF classifier of important parameters involved in the prediction of floor level within an educational building, which stays naturally non-air-conditioned. Fans, humidity (%), and temperature (°C) were the most contributory factors, having maximum mean SHAP values of 0.5, 0.4, and 0.25, respectively. This highlights their key role in shaping floor level perceptions. Humidity and temperature are essential environmental factors that affect indoor air quality at the floor level. In contrast, the number of fans is an important layout factor that greatly impacts airflow perception. The number of lights and the number of windows also had moderate effects, with SHAP values around 0.20–0.22, underscoring the importance of layout-related factors.

This ranking suggests that in the studied NVEB, layout-related features, particularly fan distribution, exert a measurable impact on the prediction of floor level. From a physical perspective, the number of fans may be a strong proxy for design and operational differences between floors, as fan placement often varies according to ceiling height, expected occupancy density, and budget constraints during construction.

Humidity and temperature, both environmental parameters, strengthen the impact of vertical stratification concerning IEQ. In tropical locations like Dhaka, the upper floors of the buildings experience higher temperatures and lower humidity in the afternoon due to increased solar gain and wind. Also, the moderate spread of importance scores from the RF model suggests that its strength lies in capturing complex, nonlinear interactions and is not dependent on a single dominant feature.

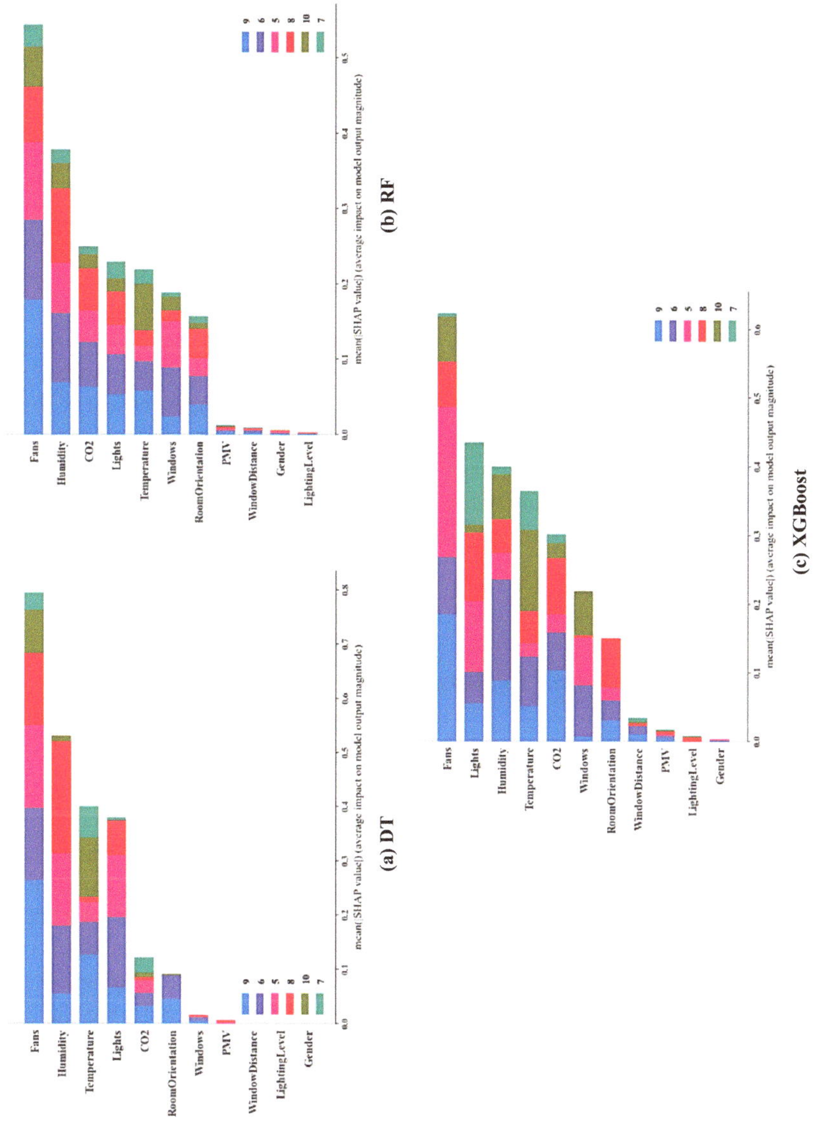

Fig. 2 SHAP diagram for DT (a), RF (b), and XGBoost (c)

The SHAP values of the secondary features, like the Number of Lights and Number of Windows, which were in the range of values between 0.20 and 0.22, demonstrate that the changes to the layout of the space and the amount of daylight coming inside the space influence the occupants' perception and the measured IEQ values.

XGBoost. Analyzing the SHAP values for XGBoost classifier, as represented in Fig. 2c, reveals how certain factors contribute to predicting the floor level in the ventilated educational building. Cumulatively, the number of fans stands out as the most significant feature, surpassing 0.6 in mean SHAP value. This underscores its important role in airflow perception and its relationship to vertical zoning in the utilized spaces. The number of lights and humidity percentage follow closely, with SHAP values around 0.45 and 0.4, respectively. This also demonstrates that both layout-related and environmental factors are significant predictors.

Along with these, temperature in Celsius and the level of CO_{22} in ppm contribute to SHAP values of roughly 0.35 and 0.3, respectively. These factors indicate the role they fulfill in perceived comfort and indoor climate across different floors.

The second and third most significant features, humidity level and lights, respectively, with SHAP values of 0.4 and 0.45, confirm that both spatial layout and environmental parameters are significant. This is likely because of the electrical performance within a building that can change with an increasing number of floors, versus floor-to-ceiling daylight impacting the electrical layout or window-to-wall ratios, or zoning architecturally. SHAP values for environmental factors of temperature and CO_2 level reaching 0.35 and 0.3, respectively, suggesting that upper levels might endure as well.

4.2 *Model Accuracy and Performance*

Accuracy. In order to predict floor level in naturally ventilated educational buildings, this study assesses the effectiveness of three interpretable machine learning classifiers: Random Forest (RF), Decision Tree (DT), and XGBoost. Based on their individual confusion matrices, all models demonstrated high predictive accuracy, as indicated in Table 3. The confusion matrix generally shows the performance of these classification models for seven different floors: 4th, 5th, 6th, 7th, 8th, 9th, and 10th. Figure 3 shows the confusion matrices representing the performance of the level predicting model using three machine learning models. The best-performing model was XGBoost (Accuracy: 0.98), followed by RF (0.97) and DT (0.96). SHapley Additive exPlanations (SHAP) analysis showed that, despite similar performance, the models differed in feature sensitivity and interpretability. The accuracy of all three models is very high, as the accuracy value exceeded 96%. However, this may have occurred because we used three types of parameters in this study: layout-related, environmental, and occupant perception-related parameters, where the layout-related parameters are the same or constant for each floor. Although the SHAP analysis clearly demonstrated that environmental parameters and occupant perception-related parameters also have a significant impact on the accuracy calculation.

Table 3 Model accuracy and key SHAP drivers

Model	Accuracy	F1	Top SHAP features (impact range)
RF	0.97	0.96	Number of fans 0.5, humidity 0.4, and temperature 0.25
DT	0.96	0.95	Number of fans 0.85, humidity 0.60, and temperature 0.45
XGBoost	0.98	0.97	Number of fans 0.6, number of lights 0.45, and humidity 0.4

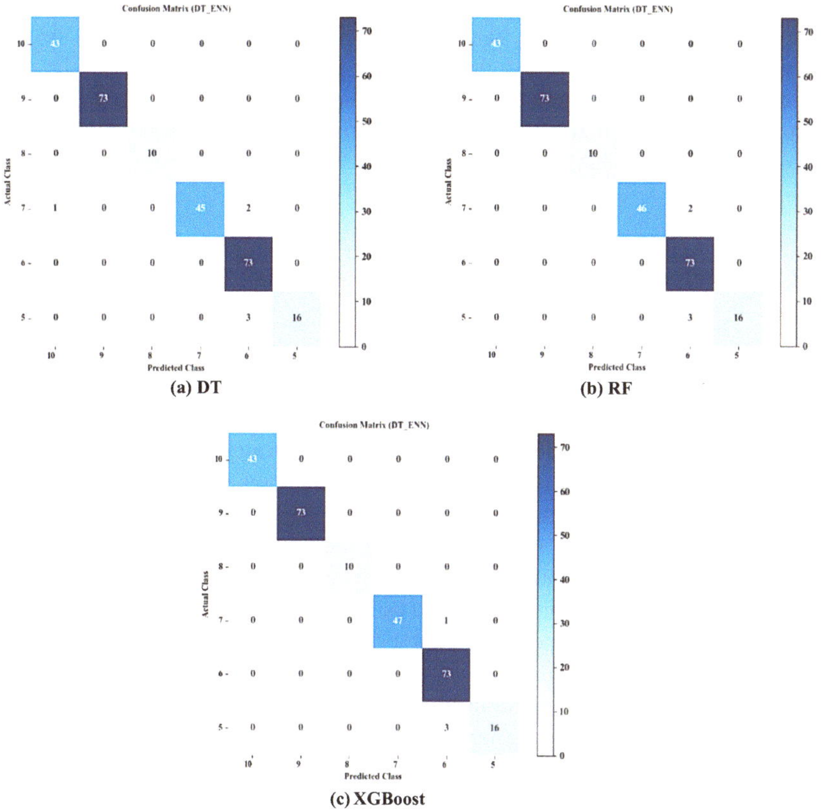

Fig. 3 Confusion matrix for DT (**a**), RF (**b**), and XGBoost (**c**)

The XGBoost classifier, with its boosting framework, captured strong nonlinear relationships and found that the most influential features were the number of fans (SHAP: 0.6), the number of lights (0.45), and humidity (0.4). The results are useful because they offer a range of thresholds that can be applied at different scales in interventions, within the framework of designing (e.g., increasing lighting on upper floors). Using a simple, rule-based architectural framework made it clear for the DT classifier that the number of fans (0.85), humidity (0.60), and temperature (0.45) were potential splitters. In contrast, the RF model spread importance more broadly with top features temperature (0.25), humidity (0.4), and number of fans (0.5), despite its high level of accuracy as well. Although it is less interpretable, this shows how well RF models subtle, nonlinear dependencies.

Cross-model interpretations. All three classifiers agreed that the number of fans, temperature, and humidity were important variables, despite variations in modeling techniques. The strength of these characteristics as indicators of vertical IEQ variation in NVEBs is supported by this agreement. While it was less prominent in RF, the number of lights became a significant secondary feature in XGBoost and DT. This implies that in models that are susceptible to direct feature thresholds or discrete layout changes, the number of lights might be more important. In a similar vein, CO_2 concentration performed better in XGBoost, most likely due to the ability of boosting algorithms to better capture the interaction effects between ventilation rates and occupancy patterns across floors.

5 Conclusion

This study develops a predictive model specially for floor level classification in a naturally ventilated educational building (NVEBs) which is located in a low-income economy with a hot and humid climate. To reach the goal of this study, a total of 1087 data samples were collected from an NVEB, which is located in Dhaka, Bangladesh. From this study, it is clear that the AI-based classification models, such as RF, DT, and XGBoost, are very effective and potential in predicting floor level classification within NVEBs classrooms. After performing the necessary data cleaning, model selection and model preparation, the analysis was done. For effective and optimal performance, cross-validation was applied using GridSearchCV. This study achieved superior prediction accuracy of more than 96% for three powerful machine learning algorithms (RF, DT, and XGBoost), indicating that these models are capable of being applied in complex indoor environmental data. While the XGBoost model achieved the highest accuracy of 98%. The results are more interpretable as SHAP gives an explanation to each of the classification results, pointing out key features. The three most important features were the number of fans, humidity, and temperature. Additionally, two features had a noticeable impact. These five features are related to environmental parameters and layout-related data. This classification marks a critical separation between indoor environmental conditions and the layout of the building in terms of the indoor environmental experience. Humidity and temperature are key to occupant health and perception, along with productivity. At the same time, the number of fans, the number of lights, and the number of windows also significantly affect the indoor environment and the perceived air quality of the space. These aspects and other similar spatial and environmental independent factors highlight that the quality of indoor environments, as well as the variations in human perception across floor levels, result from the features of building designs and the complex environmental dynamics and conditions in heavily populated and climate-sensitive areas.

The results strongly suggest that, rather than a flat floor and uniformity assumption throughout a building, architects, engineers, facility managers, and policymakers take into account the differences in air quality, occupant perception (thermal and visual comfort), and general indoor environmental conditions that occur across different floor levels. In cities that are both densely populated and developing, like Dhaka, where ventilation patterns, building forms, and surrounding climates intertwine in multifaceted ways, optimal and harmonized indoor conditions across all floors must be accounted for. In addition to environmental parameters like temperature, humidity, and CO_2, design elements or layout-related parameters like the number of fans, windows, and lights are considered. Oversights in this regard lead to inefficient energy utilization, lost productivity, and discomfort. Thus, design approaches must focus on the specific comfort challenges of each floor, targeting solutions such as adaptive control of lighting, strategic window placement, localized ventilation enhancement, and the optimized distribution of fans. Hybrid modeling methods: Future studies may consider combined energy simulation and machine learning approaches for building performance prediction. Also, environmental parameters, like wind velocity, and layout-related parameters, like floor area, could be added from different places, like the lobby, corridor, and stairs. In this way, the benefits of both fields (machine learning's real-world pattern-recognition capabilities and simulation's hypothetical modeling abilities) would be extended to each other to create more accurate, reliable, and practical predictive tools. This study finally shows that suitable explainable-AI frameworks, as mentioned, e.g., those in the form of SHAP analysis, can provide useful help for improving sustainable building design, energy efficiency maximization, and indoor comfort improvement. By embracing these tactics, foregrounding those specifically aimed at the twin pressures of rapid population growth and environmental stress, urban environments can become smarter, healthier, body-friendly, and more adaptable for building occupants both now and in the future.

Acknowledgements This research did not receive any specific grant from funding agencies in the public, commercial, or not-for-profit sectors.

Data Availability Data will be made available on request.

AI Use Disclosure Statement During the preparation of this manuscript, the author utilized pre-trained large language models to assist with grammar refinement and clarity enhancement. The author subsequently reviewed and edited the text thoroughly and assumes full responsibility for the final published version.

References

1. Sullivan AM (1991) Tall buildings on cheap land: building heights and intrabuilding travel costs. J Urban Econ 29. https://doi.org/10.1016/0094-1190(91)90004-Q
2. Zhang J (2020) Integrating IAQ control strategies to reduce the risk of asymptomatic SARS CoV-2 infections in classrooms and open plan offices. Sci Technol Built Environ 26:1013–1018. https://doi.org/10.1080/23744731.2020.1794499

3. Ma G, Pan X (2021) Research on a visual comfort model based on individual preference in China through machine learning algorithm. Sustainability 13:7602. https://doi.org/10.3390/su13147602
4. Davoodi A, Johansson P, Aries M (2020) The use of lighting simulation in the evidence-based design process: a case study approach using visual comfort analysis in offices. Build Simul 13:141–153. https://doi.org/10.1007/s12273-019-0578-5
5. Shah R, Pandit RK, Gaur MK (2022) Thermal comfort analysis through development of artificial neural network models: an experimental study in Cwa climate. Mater Today Proc 57:2018–2025. https://doi.org/10.1016/j.matpr.2021.11.139
6. Uddin MN, Wang Q, Wei HH et al (2021) Building information modeling (BIM), system dynamics (SD), and agent-based modeling (ABM): towards an integrated approach. Ain Shams Eng J 12:4261–4274. https://doi.org/10.1016/j.asej.2021.04.015
7. Ngarambe J, Irakoze A, Yun GY, Kim G (2020) Comparative performance of machine learning algorithms in the prediction of indoor daylight illuminances. Sustainability 12:4471. https://doi.org/10.3390/su12114471
8. Han Y, Shen L, Sun C (2021) Developing a parametric morphable annual daylight prediction model with improved generalization capability for the early stages of office building design. Build Environ 200:107932. https://doi.org/10.1016/j.buildenv.2021.107932
9. Sadrizadeh S, Yao R, Yuan F et al (2022) Indoor air quality and health in schools: a critical review for developing the roadmap for the future school environment. J Build Eng 57. https://doi.org/10.1016/j.jobe.2022.104908
10. Brink HW, Loomans MGLC, Mobach MP, Kort HSM (2021) Classrooms' indoor environmental conditions affecting the academic achievement of students and teachers in higher education: a systematic literature review. Indoor Air 31. https://doi.org/10.1111/ina.12745
11. Rosbach JT, Vonk M, Duijm F et al (2013) A ventilation intervention study in classrooms to improve indoor air quality: the FRESH study. Environ Health 12. https://doi.org/10.1186/1476-069X-12-110
12. Satish U, Mendell MJ, Shekhar K et al (2012) Is CO_2 an indoor pollutant? Direct effects of low-to-moderate CO_2 concentrations on human decision-making performance. Environ Health Perspect 120:1671–1677. https://doi.org/10.1289/ehp.1104789
13. Mba EJ, Oforji PI, Okeke FO et al (2025) Assessment of floor-level impact on natural ventilation and indoor thermal environment in hot-humid climates: a case study of a mid-rise educational building. Buildings 15:686. https://doi.org/10.3390/buildings15050686
14. Dahlblom M, Tech L, Jensen L. Vertical temperature increase in multi-storey buildings
15. Chatzidiakou L, Archer R, Beale V et al (2023) Schools' air quality monitoring for health and education: methods and protocols of the SAMHE initiative and project. Dev Built Environ 16. https://doi.org/10.1016/j.dibe.2023.100266
16. Nicol F, Humphreys M, Roaf S (2012) Adaptive thermal comfort: principles and practice. Routledge
17. Afifi S, Kamel T, Ezzeldin S (2025) Indoor environmental quality assessment of naturally-ventilated school classrooms within a dense arid urban setting of Cairo. Egypt Sci Rep 15:16245. https://doi.org/10.1038/s41598-025-98602-y
18. Romero P, Valero-Amaro V, Rubio S, Miranda MT (2024) An analysis of thermal comfort as an influencing factor on the academic performance of university students. Educ Sci (Basel) 14:1340. https://doi.org/10.3390/educsci14121340
19. Singh MK, Ooka R, Rijal HB et al (2019) Progress in thermal comfort studies in classrooms over last 50 years and way forward. Energy Build 188–189
20. Ma J, Yu Z, Qu Y et al (2020) Application of the xgboost machine learning method in PM2.5 prediction: a case study of shanghai. Aerosol Air Qual Res 20. https://doi.org/10.4209/aaqr.2019.08.0408
21. Uddin MN, Wei H-H, Chi H-L et al (2022) Building layout influence on occupant's energy consumption behaviour: an agent-based modeling approach. https://doi.org/10.3390/environsciproc2022015022

22. Apostolopoulos ID, Dovrou E, Androulakis S et al (2025) Monitoring of indoor air quality in a classroom combining a low-cost sensor system and machine learning. Chemosensors 13:148. https://doi.org/10.3390/chemosensors13040148
23. Zimmerman N, Presto AA, Kumar SPN et al (2018) A machine learning calibration model using random forests to improve sensor performance for lower-cost air quality monitoring. Atmos Meas Tech 11:291–313. https://doi.org/10.5194/amt-11-291-2018
24. Breiman L (2001) Random forests. Mach Learn 45:5–32. https://doi.org/10.1023/A:1010933404324
25. Chen T, Guestrin C (2016) XGBoost. In: Proceedings of the 22nd ACM SIGKDD international conference on knowledge discovery and data mining. ACM, New York, NY, USA, pp 785–794
26. Safavian SR, Landgrebe D (1991) A survey of decision tree classifier methodology. IEEE Trans Syst Man Cybern 21:660–674. https://doi.org/10.1109/21.97458
27. Lundberg SM, Lee SI (2017) A unified approach to interpreting model predictions. In: Advances in neural information processing systems
28. Fayyaz M, Farhan AA, Javed AR (2022) Thermal comfort model for HVAC buildings using machine learning. Arab J Sci Eng 47:2045–2060. https://doi.org/10.1007/s13369-021-06156-8
29. Uddin MN, Lee M, Cui X et al (2024) Thermal and visual comforts of occupants for a naturally ventilated educational building in low-income economies: a machine learning approach. J Build Eng 94:110015. https://doi.org/10.1016/j.jobe.2024.110015
30. Ke H, Gong S, He J et al (2022) A hybrid XGBoost-SMOTE model for optimization of operational air quality numerical model forecasts. Front Environ Sci 10. https://doi.org/10.3389/fenvs.2022.1007530
31. Babich F, Torriani G, Corona J, Lara-Ibeas I (2023) Comparison of indoor air quality and thermal comfort standards and variations in exceedance for school buildings. J Build Eng 71:106405. https://doi.org/10.1016/j.jobe.2023.106405
32. Korsavi SS, Montazami A, Mumovic D (2020) The impact of indoor environment quality (IEQ) on school children's overall comfort in the UK; a regression approach. Build Environ 185:107309. https://doi.org/10.1016/j.buildenv.2020.107309
33. Chai Q, Wang H, Zhai Y, Yang L (2020) Using machine learning algorithms to predict occupants' thermal comfort in naturally ventilated residential buildings. Energy Build 217:109937. https://doi.org/10.1016/j.enbuild.2020.109937
34. Frontczak M, Wargocki P (2011) Literature survey on how different factors influence human comfort in indoor environments. Build Environ 46:922–937. https://doi.org/10.1016/j.buildenv.2010.10.021

A Novel Approach to Solve Triangular Fuzzy Multi Objective Linear Programming Problem Using Novel Similarity Measure

Pallabi Pal, Sayanta Chakraborty, and Apu Kumar Saha

Abstract Multiple Objective Linear Programming Problem (MOLPP) deals with the problems where more than one objective are present. In decision making problems of real life, there exists many scenarios where the information regarding the problem is full of uncertainty and vagueness. In such cases, crisp MOLPP fails to solve these problems because of imprecise data. To overcome this difficulty, Fuzzy Multiple Objective Linear Programming Problem (FMOLPP) is used. On the other side, similarity measure (SMr) indicates how similar two objects are. The aim of the present study is to introduce a new method to solve Triangular Fuzzy Multi Objective Linear Programming Problem (TFMOLPP) with the help of a novel SMr. To carry out the present research, firstly, a novel SMr has been introduced. All the salient traits of an ideal SMr has been examined theoretically and also validated through numerical examples. Secondly, a closeness degree has been formulated to maximize the similarity of each of the objective value with the corresponding ideal value, regarded as the optimistic value (OV) of that objective. Here the aim of the present research is to find a solution which provides the objective values nearer to the ideal objective values. The proposed method has been illustrated by modelling and solving a numerical example. It could be observed that the proposed SMr based solution process elicits an efficient solution which provides the objective values nearer to their ideal values. To check the reliability of the proposed model, the results have been compared with the solutions obtained by an existing method.

Keywords Similarity measure · Triangular fuzzy number · Multi-objective linear programming problem · Optimistic objective value

P. Pal (✉) · S. Chakraborty · A. K. Saha
Department of Mathematics, National Institute of Technology Agartala, Agartala, Tripura, India
e-mail: pallabipal706@gmail.com

A. Azizi et al. (eds.), *Advanced Decision-Making Under Uncertainty*, Emerging Trends in Mechatronics, https://doi.org/10.1007/978-981-95-8696-7_5

1 Introduction

In real life decision making problems where multiple objectives are associated, MOLPP can be used to solve them. MOLPP can deal with the problems in which multiple objectives are involved, where objectives are conflicting in nature. In this case, one cannot find an optimal solution which can simultaneously optimize all the objectives. So, one has to search for a compromise solution or trade-off solutions or efficient solution or non-dominated solution. To deal with multiple objectives, there are different methods like (i) Ideal Multi Objective procedure, (ii) Preference based approach. In the ideal multi objective procedure, at first multiple trade-off solutions are computed. Then among the multiple trade-off solutions, one trade-off solution is taken by using higher level information. In the preference based approach, by assigning preferences to each of the objective based on the decision maker's choice, a composite objective function is formulated to convert the multiple objectives to the single objective function and then the composite objective function is optimized. It has various applications in business, finance, logistics, supply chain and in the different sectors like agriculture, healthcare etc.

But, there are certain situations, where uncertainty may associate in real life decision making. In that case, crisp set theory fails to deal with such situation. So, to handle such ambiguity, Zadeh introduced Fuzzy Set (FS) [1], which is an ordered pair, consisting of elements and its corresponding membership degrees. Membership degree represents the belongingness of the element in the set. Generally, to quantify the fuzziness in problem solving, they can be represented by fuzzy numbers (FNs) [2]. FNs are convex, normalized fuzzy sets, where for exactly one element in the fuzzy set, the corresponding membership value will be 1 and the membership function should be piecewise continuous. Triangular Fuzzy Number (TFN) [2] is one of the efficient FN, which has been widely used to solve real life decision-making problems. FS and hierarchical extension such as TFNs gives strong representation of uncertainty. The crisp MOLPP cannot deal with the situations where the uncertainty exists, so, to overcome this, FMOLPP [2] is used to solve the problem. FMOLPP is an extension of crisp MOLPP by incorporating FNs, which is able to handle vagueness and uncertainty in the real life decision making problems. FMOLPP is useful in different areas like production management, finance, supply chain management etc. But, there exists no general method to solve FMOLPP. So, our aim to develop a mathematical model to solve FMOLPP, which can provide a solution eliciting objective values nearer to ideal values.

On the other hand, SMr is a mathematical tool used to measure the similarity between two objects, points or sets. SMr is widely used in pattern recognition, image processing, multi-criteria decision making, data mining, clustering etc. In FMOLPP, SMr can be used to compare different solutions to find out which solution is closer to the ideal solution.

The aim of the present study is to develop a novel SMr based closeness degree, which will provide a solution yielding the objective values closer to the corresponding ideal objective values. The maximization of the closeness degree will provide a solution, obtaining objective values nearer to the corresponding ideal values. To validate the applicability of the proposed SMr based approach to solve FMOLPP, a numerical example has been structured and solved and the results have been compared with an existing method.

1.1 Literature Review

In this section, a brief literature review on the approaches used to solve fuzzy MOLPPs have been discussed below.

Dong and Wan [3] proposed a two stage method to solve FMOLPP with the help interval ranking and TOPSIS. Hashemi et al. [4] proposed a modified multi objective linear programming model for project management. Premkumar [5] proposed a multi objective arithmetic optimization algorithm for solving real life multi objective optimization problem. Deng et al. [6] developed a fuzzy random multi objective portfolio model with the help of entropy measures and a method to solve the developed model. Wang et al. [7] solved multi objective fuzzy job scheduling problem by using a hybrid adaptive differential evolution algorithm. Gulia et al. [8] evaluated the efficiencies, challenges and applications of the techniques to solve FMOLPP. Yuvashri and Saraswathi [9] proposed an approach to solve spherical fuzzy multi objective linear programming problem with the help of score function and Zimmermann's approach. Ali and Javaid [10] presented a mathematical model of traditional transportation problem with the help of Fermatean fuzzy parameters and transformed it into crisp form by using a new score function. Ali et al. [11] proposed fuzzy programming approach, weighted sum approach and neutrosophic goal programming approach to solve multi objective linear plus linear fractional programming problem. Malik and Gupta [12] proposed a method to solve multi objective linear fractional programming problems with the help of linearization, accuracy function, membership function and non-membership function.

1.2 Research Gap and Novelty

From the literature review, it could be observed that there is hardly any research where SMr based approach has been developed to solve FMOLPP. Moreover, very few researches focused on developing solution processes for MOLPPs, which can provide a solution to obtain an objective value nearer to the ideal solution. So, it is pertinent to develop a solution model of solving MOLPP, which can elicit a solution providing an objective value nearer to the ideal value, with the aid of an appropriate SMr. Thus the present study has focused on the following aspects to develop such an efficient model to solve MOLPP under fuzzy environment.

- A new SMr has been proposed, which follows all the significant properties of an ideal SMr.
- A novel closeness degree has been developed with the help of the proposed SMr, which will signify how closer the derived objective value with that of the ideal objective value of a FMOLPP.
- By using the proposed SMr and proposed closeness degree, a new method has been introduced to solve FMOLPP under triangular fuzzy environment, which will identify a solution providing an objective value closer to the ideal one.

1.3 Contributing Work

Along with all the novel aids, the present study has contributed the followings also:

- To find the similarity between objectives and their corresponding OVs and pessimistic values (PVs), based on which the closeness degree has been computed.
- To validate the applicability of the proposed model, a numerical example has been modelled and solved.
- To check the reliability of the results obtained by the proposed method, a comparative study has been executed with an existing method.

The remaining paper is organized as:

Section 2 consists of the basic concepts related to the present work. In Sect. 3, the proposed work is discussed. In Sect. 4, the application of proposed work is discussed by solving a numerical example. Result analysis has been done in Sect. 5. The conclusion and the future scope of present work have been described in Sect. 6.

2 Preliminaries

In this section, the basic concepts related to fuzzy MOLPP and SMr have been discussed.

2.1 Multi Objective Linear Programming Problem (MOLPP)

The MOLPP model [13, 14] can be described as

$$\max\{Z_s(x) = C_s^T x, s = 1, 2, .., k\}$$
$$\text{Subject to } \begin{cases} Ax \le b \\ x \ge 0 \end{cases} \tag{1}$$

where Z_s is sth objective function, $C_s^T = (C_{s1}, C_{s2}, ..., C_{sw,})$ is the cost vector corresponding to the sth objective function, $x = (x_1, x_2, .., x_w)^T$ is the vector of decision variable, $A = (a_{ij})_{v\times w}$ is the coefficient matrix, $b = (b_1, b_2, ..., b_v)^T$ is the resource vector.

2.2 *Efficient Solution*

$\overline{x}$ is called efficient solution [2] of model (1) if there exist no x' such that $Z_s(x') \geq Z_s(\overline{x})$, $s = 1, 2, .., k$ and $Z_s(x') > Z_s(\overline{x})$ for at least one $s = 1, 2, .., k$.

2.3 *Ideal Solution*

Let x_s^0 be the optimal solution to the sth objective of model (1), $s = 1, 2, ..., k$. Then the ideal solution [15] of model (1) is the value of the objective function $(Z_1(x_1^0), Z_2(x_2^0), ..., Z_k(x_k^0))$.

2.4 *Best Compromise Solution*

Best compromise solution [15] of model (1) is an efficient solution x^* of model (1) such that distance between the ideal objective values of model (1) and the objective value at x^*, that is, $(Z_1(x^*), Z_2(x^*), ..., Z_k(x^*))$ is minimum among the efficient solutions of the model (1).

2.5 *Fuzzy Set (FS)*

Let X be the universal set, $\tilde{A} = \{(x,\ \delta_{\tilde{A}}(x))|x \in X\}$ is called FS [2] in X, where $\delta_{\tilde{A}}(x)$ is called the membership degree.

2.6 *Fuzzy Number (FN)*

A FS $\widetilde{\mathcal{N}} = (x,\ \delta_{\widetilde{\mathcal{N}}}(x))$ is called a FN [2], if $\widetilde{\mathcal{N}}$ is convex normalized set of real line R such that

(a) There exists exactly one $x_0 \in R$ with $\delta_{\widetilde{\mathcal{N}}}(x_0) = 1$.
(b) $\delta_{\widetilde{\mathcal{N}}}(x)$ is piecewise continuous.

2.7 Triangular Fuzzy Number (TFN)

A TFN [2] can be defined as $\widetilde{\mathcal{T}} = (\bar{a}_1, \bar{a}_2, \bar{a}_3)$, where the membership function $\delta_{\widetilde{\mathcal{T}}}(x)$ of $\widetilde{\mathcal{T}}$ is defined as

$$\delta_{\widetilde{\mathcal{T}}}(x) = \begin{cases} 0, \bar{a}_1 \leq x \\ \frac{x-\bar{a}_1}{\bar{a}_2-\bar{a}_1}, \bar{a}_1 \leq x \leq \bar{a}_2 \\ 1, x = \bar{a}_2 \\ \frac{\bar{a}_3-x}{\bar{a}_3-\bar{a}_2}, \bar{a}_2 \leq x \leq \bar{a}_3 \\ 0, x \geq \bar{a}_3 \end{cases}$$

Here $\bar{a}_2$ is the mean value and $\bar{a}_1$ and $\bar{a}_3$ are the left and right spreads respectively.

2.8 Operation of TFN [16]

Let $\tilde{T}' = (t_1, t_2, t_3)$ and $\tilde{G}' = (g_1, g_2, g_3)$ be two TFNs. Then.

(i) Addition

$$\tilde{T}' + \tilde{G}' = (t_1, t_2, t_3) + (g_1, g_2, g_3) = (t_1 + g_1, t_2 + g_2, t_3 + g_3)$$

(ii) Subtraction

$$\tilde{T}' - \tilde{G}' = (t_1, t_2, t_3) - (g_1, g_2, g_3) = (t_1 - g_3, t_2 - g_2, t_3 - g_1)$$

(iii) Multiplication

$$\tilde{T}' \times \tilde{G}' = (\min(t_1g_1, t_1g_3, t_3g_1, t_3g_3), t_2g_2, \max(t_1g_1, t_1g_3, t_3g_1, t_3g_3))$$

(iv) Division

$$\tilde{T}'/\tilde{G}' = (\min(t_1/g_1, t_1/g_3, t_3/g_1, t_3/g_3), t_2g_2, \max(t_1/g_1, t_1/g_3, t_3/g_1, t_3/g_3))$$

(v) Symmetric image

$$-\left(\tilde{T}'\right) = (t_3, t_2, t_1)$$

2.9 Defuzzification Method of TFN

Let $\widetilde{\mathcal{T}} = (\bar{a}_1, \bar{a}_2, \bar{a}_3)$ be a TFN. Then the defuzzified crisp value of $\widetilde{\mathcal{T}}$[17]

$$\mathcal{T} = \frac{\bar{a}_1 + \bar{a}_2 + \bar{a}_3}{3} \tag{2}$$

2.10 Similarity Measure (SMr)

Let X be any non-empty set. A function $\acute{S}: X \times X \rightarrow R$ is a SMr [18], if for any $x, v \in X$, the following properties are satisfied.

$$\acute{S}(x, v) \geq 0$$

$$\acute{S}(x, v) = \acute{S}(v, x)$$

$$\acute{S}(x, x) > \acute{S}(x, v)$$

3 Proposed SMr Based Approach to Solve TFMOLPP

In this section, a novel SMr has been proposed and its salient features have been analyzed both theoretically as well as with numerical example. Based on the proposed SMr, a closeness degree has also been formulated, which can elicit an efficient solution of the TFMOLPP closer to the ideal objective value. Thereafter, a new method has been developed with the aid of the proposed SMr to solve TFMOLPP.

3.1 Proposed Novel SMr

Let R^n be the set of n-tuples of real numbers and $V = (v_1, v_2, \ldots, v_n)$ and $R = (r_1, r_2, \ldots, r_n)$ be two points of R^n.

For any $V, R \in R^n$, the proposed SMr, $S: R^n \times R^n \rightarrow R$ is defined as

$$S(V, R) = \frac{\sum_{i=1}^{n} \left|(v_i - r_i)^2 - 1\right|}{\sum_{i=1}^{n} \left|(v_i - r_i)^2 + 1\right|} \tag{3}$$

The prominent traits of the proposed SMr have been discussed below:

Prop. 1 $S(V, R) \geq 0$.

Proof. $\sum_{i=1}^{n} \left|(v_i - r_i)^2 - 1\right| \geq 0$ and $\sum_{i=1}^{n} \left|(v_i - r_i)^2 + 1\right| \geq 1$.

Table 1 Coordinates of the points

P	Ń	Ģ	Ĥ	Ŭ	Ľ	Į	Ɏ
(1,5)	(2,3)	(−5,4)	(7,11)	(−6,−13)	(−9,−8)	(10,0)	(−15,−1)

So, $\frac{\sum_{i=1}^{n}\left|(v_i - r_i)^2 - 1\right|}{\sum_{i=1}^{n}\left|(v_i - r_i)^2 + 1\right|} \geq 0.$

Therefore, $S(V, R) \geq 0$.

Prop. 2 $S(V, R) = S(R, V)$.

Proof.

$$S(V, R) = \frac{\sum_{i=1}^{n}\left|(v_i - r_i)^2 - 1\right|}{\sum_{i=1}^{n}\left|(v_i - r_i)^2 + 1\right|} = \frac{\sum_{i=1}^{n}\left|(r_i - v_i)^2 - 1\right|}{\sum_{i=1}^{n}\left|(r_i - v_i)^2 + 1\right|} = S(R, V)$$

Prop. 3 $S(V, V) > S(V, R)$.

$$S(V, V) = \frac{\sum_{i=1}^{n}\left|(v_i - v_i)^2 - 1\right|}{\sum_{i=1}^{n}\left|(v_i - v_i)^2 + 1\right|} = 1$$

Clearly, $S(V, R) = \frac{\sum_{i=1}^{n}\left|(v_i - r_i)^2 - 1\right|}{\sum_{i=1}^{n}\left|(v_i - r_i)^2 + 1\right|} < 1.$

Therefore, $S(V, V) > S(V, R)$.

3.1.1 Validation of the Proposed SMr by Numerical Examples

To authenticate the integrity of the proposed SMr, all the properties have been validated through the following numerical example.

Let (P, Ń),(Ģ, Ĥ),(Ŭ, Ľ)and (Į, Ɏ) be four pairs of points in R^2. The coordinates of the points have been presented in Table 1. Prop.1 to Prop.3 have been validated through these points and the results have been presented in Table 2.

From Table 2, it could be observed that the properties have been satisfied for the given points, which ensures the authenticity of the proposed SMr.

3.1.2 Proposed Closeness Degree

The proposed closeness degree is defined as

Table 2 Verification of the properties of SMr

SL no.	Pairs of points	Similarity (*S*)	Prop.1	Prop.2	Prop.3
1	(P, Ń)	0.43	Satisfied	Satisfied	Satisfied
	(P, P)	1			
	(Ń, P)	0.43			
2	(Ģ, Ĥ)	0.98	Satisfied	Satisfied	Satisfied
	(Ģ, Ģ)	1			
	(Ĥ, Ģ)	0.98			
3	(Ŭ, Ľ)	0.89	Satisfied	Satisfied	Satisfied
	(Ŭ, Ŭ)	1			
	(Ľ, Ŭ)	0.89			
4	(Į, Ɏ)	0.99	Satisfied	Satisfied	Satisfied
	(Į, Į)	1			
	(Ɏ, Į)	0.99			

$$n(x) = \frac{S^{opt}(x)}{S^{opt}(x) + S^{pess}(x)} \tag{4}$$

Here $S^{opt}(x)$ and $S^{pess}(x)$ denote the similarities between the objective values with their corresponding OVs and PVs respectively. This degree measures the similarity between the objective values with their ideal values corresponding to the total similarity from both OVs and PVs.

3.2 Proposed Approach to Solve TFMOLPP

In this sub-section, the new method has been developed to solve fuzzy MOLPP with the aid of the proposed SMr and closeness coefficient. The aim is to find a solution of the problem which will elicit an ideal value nearer to the ideal objective value. So, SMr has been considered here based on which a closeness coefficient has been developed and this measure will provide values of objectives nearer to ideal values. Here, using the closeness degree, the MOLPP has been converted into single objective LPP. The algorithm of the proposed approach has been discussed below:

Step 1. Construct the TFMOLPP a

$$\begin{aligned} &\max\left\{\tilde{z}_k(x) = \tilde{C}_k^T x\ ,\ k = 1, 2, .., q\right\} \\ &\text{s.t}\ \begin{cases} \tilde{A}x \le \tilde{b} \\ x \ge 0 \end{cases} \end{aligned} \tag{5}$$

Here $\tilde{z}_k$, $\tilde{C}_k^T$, $\tilde{A}$ and $\tilde{b}$ are the triangular fuzzy representation of z_k, C_k^T, A and b respectively.

Step 2. Using Eq. (2), convert all the TFNs in model (5) to crisp numbers and transform the TFMOLPP into crisp MOLPP. Therefore the model changes to

$$\max \left\{ z_k(x) = C_k^T x, k = 1, 2, .., q \right\}$$
$$\text{s.t} \begin{cases} Ax \le b \\ x \ge 0 \end{cases} \tag{6}$$

Step 3. Determine the OV and PV for each objective.

I. **Calculation of OV for the** ith **Objective**

To find out the OV for the ith objective in model (6), consider the ith objective along with all the constraints and solve it. For maximizing objective, the maximum value will be the OV and for minimizing objective, its minimum value will be the OV. Suppose, the OV for the ith objective be Z_i^{opt}.

II. **Calculation of PV for the** ith **objective**

To find out the PV for the ith objective, put the optimal solution of the other objectives to the ith objective function. Among those values, choose the highest value as the PV of the ith objective. Suppose, that value be Z_i^{pess}. Continue the process for all the objectives to calculate their OVs and PVs.

Step 4. Compute the closeness degree n(x) by using the Eq. (4).

Step 5. Maximize n(x) with respect to the following constraints.

$$\max \ n(x)$$
$$\text{Subject to} \begin{cases} Ax \le b \\ x \ge 0 \end{cases} \tag{7}$$

Step 6. Solve model (7) using usual simplex method.

4 Application of the Proposed Approach

To validate the applicability of the proposed SMr based model of solving TFMOLPP, a numerical example has been modelled and solved using the proposed method, which has been discussed below.

4.1 A Numerical Example

To check the applicability of the proposed SMr based FMOLPP, a numerical example [3] has been structured and solved. The application of the proposed approach to solve the example has been discussed below:

Step 1. The TFMOLPP has been given below:

$$\text{Max } \tilde{z}_1(x) = (40, 50, 80)x_1 + 100\ x_2 + 17.5\ x_3$$

$$\text{Max } \tilde{z}_2(x) = 92x_1 + (70, 75, 90)x_2 + 50\ x_3$$

$$\text{Max } \tilde{z}_3(x) = (10, 20, 70)x_1 + 100\ x_2 + 75x_3$$

$$\text{Such that} \begin{cases} (6, 12, 14)x_1 + 17x_2 \le 1400 \\ 3x_1+9\ x_2 + (3, 8, 10)x_3 \le 1000 \\ 10x_1 + (7, 13, 15)x_2 + 15x_3 \le 1750 \\ (4, 6, 8)x_1 + 16x_3 \le 1325 \\ (7, 12, 19)x_2 + 7x_3 \le 900 \\ 9.5\ x_1 + (3.5, 9.5, 11.5)x_2 + 4x_3 \le 1075 \end{cases} \tag{8}$$

$$x_t \ge 0(t = 1, 2, 3)$$

Step 2. Using the Eq. (2), all the TFNs of model (8) have been converted to crisp numbers and accordingly, the TFMOLPP has been converted to crisp MOLPP, shown in model (9).

The crisp MOLPP is

$$\text{Max } z_1(x) = 56.67x_1 + 100x_2 + 17.5x_3$$

$$\text{Max } z_2(x) = 92x_1 + 78.33x_2 + 50x_3$$

$$\text{Max } z_3(x) = 33.33x_1 + 100x_2 + 75x_3$$

$$\text{Such that} \begin{cases} 10.67\ x_1 + 17x_2 \le 1400 \\ 3x_1 + 9x_2 + 7x_3 \le 1000 \\ 10x_1 + 16.67x_2 + 15x_3 \le 1750 \\ 6x_1 + 16x_3 \le 1325 \\ 12.67x_2 + 7x_3 \le 900 \\ 9.5x_1 + 8.17x_2 + 4x_3 \le 1075 \end{cases} \tag{9}$$

$$x_t \ge 0(t = 1, 2, 3)$$

Step 3. By the proposed method, the optimistic and pessimistic values have been calculated for all the objectives and represented in Table 3.

Step 4. Using Eqn. (3), similarity to OV and PV of the objectives have been computed and presented below:

$$s^{opt}(x) = \frac{\left|(56.67x_1 + 100x_2 + 17.5x_3 - 8446.08)^2 - 1\right| + \left|(92x_1 + 78.33x_2 + 50x_3 - 10979.49)^2 - 1\right| + \left|(33.33x_1 + 100x_2 + 75x_3 - 9025.63)^2 - 1\right|}{\left|(56.67x_1 + 100x_2 + 17.5x_3 - 8446.08)^2 + 1\right| + \left|(92x_1 + 78.33x_2 + 50x_3 - 10979.49)^2 + 1\right| + \left|(33.33x_1 + 100x_2 + 75x_3 - 9025.63)^2 + 1\right|}$$

$$s^{pess}(x) = \frac{\left|(56.67x_1 + 100x_2 + 17.5x_3 - 6511.78)^2 - 1\right| + \left|(92x_1 + 78.33x_2 + 50x_3 - 9540.13)^2 - 1\right| + \left|(33.33x_1 + 100x_2 + 75x_3 - 9000.95)^2 - 1\right|}{\left|(56.67x_1 + 100x_2 + 17.5x_3 - 6511.78)^2 + 1\right| + \left|(92x_1 + 78.33x_2 + 50x_3 - 9540.13)^2 + 1\right| + \left|(33.33x_1 + 100x_2 + 75x_3 - 9000.95)^2 + 1\right|}$$

Step 5. Using Eq. (4), model (9) has been converted to single objective LPP as. follows;

$$\begin{aligned} &\text{Max } n(x) \\ &\text{Subject to} \begin{cases} 10.67x_1 + 17x_2 \le 1400 \\ 3x_1 + 9x_2 + 7x_3 \le 1000 \\ 10x_1 + 16.67x_2 + 15x_3 \le 1750 \\ 6x_1 + 16x_3 \le 1325 \\ 12.67x_2 + 7x_3 \le 900 \\ 9.5x_1 + 8.17x_2 + 4x_3 \le 1075 \end{cases} \\ &x_t \ge 0 (t = 1, 2, 3) \end{aligned} \tag{10}$$

Step 6. Model (10) has been solved using Lingo 20.0 software and the solution has been presented in Table 4.

From the Tables 3 and 4 it could be observed that the objective values obtained by the proposed method are closer to the optimistic values.

5 Result Analysis

To check the reliability of the proposed approach, the results obtained by the proposed method have been compared to the results obtained by Dong and Wan [3] and have been shown in Table 5.

Table 3 The OVs and PVs of the objectives

Solution			Values of objectives			OV			PV		
x_1	x_2	x_3	z_1	z_2	z_3	z_1	z_2	z_3	z_1	z_2	z_3
41.29	56.44	26.41	8446.08	9540.13	9000.95	8446.08	10,979.49	9025.63	6511.78	9540.13	9000.95
84.51	8.28	51.12	6511.78	10,979.49	7478.72						
11.03	27.57	78.68	4758.97	7108.32	9025.63						

Table 4 Values of objectives attained by the proposed approach

Value of proposed closeness degree	Solution			Values of objectives		
0.6099	x_1	x_2	x_3	z_1	z_2	z_3
	62.30	28.29	43.68	7124.93	10,132.61	8182.55

Table 5 Comparison with the existing method

	Solution			Values of objectives		
	x_1	x_2	x_3	z_1	z_2	z_3
Proposed method	62.30	28.29	43.68	7124.93	10,132.61	8182.55
Dong and Wan [3]	52.67	36.71	47.29	7483.43	10,085.87	8973.15

From Table 5, it could be observed that the proposed method gives the solution closer to the optimistic value for the second objective, which is better than the value obtained by the existing approach. For the other two objectives, the results are closer to the existing approach. So it could be concluded that the results obtained by the proposed approach are reliable.

6 Conclusion

MOLPP can handle the situations where several conflicting objectives are connected. But MOLPP can not deal with imprecise information, so the FMOLPP is used to solve these types of problems. FMOLPP is widely used in many areas like business, transportation and logistics, health care, resource allocation etc. SMr is an effective metric to find the similarity between two sets or objects. In view of this characteristic of SMr, the present treatise aims to develop an appropriate SMr to solve FMOLPP by finding the closeness of the objective values with respect to their OVs and PVs by maximizing the closeness degree. To do so, a novel SMr has been developed and its prominent features have been investigated. A new method to solve MOLPP under triangular fuzzy environment has been introduced with the help of a novel SMr and a novel closeness degree. With the aid of the proposed measure, a closeness degree has been computed and maximized to get a solution of an FMOLPP, which can elicit objective values nearer to its ideal values. Here, all the parameter involved in the MOLPP have been considered as TFNs. To validate the applicability of the proposed approach, a numerical example has been solved. To check the reliability of the proposed method, the results have been compared with the results obtained by the existing method.

The key advantages of the proposed work are twofold. Firstly, the proposed novel SMr is an appropriate SMr, as it satisfies all the axiomatic traits of an ideal SMr.

The measure has been validated through numerical examples also. Secondly, the proposed SMr based approach can elicit a solution which can provide the values of the objectives nearer to their ideal values. However, the present work has considered only maximizing type of objectives, so our next endeavour is to develop a general type of model, where both maximizing and minimizing type of objectives will be considered. Moreover, instead of taking only TFNs, the parameters can be chosen as orthopair or orthotriplet fuzzy numbers, for which q-Rung Orthopair fuzzy sets and Spherical fuzzy sets may be taken into consideration.

AI Use Disclosure Statement The authors used generative AI technologies only for proofreading. This includes minor corrections to grammar, spelling, and syntax. No AI was used for ideation, data analysis, or writing original content.

References

1. Zadeh LA (1965) Fuzzy sets. Inf Control 8:338–353
2. Zimmermann HJ (2011) Fuzzy set theory—and its applications. Springer Science & Business Media
3. Dong J, Wan S (2019) A new method for solving fuzzy multi-objective linear programming problems. Iran J Fuzzy Syst 16(3):145–159
4. Hashemi SM, Mousavi SM, Patoghi A (2021) A fuzzy multi-objective mathematical programming model for project management decisions considering quality and contractual reward and penalty costs in a project network. Arab J Sci Eng 46(2):1617–1629
5. Premkumar M, Jangir P, Kumar BS, Sowmya R, Alhelou HH, Abualigah L, Mirjalili S (2021) A new arithmetic optimization algorithm for solving real-world multi-objective CEC-2021 constrained optimization problems: diversity analysis and validations. IEEE Access 9:84263–84295
6. Deng X, He X, Huang C (2022) A new fuzzy random multi-objective portfolio model with different entropy measures using fuzzy programming based on artificial bee colony algorithm. Eng Comput 39(2):627–649
7. Wang GG, Gao D, Pedrycz W (2022) Solving multiobjective fuzzy job-shop scheduling problem by a hybrid adaptive differential evolution algorithm. IEEE Trans Industr Inf 18(12):8519–8528
8. Gulia P, Kumar R, Viriyasitavat W, Aledaily AN, Yadav K, Kaur A, Dhiman G (2023) A systematic review on fuzzy-based multi-objective linear programming methodologies: concepts, challenges and applications. Arch Comput Methods Eng 30(8):4983–5022
9. Yuvashri P, Saraswathi A (2024) A novel approach for multi-objective linear programming model under spherical fuzzy environment and its application. J Intell Fuzzy Syst 46(2):3259–3280
10. Ali W, Javaid S (2025) A solution of mathematical multi-objective transportation problems using the fermatean fuzzy programming approach. Int J Syst Assur Eng Manag 1–19
11. Ali W, Sadia S, Javaid S, Abidi N, Nabeel M, Javaid MS, Nusrat A (2025) Solving the multi-objective linear plus linear fractional programming problem using different mathematical programming approaches. Palest J Math 14(1)
12. Malik M, Gupta SK (2025) On optimistic, pessimistic and mixed fuzzy-programming based approaches to solve multi-objective fully intuitionistic fuzzy linear fractional programming problems. Ann Oper Res 346(2):13

13. Bhattacharya D, Chakraborty S (2018) Solution of the general multi-objective De-Novo programming problem using compensatory operator under fuzzy environment. In: Journal of physics: conference series, vol 1039, no 1, p 012012. IOP Publishing
14. Chakraborty S, Bhattacharya D (2022) Fuzzy approach to solve general de-novo programming problem. In: Advances in mathematical modeling, applied analysis and computation: proceedings of ICMMAAC 2021. Springer Nature Singapore, Singapore, 181–192
15. Pandian P, Jayalakshmi M (2013) Determining efficient solutions to multiple objective linear programming problems. Appl Math Sci 7(26):1275–1282
16. Gani AN, Assarudeen SM (2012) A new operation on triangular fuzzy number for solving fuzzy linear programming problem. Appl Math Sci 6(11):525–532
17. Voskoglou M (2015) Use of the triangular fuzzy numbers for student assessment. arXiv:1507.03257
18. Bădică A, Bădică C, Logofătu D, Neremzoiu ID (2025) Generalizations and properties of normalized similarity measures for Boolean models. Mathematics 13(3):384

Intelligent Prediction of Mechanical Properties of Fiber-Reinforced Soils: A Comprehensive Review

Leyla Bouaricha, Jitendra Khatti, and Benathmane Baghdir

Abstract The use of AI for the characterization of fiber-reinforced soils (FRS) has dramatically improved geotechnical engineering by offering data-driven alternatives to traditional experimental methods. This review provides a comprehensive appraisal of AI methods, including machine learning (ML), deep learning (DL), and hybrid learning (HL) techniques, for the prediction of im-portant mechanical properties (unconfined compressive strength, shear strength, tensile strength, and deformation behavior) of FRS. In general, studies have shown that artificial neural networks (ANNs) and support vector machines (SVMs) tend to have better prediction capacities with coefficient of determination (R^2) values of greater than 0.99, suggesting they provide a more valid representation of nonlinear soil-fiber interactions than classical empirical models. Hybrid models such as GA-ANN and WFLSSVR also demonstrated improved prediction accuracy due to optimized metaheuristic algorithms with observed mean absolute percentage errors (MAPE) of less than 3%. Important complexities in model prediction can arise through fiber type (natural vs. synthetic), fiber length, and fiber and soil properties. Challenges still remain with regard to data availability for natural fibers, interpretability of models, and transferability of results to field scale. Future developments could be boosted by implementing approaches such as physics-informed machine learning, explainable AI (XAI), and IoT-based monitoring of site factors to link numerical models with real-world observations. Overall, AI presents transformative potential for optimizing FRS design and deployment, supporting sustainable geotechnical solutions through the use of standardized datasets and hybrid experimental-computational frameworks.

L. Bouaricha (✉)
Department of Civil Engineering and Architecture, Faculty of Sciences and Technology, University of Khemis Miliana, Miliana, Algeria
e-mail: l.bouaricha@univ-dbkm.dz

J. Khatti
Department of Civil Engineering, Rajasthan Technical University, Kota, Rajasthan, India
e-mail: jitendrakhatti197@gmail.com

B. Baghdir
Laboratory of Numerical and Experimental Investigation in Geotechnical Engineering, University of Saad Dahlab, Blida, Algeria
e-mail: othmanegcv@gmail.com

A. Azizi et al. (eds.), *Advanced Decision-Making Under Uncertainty*, Emerging Trends in Mechatronics, https://doi.org/10.1007/978-981-95-8696-7_6

Keywords Fiber-reinforced soil · Artificial intelligence · Machine learning · Mechanical properties · Geotechnical engineering · Predictive modeling

1 Introduction

Soil stabilization is the method of improving the engineering characteristics of the subsoil before any construction using different types of techniques such as preloading, sand drains, prefabricated vertical drains, and chemical additives [1–3]. In contrast to stabilization, reinforcement involves the incorporation of materials with desirable properties into another material to compensate for its deficiencies [4]. In geotechnical engineering, soil reinforcement is a commonly used practice to improve important soil properties, i.e., shear strength, compressibility, density, and hydraulic conductivity [5–8]. Earth is reinforced by alternating layers of compacted backfill and synthetic reinforcement, and the primary app, to improve stability, increase bearing capacity, improve settlement, and improve lateral deformations [9–11]. There have been a number of investigations reported, examining the advancement of the use of natural and synthetic fibers in soil reinforcing with the waste calcareous materials including bamboo, sugarcane, coconut fibers [12], jute [13], coir [14–16], lime [17–20], in addition to recycled materials such as tire cord waste and polypropylene (PP) fiber [19, 21–23].

Fibers enhance soil strength by forming a three-dimensional reinforcement network, which increases shear resistance and mitigates post-peak strength loss [24, 25]. According to studies, PP fibers are known to significantly increase the unconfined compressive strength (UCS) of clayey soils while improving their swelling potential due to their high acid-alkali resistance [26–28]. Likewise, glass fibers have shown suitability for improving the shear strength and permeability of fine-grained soils for subgrade stabilization [29, 30]. Fibers, unlike natural and synthetic fibers, experience less significant degradation under wet-dry cycles, which is crucial to ensuring their long-term performance without additional stabilization treatment [31]. In addition, the random fiber distribution in soil provides isotropic strength enhancement and does not induce weak planes which can occur to oriented reinforcements [32].

Laboratory methods to evaluate the mechanical properties of reinforced soil, although essential, are both lengthy to conduct and require human resources. Because of this time requirement, various artificial intelligence (AI) methods, including artificial neural networks (ANN), support vector machines (SVM), and gradient boosting (GB), have been used to evaluate the mechanical properties of treated soils, and hybrid methods such as gradient-boosted ANN and wavelet-bagged ANN have been shown to have positive influence on predicting the mechanical properties of reinforced soils [33]. Nevertheless, assessing the type, content, and size of fiber using trial and error ultimately requires significant time, which is a key driver of the AI-method use in soil improvement [34, 35]. However, optimizing fiber type, content, and size through trial and error remains time-consuming, underscoring the need for AI-driven solutions in soil improvement research [36, 38].

Recent studies have demonstrated the trustworthiness of artificial intelligence (AI) techniques in predicting the mechanical properties of fiber-reinforced soils [36]. Artificial Neural Networks (ANNs) were employed to model the unconfined compressive strength (UCS) and splitting tensile strength of polypropylene fiber–reinforced soil, achieving an R^2 of 0.95. Fiber content, curing conditions, and freeze–thaw cycles were identified as key input variables. Similarly, Wang et al. [39] used linear regression (LR) to predict the splitting tensile strength of basalt fiber (BF)-stabilized soil, attaining an R^2 of 0.99 by considering BF length, content, and curing time. The support vector regression (SVR) and ensemble methods demonstrated high accuracy [40, 41] with R^2 values of 0.98 and 0.99, respectively, for predicting the properties of soils reinforced with diverse fibers (e.g., polypropylene, steel, and coir). Meanwhile, Yu et al. [42] compared ANNs, SVM, and extreme learning machines (ELM) for UCS estimation, with ANNs yielding an R^2 of 0.94. Kumar and Harinder [43] further highlighted the versatility of ANN and SVM in evaluating glass fiber-reinforced soils, achieving an R^2 of 0.90 for the California Bearing Ratio (CBR) test. Overall, these investigations collectively highlight the promise of AI for fiber-reinforced soil design to limit expensive experimental trials. Over the years, AI techniques (i.e., ANNs, SVM, and fuzzy logics (FL)) have been successfully used to estimate or predict the mechanical properties of fiber-reinforced soils. For instance, ANNs have been used successfully to predict axial compressive strength-based soil moisture content with R values above 0.86 [44]. In another study, FL outperformed the other AI techniques when predicting the maximum deviator stress and deformation to failure, demonstrating their potential to optimize soil reinforcement approaches [45].

Moreover, Tinoco et al. [46] showed that ANNs estimated UCS and tensile strengths in soils mixed with natural (sisal) and synthetic (polypropylene) fibers with $R^2 > 0.95$. Similarly, Houcine et al. [47] used ANNs to predict compressive and tensile strength in compressed earth blends with natural fibers (palm, typha, millet waste) showing R value of 0.9773. Chou et al. [40] also obtained improved R values with ensemble and meta-ensemble techniques used for estimating the peak shear strength (i.e. $R = 0.94$; MAPE $\approx$ 3–5%). Ndepete et al. [45] showed fuzzy logic models were effective in predicting the deformation to failure ($R^2 = 0.938$) in silty soils reinforced with basalt fibers. Other studies highlight the effectiveness of support vector machines and random forests, with critical input variables including fiber type, length (12–76 mm), content (0.2–2%), and binder composition playing a key role in model performance [40, 45–47].

Recent advancements in artificial intelligence (AI), particularly machine learning (ML), deep learning (DL), and hybrid models, have demonstrated significant potential in predicting the mechanical properties of fiber-reinforced soils, offering a data-driven alternative to traditional experimental methods. Studies utilizing ML techniques, such as ANNs and random forest (RF), have achieved high predictive accuracy ($R^2 \geq 0.93$) for elastic modulus, compressive strength, and shear strength in fiber-reinforced composites [40, 48, 49]. Machine learning approaches, including ANNs, have further enhanced prediction capabilities by modeling complex nonlinear interactions within soil-fiber matrices [47, 50]. Hybrid AI models that integrate micromechanical principles with ML algorithms have improved generalization and parametric analysis for high-performance fiber-reinforced materials [51]. However, challenges persist regarding data quality, uncertainty quantification, and the need for comprehensive models that incorporate diverse fiber types, soil compositions, and environmental factors [52, 53].

This review of the literature intends to synthesize research on the use of AI in the prediction and optimisation of the mechanical characteristics of fiber-reinforced soils (FRS) (i.e., shear strength, compressibility, and durability). The review compares AI models (e.g., ANN, SVM, and hybrid) with traditional physical testing methods and summarizes the prediction capabilities of the models (e.g., $R^2 > 0.90$ when predicting strength) and the computational efficiency of the models (e.g., run time). The published literature has identified the influencing factors that can alter the behavior of FRS (e.g., fiber [composition and type], quantity of fibers, type of soil) and the predictive/computational aspects of the machine learning process. The literature also indicates some limitations that are important in practical application. For example, limitations include the lack of data on naturally derived fibers and the inability to assess the long-term durability of fiber-reinforced soil (FRS). This literature review aims to assist geotechnical researchers in advancing AI-driven geotechnical solutions and to identify potential future research directions, such as physics-informed machine learning and multi-objective optimization, with the overall goal of achieving sustainable soil reinforcement techniques.

2 Mechanical Properties of Fiber-Reinforced Soil

The FRS has been recognized as a very efficient method in geotechnical engineering for improving the mechanical behavior of soils under different types of loading conditions. The use of synthetic or natural fibers in soil systems leads to a notable improvement of essential engineering properties of soil such as shear strength, tensile strength, ductility, and deformation resistance [10, 54–58]. Several studies on synthetic fibers (polypropylene (PP), polyethylene terephthalate (PET), glass, basalt, and carbon fibers) highlight their high mechanical properties, durability, and resistance to environmental degradation [21, 59–62]. Natural fibers, such as coconut, sisal, and oil palm empty fruit bunch (OPEFB), are low-cost and biodegradable materials; however, their hydrophilic nature and susceptibility to decay limit their durability. [10, 56, 58]. The mechanical performance of FRS is influenced by multiple factors including fiber type, fiber content, aspect ratio of fiber, length of fiber, and soil matrix [63, 64]. Experimental studies demonstrated an optimal fiber content, typically between 0.2 and 0.75% by weight, which leads to yield diminishing returns or deleterious effects beyond those percentages [61, 65]. Fiber reinforcement enhances soil cohesion but does not affect the internal friction hence provides stability without acid altering the frictional behavior of the soil [54, 57, 59].

The method of fiber distribution—whether random (RDFS) or oriented (ODFS) (Fig. 1)—is also an important factor influencing the mechanical behavior of reinforced soil. Randomly distributed fibers provide an isotropic strength enhancement by

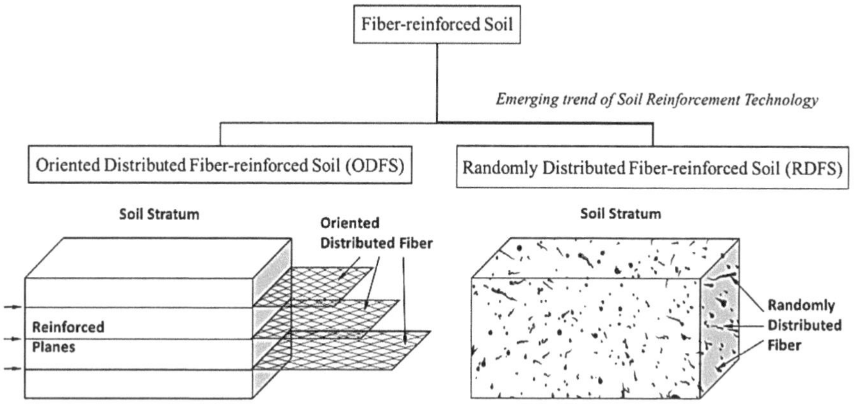

Fig. 1 Classification and mechanisms of fiber-reinforced soil [32]

uniformly mobilizing tensile resistance throughout a soil matrix, and follow natural reinforcement mecha-nisms of roots [32, 66, 67]. In contrast, oriented fiber systems, such as a woven or layered geotextile, provide strength enhancement along planes but may leave zones unreinforced and subject to localized failure [68, 69]. Both triaxial and direct shear tests have shown that inclusion of fibers increases peak shear strength, reduces the loss of strength after peak strength, and improves ductility of soils. In cases where fibers are used, the brittle behavior of soils change to a more ductile, energy-absorbing material behavior [55, 70–72]. In addition, fiber-reinforced soils are more resistant to cracking, erosion, and moisture-induced weakening/rendering, making these fibrous materials ideal for slope stability, embankment areas, or as subgrade materials [32, 65, 73].

3 Analytical Models for Predicting Shear Strength in Fiber-Reinforced Soil Systems

Researchers have introduced simplified models to predict the behavior of fiber-reinforced soil, based on experimental studies that employed various proce-dures (force equilibrium/mechanistic, energy dissipation, statistical, and superimposition of soil and fiber effects) [74–77]. Waldron [78] first proposed a simple force-equilibrium model to estimate the strength enhancement of soil reinforced by non-rigid plant roots. However, it was observed that the model focused solely on the tension created in the root reinforcement and included it in Mohr-Coulomb's in the following modified form:

$$S_R = S + \Delta S = c + \sigma \tan\phi + \Delta S \tag{1}$$

where S_R represents the shear strength of reinforced soil, S represents the shear strength of unreinforced soil, and ΔS refers to the increase in shear strength due to plant root reinforcement. The parameter σ denotes the total normal confining stress on the shear or failure plane, and c and ϕ are the cohesion intercept and angle of shearing resistance, respectively, of the unreinforced soil.

Gray and Ohashi [79] extended the model by Waldron [78], demonstrating the deformation and failure mechanisms of fiber-reinforced cohesionless soil. The extension also predicted how much fiber reinforcement will increase the shear strength of the soil. It was suggested that soil could fail because of soil-fiber slip, dilated soil, and fiber rupture. This model, which remains the basis for many current models, provided a fundamental understanding of fiber reinforcement, thereby increasing the shear strength of the soil. The following formulas are used to estimate increased shear strength ΔS due to the fiber reinforcement in sand:

$$\Delta S = \sigma_R(\sin\theta + \cos\theta\ \tan\phi) \tag{2}$$

For fibers that are initially oriented perpendicular to the shear plane, and

$$\begin{aligned}\Delta S &= \sigma_R\left[\mathrm{sinsin}(90^\circ - \psi) + \cos(90^\circ - \psi)\mathrm{tantan}\phi\right] \\ &= \sigma_R(\cos\psi + c\sin\psi\tan\phi)\end{aligned} \tag{3}$$

For fibers that are initially oriented at an arbitrary angle i with the horizontal,

$$\psi = \tan^{-1}\left(\frac{z}{x + x'}\right) = \tan^{-1}\left(\frac{1}{k + \mathrm{cotcot}i}\right) \tag{4}$$

where, σ_R represents the mobilized tensile strength of the fibers per unit area of the fiber-reinforced granular soil, with the majority of the strength coming from the fibers themselves. The angle of internal friction of unreinforced soil is represented by ϕ, while θ represents the angle of shear distortion. x is the horizontal shear displacement, z is the thickness of the shear zone, i is the initial orientation angle of the fiber with respect to the shear surface, and $k = x/z$ is the shear distortion ratio.

In direct shear tests, Jewell and Wroth [80] described a symmetric placement of stiff reinforcement. This way, the force from the reinforcement on each side of the shear plane was decomposed into normal and tangential components. Their model for the increase in shear strength is consistent with the models presented by Waldron [78] and Gray and Ohash [79]. Based on data from triaxial compression tests and statistical analysis, Maher and Gray [81] proposed a strength behavior model for randomly distributed fiber-reinforced sand subjected to static loads. Their model estimates the increase in shear strength due to the fiber reinforcement, using the following expressions:

$$\Delta S = N_S\left(\frac{\pi D^2}{4}\right)(2a_r\sigma_{3\mathrm{av}}\tan\delta)(\sin\theta + \cos\theta\tan\phi)\eta \tag{5}$$

For $0 < \sigma_{3av} < \sigma_{3crit}$, and

$$\Delta S = N_S\left(\frac{\pi D^2}{4}\right)(2a_r\sigma_{3crit}\tan\delta)(\sin\theta + \cos\theta\tan\phi)\eta \quad (6)$$

For $\sigma_{3av} > \sigma_{3crit}$, where σ_{3av} is the average confining stress, N_S is the average number of fibers intersecting a unit area of the shear plane, a_r is the aspect ratio of fibers, δ is the fiber-skin friction angle, and η is an empirical coefficient depending on the soil characteristics (D_{50}, particle sphericity S_S, and the coefficient of uniformity C_U) and the fiber parameters (aspect ratio and skin friction). The value of σ_{3crit} is determined experimentally from experimental values and is dependent on soil characteristics and fiber characteristics. Other researchers have developed alternative models and performed numerical studies. Ranjan et al. [74] described a Ranjan, Vasan, and Charan (RVC) model where they ran statistical and regression analysis of data taken from using more than 500 triaxial compression tests on randomly distributed discrete FRS. The RVC model is used to examine the effect of fiber properties, soil properties, and confining stress on the shear strength of reinforced soil. The RVC model was validated by laboratory tests using varying soils and fibers, showing a strong correlation with experimental data. Zornberg [76] developed a model using a discrete approach to predict the shear strength of FRS. The model was based on the properties of the fibers and the soil, including fiber content, fiber aspect ratio, and the shear strength of the unreinforced soil. The model assumes that the fibers contributed to the enhancement of shear strength by mobilizing tensile stresses as the shear plane develops.

$$\phi_f = \tan^{-1}\left(\left(1 + \alpha \cdot \eta_f \cdot \chi_f \cdot c_{i,\phi}\right) \cdot \tan\phi\right) \quad (7)$$

where α is an empirical coefficient that accounts for the orientation of fibers. For the case of randomly oriented fibers, α equals 1. For any preferred orientation, α should be selected suitably between 0 and 1.

Michalowski and Čermák [77] defined an energy dissipation model for estimating the ultimate shear strength of discrete random installations of fiber-reinforced axisymmetric deformation processes. The model developed a "macroscopic or equivalent" internal friction angle, ϕ, that allows the designer to practically substitute the FRS with an equivalent granular material with a higher internal friction angle than the unreinforced material ϕ_f.

$$\phi_f = 2\tan^{-1}\sqrt{\frac{\eta_f \cdot \chi_f \cdot M \times \tan(\delta) + 6\tan^2(45^\circ + \phi/2)}{6 - \eta_f \cdot \chi_f \cdot M \times \tan(\delta)}} - \frac{\pi}{2} \quad (8)$$

where

$$M = \tan^2(45^\circ + \phi/2) \cdot \sin\tan^{-1}\sqrt{\tan^2(45^\circ + \phi/2)/2} \quad (9)$$

Ding and Hargrove [82] developed a nonlinear stress–strain relationship for a flexible fiber-reinforced soil model, considering both the energy in the soil and the energy in the fibers. Their model links the shear modulus of the fiber-reinforced soil to the fiber content, fiber distribution, geometrical features of the fiber, and soil-fiber interaction.

Lastly, it is worth considering the use of an axisymmetric model as the basis for predicting the friction angle for all test types, whether triaxial, direct shear, or ring shear, as this may have influenced the representations of the bias factor statistics for the energy-based model. The energy-based model, first described in Michalowski [83] and subsequently presented in conjunction with a planar deformation process, has been used to circumvent this deficiency. The friction angle for the reinforced sand in the planar strain energy-based model is obtained by the expression:

$$\phi_f = \text{arc}\sin\left(\sin\phi + \frac{N \cdot \eta_f \cdot \chi_f \tan\delta}{3}\right) \tag{10}$$

where

$$N = \frac{\cos\phi}{\pi} + \left(\frac{1}{2} + \frac{\phi}{\pi}\right)\sin\phi \tag{11}$$

Moreover, Diambra et al. [66] formulated a model to couple the effects of fibers with the stress–strain relation of unreinforced soil based on the rules of mixtures. They also presented a constitutive model for the fiber-reinforced sand mass by using the following expressions:

$$\Delta\sigma = \Delta\sigma' + p_{vf}\Delta\sigma_f = [M_m]\Delta\varepsilon + p_{vf}\left[M_f\right]\Delta\varepsilon \tag{12}$$

where $\Delta\sigma$ is the total stress increase, $\Delta\sigma'$ is the stress increase in sand particles, p_{vf} is the volumetric fiber content, $\Delta\sigma_f$ is the stress increase in fibers, M_m is the stiffness matrix for the sand, M_f is the stiffness matrix for the fibers, and $\Delta\varepsilon$ is the strain increase in sand particles and fibers, caused by stress increase $\Delta\sigma$.

Najjar et al. [84] evaluated the predictive uncertainty of shear strength models for fiber-reinforced sand, finding that the energy-based model underestimated strength by 10% (COV = 0.20). In comparison, the discrete model overestimated it by 6% (COV = 0.17). They attributed these biases to unaccounted factors, such as fiber-grain size interactions, scale effects, and nonlinear relationships between fiber content and shear strength, highlighting the need for improved models that incorporate these parameters. Shukla [62] predicts the shear strength of fiber-reinforced granular soils under high confining stresses, where fiber pullout is negligible. The model introduces an apparent cohesion (c_R) and enhanced normal stress (σ_{RS}) due to fiber reinforcement, with the total shear strength expressed as follows:

$$S_R = c_R + \sigma_{\mathrm{RS}}\tan\phi \tag{13}$$

where c_R and σ_{RS} depend on fiber content (p_f), aspect ratio ($a_r = L/D$), and soil-fiber friction (ϕ_i). A key output is the Shear Strength Ratio (SSR), quantifying improvement over unreinforced soil:

$$\text{SSR} = 1 + 2a_r \tan\phi_i \sin i\left(\sin\psi + \frac{\cos\psi}{\tan\phi}\right)\left[\frac{1 - \sin\phi\sin(\phi - 2i)}{\cos^2\phi}\right]\left[\frac{p_f\left\{\frac{G}{G_f(1+e_s)}\right\}}{1 + p_f\left\{\frac{G}{G_f(1+e_s)}\right\}}\right] \quad (14)$$

The model is based on the full mobilization of fiber tension, so it is only appropriate to apply under high stress; low confinement conditions will disregard pullout effects. The approaches typically involve formulating mathematical equations and relationships based on several main parameters: characterizing soil and fiber properties, assessing interface properties, and estimating the stress state of the fiber-reinforced soil (FRS). Although the proposed methods can produce a shear strength for FRS, they may not be broadly accepted within the academic community due to difficulty involved with reliably quantifying the soil-/fiber-interaction parameters. Recently, soil and geotechnical researchers have utilized data mining and machine learning techniques to improve their understanding of FRS shear strength and analyze performance level databases of FRS test data while simultaneously isolating the most significant soil and fiber properties from a much larger dataset.

4 AI Techniques in Geotechnical Engineering Applications

The distribution of AI methods across nine geotechnical engineering domains shows clear trends in the adoption of different computational techniques. Traditional AI models such as ANNs and SVMs are used more often as they are robust approaches for addressing nonlinear geotechnical data. Hybrid models such as Adaptive Neuro-Fuzzy Inference Systems (ANFIS) are in moderate use, particularly in the fields of foundation engineering and unsaturated soils, as they balance accuracy and interpretability of results. In slope stability analysis and tunneling applications, deep learning methods, such as convolutional neural networks (CNN) and long short-term memory (LSTM) methods, are increasingly being used as they can analyze spatial and temporal data. There is little use of Generative Adversarial Networks (GANs) as their use demand the generation of synthetic data and likely niche applications. For field application, there is a preference to employ greater AI methods in landslide risk assessment and rock mechanics, likely an opportunity for wider AI applications in lesser researched areas (e.g., frozen soils) [85]. Among the various AI methods applied in geotechnical engineering, the most widely used technique is ANN ($\approx$35%), followed by random forest (RF, $\approx$19%) and SVM ($\approx$17%) with the remaining ($\approx$29%) consisting of new techniques such as CNN, LSTM, GANs, and BN [86, 87]. This further validates the findings of earlier studies which recognized ANN, SVM, and tree-based techniques (i.e., RF) to be the major techniques utilized in geotechnical applications. As a result, deep learning has started to become used where specialized tasks require development, and specially in that area where ANN or other techniques are challenged, such as in image analysis (CNN) or sequential data modeling (LSTM). Nonetheless, it has shown promise in other areas [87].

5 Artificial Intelligence in FRS Assessment

The use of artificial intelligence (AI) as an alternative to experimental approaches in predicting the mechanical characterization of soil-fiber composites is a prospective research field, as it has the potential to be more efficient and accurate. AI models, and more specifically, machine learning algorithms, can be used to predict various complex mechanical properties, such as shear strength, elastic modulus, and compressive strength, of reinforced soils with fibers by observing the continuous interactions between the soil and fiber in the composites [88, 89]. Once trained with experimental data, AI models can then be used to make general predictions in unpredictable variable scenarios regarding fiber and soil composition. The subsequent sections outline the primary reasons for excitement regarding this research area.

Recent developments in AI have advanced the field of geotechnical engineering, especially in liquefaction and slope stability assessment, based on solid scientific approaches. For liquefaction prediction, machine learning (ML), and deep learning (DL) methods, including long-short term memory (LSTM) [90], gated recurrent unit (GRU) [91], multi-gene genetic programming (MGGP) [92], and neural networks [93], have shown high prediction accuracy (up to 98%) in comparison to traditional approaches by learning complex behavior of soils and temporal dependencies. The use of fiber-reinforced soils (FRS) compliments the use of AI by providing physical mitigation techniques that can be tested in the laboratory and field [94]. In slope stability, hybrid AI models (e.g., ANN-PSO, ANFIS-GA) and AutoML ensembles achieved an accuracy of over 97% by optimizing important slope stability parameters (e.g., slope angle, cohesion) [95–98]. These studies also used robust statistical validation techniques (e.g., ROC analysis, Nash–Sutcliffe efficiency), sensitivity analysis and multi-objective optimization techniques (e.g., NSGA-II, SPEA-II) to support findings.

The integration of artificial intelligence (AI) in evaluating fiber-reinforced soils (FRS) has revolutionized geotechnical engineering by enhancing the prediction of mechanical properties. AI approaches (machine learning (ML), deep learning (DL), and hybrid AI approaches). However, they do not replace traditional experimental approaches; they are more accurate and reliable than traditional approaches to measuring FRS properties. Using our research articles, we will classify AI approaches to FRS into machine learning approaches, deep learning approaches, and hybrid AI approaches while also discussing inherent mechanical properties such as unconfined compressive strength (UCS), shear strength parameters (c, φ), flexural and tensile strength, and stress–strain behavior.

5.1 Machine Learning Techniques

Machine learning (ML) methods (i.e., support vector machines (SVM) and random forests (RF)) are commonly used for predicting the mechanical performance of fiber-reinforced soils (FRS), comprising 35% and 19% of the applications reviewed, respectively [86]. Artificial neural networks (ANNs) can help relate the input parameters of fiber content, soil type, and moisture to mechanical outputs, with good performance, for instance, in predicting tensile and compressive strength [99]. Moreover, SVM and RF have been performed for classification and regression tasks and R^2 values have been reported at >0.93 for predicting elastic modulus and shear strength [100]. In predicting the unconfined compressive strength (UCS) of FRS, ensemble learning methods which combine multiple machine learning models have outperformed any individual classifier [48]. Notable applications for ML use relate to predicting UCS using relationships between fiber dosage and soil properties [101], modeling shear strength parameters of different FRS [73], predicting flexural/tensile strength using neural networks [65], and replicate stress–strain behavior [102].

5.2 Deep Learning Approaches

Given their capability to model complex relationships from large datasets, deep learning (DL) methods are ideal for modeling behavior of FRS. While DL approaches are not as common as hybrid approaches, they have begun to be incorporated in models because they represent nonlinear relationships in soil behavior and enhance predictions [86]. Convolutional neural networks (CNNs) have been used to model failure mechanisms using images of microstructures and Recurrent Neural Networks (RNNs) have been utilized to model time dependent deformations (creep and consolidation) in soils [50]. The use of DL has contributed to a number of significant advances in predictions of UCS from experimental data [103], enhanced estimation of shear strength parameters through improved identification of fiber-soil interactions [99], and have also provided critical insights into nonlinear deformation associated with imposed loading through stress–strain modeling [50].

5.3 Hybrid Learning Models

Hybrid AI models merge various approaches in a format that enhances the evaluation of FRS, with Neuro-Fuzzy Systems allowing fuzzy logics to integrate with ANNs for uncertain handling and ensemble learning, which, along with ensemble learning, improves robustness of predictions by averaging them [95]. Techniques such as weighted-feature least squares support vector regression (WFLSSVR), optimized with metaheuristic algorithms, have displayed higher predictive capabilities in shear strength, compared to traditionally utilized methodologies [104]. Successful applications have included LSSVM-CSO hybrids for predicting shear strength ($R^2 = 0.885$) [105], GA-ANN for cohesion predictions ($R^2 = 0.957$) [106], K-OPLS for dosage, resulting in 50% reduction in lime [107]. Each of these successfully optimized UCS as a response to fiber content and/or length [99], and predicted shear strength factors [108], as well as flexural and/or tensile cracking resistances [65].

The distribution of AI methods used in assessing fiber-reinforced soils (Fig. 2) indicates that artificial neural networks (ANNs) are the leading approach (37.5%), which is not surprising given their demonstrably successful utility in modeling nonlinear relationships between the characteristics of fiber and the mechanical properties of compressive and tensile strength (i.e., UCS, and tensile stress) [99, 101]. Support vector machines (SVM) are also used alongside random forest (RF) at equal rates (25% each), having specific advantages in resolving high-dimensional parameter space and delivering reliable predictions of geotechnical parameters such as shear strength [48, 100]. Adaptive neuro-fuzzy inference systems (ANFIS) also had a 12.5% share, indicating again its intended use to model uncertainty in shear strength properties of composite soils [109]. This distribution of methods, which was drawn from recent Web of Science-indexed research (2021–2024), reflects the trends observed for wider AI uptake in geotechnics while providing distinctive characteristics to fiber-reinforced soils. Notably, there has been no application of deep learning models, which may be attributed to the preponderance of smaller experimental datasets currently available compared to other geotechnical applications that employ machine learning techniques [86, 87].

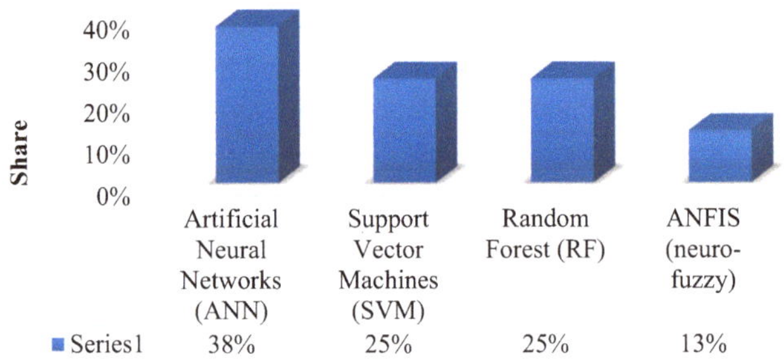

Fig. 2 Distribution of AI techniques in assessment of mechanical properties of fiber-reinforced soil

6 Comparative Analysis with Traditional Methods

The evaluation of the strength of fiber-reinforced soil (FRS) has been met with increasing interest because of its possibilities in ground improvement uses. There have been multiple models proposed, with both experimental and machine learning-based methods developed to improve strength estimates, incorporating various parameters (e.g., fiber content, fiber length, soil properties) that can help with understanding and effective use of fiber in soil engineering and geotechnical engineering settings.

More recent research has maintained a paradigm shift toward assessing the mechanical properties of fiber-reinforced soil (FRS) by introducing artificial intelligence (AI) methods in conjunction with current empirical techniques. Traditional methods are typically empirical, utilizing laboratory testing (e.g., direct shear, triaxial), and analytical models (e.g., Gray-Ohashi equations) which give a general understanding of FRS properties, but are limited by their expense, scalability, and modeling a case's fiber-soil interaction [40, 110, 111]. In comparison, machine learning (ML) algorithms, including artificial neural networks (ANNs), random forest (RF), and extreme learning machines (ELM), have demonstrated high predictive accuracy ($R^2 > 0.95$) compared to traditional methods for assessing compressive strength, tensile strength, and shear resistance, while effectively capturing the nonlinear interactions between fiber characteristics (e.g., type, length, treatment) and the mechanical performance of the composite material [37, 46, 112]. High-level methodology applications, such as Multivariate Adaptive Regression Splines (MARS) and least squares support vector regression (LSSVR), are aiding in population estimates for fiber-reinforced composite materials [113] and models that include micromechanical approaches with ML for improving prediction generalizability across soil types [51]. However, important barriers remain in: (1) addressing data standardization issues as indicated by soil mapping [114]; (2) providing model interpretation for engineering applications [115]; and (3) validating an AI prediction by measuring empirical benchmarks in a field-study context [116]. Future work should subsequently focus on creating less-rich datasets which include very diverse combinations of fiber-soil environment [117, 118], developing hybrid experimental-AI frameworks [119], and formalizing process protocols for uncertainty quantification [53] to tackle the major knowledge gaps we have in FRS mechanics.

7 Performance Evaluation of AI Models in Fiber-Reinforced Soil Analysis

The assessment of AI models' performance is summarized in Table 1. AI models are consistently outperforming empirical correlational equations when predicting the mechanical properties of fiber-reinforced soils (FRS) in models. In particular, the ML class of models, including artificial neural networks (ANNs) and support vector machines (SVMs), lead the field and provide exceptional predictive power ($R^2 >$ 0.90) for UCS, shear strength, and tensile resistance [40, 46, 47]. There is continuous evidence of significant predictive performance from ANNs ($R^2 > 0.95$) in models predicting UCS and tensile strength [46, 120]. Hybrid approaches to AI models also provide improved predictive performance, including GA-optimized ANNs and Jellyfish Search-optimized support vector regression [104, 106]. The use of ensemble approaches (e.g., random forest, XGBoost) are robust approaches to understand nonlinear fiber-soil relationships [48]. Despite advancements in machine learning methods, a challenge remains in model interpretability, and it has been confirmed by some authors' conclusions that first-order predictions do not closely resemble real-world performance [116]. Key perceived contributors to the predictive outcomes in models include type (synthetic or natural fiber), length (12–76 mm), amount (0.2–2.0%), and composition of fibers strongly contributed to the predictive outcomes of the machine learning models [52]. Further, while AI and ML models can describe results in certain forms of FRS systems, this will require greater standardization for effective predictive discussion in future experimentation [51, 112].

8 Challenges and Future Research Directions

The integration of artificial intelligence (AI) in assessing the mechanical properties of fiber-reinforced soil (FRS) represents a transformative shift in geotechnical engineering, offering data-driven solutions to overcome the limitations of traditional experimental and analytical methods. Key challenges persist, including data scarcity (especially for natural fibers), inconsistent testing protocols, and a lack of long-term durability studies [36, 52, 53]. The black-box nature of advanced ML/DL models and insufficient uncertainty quantification limit the adoption of engineering [84, 115]. Furthermore, AI models often overlook physics-based mechanisms (e.g., fiber-soil interfacial behavior) and struggle with multiscale modeling [51, 66]. While lab-scale AI predictions show promise, field validation and cost–benefit optimization remain understudied [107, 116]. Emerging opportunities include generative AI for synthetic data, explainable AI (XAI) for interpretability, and IoT integration for real-time monitoring [50, 85, 112]. Addressing these gaps will enhance the reliability and practicality of AI in FRS applications.

9 Conclusion

The integration of artificial intelligence (AI) in determining the mechanical properties of fiber-reinforced soil (FRS) represents a paradigm shift in geotechnical engineering, providing data-centric channels to address the shortcomings of conventional experimental and analytical procedures. In reviewing this study, one's key contributions include the following:

Table 1 Comparative performance of AI models in predicting mechanical properties of fiber-reinforced soils

Study	Fiber types	Property studied	Fiber effect	Best performing model and accuracy	AI methods used
[46]	Sisal, polypropylene, metallic	UCS, tensile strength	Binder and fiber properties have a significant impact on strength	ANN ($R^2 > 0.95$)	ANN, SVM, RF, MR
[45]	Basalt fiber	Max. deviator stress, failure deformation	Fiber length and cell pressure have a significant influence on strength	Fuzzy logic for failure deformation ($R^2 = 0.938$)	ANN, SVM, fuzzy logic
[40]	Polypropylene, polyester, glass, steel	Peak shear (φ)	Fiber content (at 1% optimal) and type significantly affect shear strength	Ensemble/Bagging ($R^2 = 0.94$, MAPE = 4.09%)	ANN, SVM, CART, ensemble methods
[48]	Sisal, polypropylene, metallic	Elastic modulus, UCS	Fiber and binder content are significant for elastic modulus and UCS	ANN and RF ($R^2 \geq 0.93$)	ANN, SVM, RF, MR
[121]	Geosynthetic	CBR	Geosynthetic reinforcement enhances the California Bearing Ratio (CBR)	Lazy K-Star (LKS) ($R^2 = 0.955$, MAE = 1.04)	ANN, LMSR, GPR, LKS, RF
[47]	Palm, typha, millet waste	Compressive and tensile strength	The type and content of natural fibers significantly improve strength	ANN ($R^2 = 0.977$, MSE = 0.834)	ANN, RF, MLP
[122]	Carpet (synthetic)	Shear strength, stiffness, deformation	Fiber content and soil properties influence shear strength	GA-ENN ($R^2 = 0.994$, RMSE = 3.342)	MLR, PLS, CRRF, GP, GA-ENN
[110]	Alfa (natural)	Shear strength	Fiber parameters significantly improve shear strength	ANN ($R^2 = 0.960$ overall)	ANN

(continued)

Table 1 (continued)

Study	Fiber types	Property studied	Fiber effect	Best performing model and accuracy	AI methods used
[120]	Polypropylene	Deviator stress, cohesion, φ, EAC	Fiber addition improves deviator stress, cohesion, and friction angle	ANN ($R^2 = 1.000$ training, 0.962 validation)	MLR, ANN
[104]	PP, steel, glass, natural fibers	Peak shear (φ)	Comprehensive model for shear strength using many input features	JS-WFLSSVR (MAPE = 1.265%, $R^2 = 0.989$)	WFLSSVR-Jellyfish Search
[89]	Glass fibers	UCS, failure strain	Fibers improve compressive strength and ductility	ANN (outperformed MLR with low RMSE)	ANN
[88]	Tire crumb, buffings	Deviator stress, shear strength	Fiber-shaped tire buffings increase shear strength and ductility	ANN ($R^2 = 0.961$)	ANN
[108]	Short geofiber	Shear modulus, axial shear	Lime and curing time are the most dominant factors, but fiber content also improves strength	MNN is effective for nonlinear mapping	Multilayer NN
[123]	Polypropylene (PP); 5, 10, 15 mm; 0.25–1.0%	Shear strength	Shear Stress (SS) increases with fiber content/length and vertical stress. Moisture content has a complex, nonlinear effect on the properties of the material	XGBoost_TTA (Training: R = 1.000, Testing: R = 0.9878)	Hybrid XGBoost models optimized with: AOA, BOA, QEA, TTA, WOA
[106]	Polypropylene	Cohesion	Hybrid algorithms optimize ANN for predicting cohesion	GA-ANN ($R^2 = 0.957$, outperforms PSO-ANN)	GA-ANN, PSO-ANN
[41]	PP, steel, glass, natural fibers	Shear (φ)	Optimized model for predicting shear strength from fiber/soil properties	SIOR (MAPE = 2.86%, R = 0.958)	SIOR (LSSVR + Firefly Algorithm)
[36]	Polypropylene	UCS, STS, UPV	PP fiber improves tensile strength and durability	ANN ($R^2 \approx 0.96$ for all outputs)	ANN

- The ANNs, SVM, RF, and hybrid systems achieved very high accuracies (R^2 > 0.90) in predicting FRS properties, such as UCS, shear strength, and tensile resistance. They also demonstrated a notable degree of power compared to their empirical model counterparts.
- Machine learning reduces the time spent on conventional, labor-intensive testing methods, rapidly enables the possible distributions of fiber types and fiber contents, and can streamline the testing process once relationships are established.
- Emerging challenges, including issues of dataset distribution, model interpretability, and information gained from field trials, will need to be addressed within the geotechnical engineering field for reliable applications in practice [52].
- Hybrid physics-AI models, generative data, and IoT-enabled monitoring are productive pathways in FRS research.

AI Use Disclosure Statement The author has only used Grammarly for language improvement and proofreading in the preparation and revision of this manuscript. The author used AI-assisted engineering tools such as SCISPACE and Elicit to search and gather relevant academic materials.

References

1. Cascone E, Biondi G (2013) A case study on soil settlements induced by preloading and vertical drains. Geotext Geomembr 38:51–67. https://doi.org/10.1016/j.geotexmem.2013.05.002
2. Espinosa-Santiago AL, López-Acosta NP (2020) Performance monitoring and numerical assessment of a test embankment with preloading and vertical drains on Texcoco lacustrine soft clays. Geotext Geomembr 48(4):546–560. https://doi.org/10.1016/j.geotexmem.2020.03.001
3. Verma H, Ray A, Rai R, Gupta T, Mehta N (2021) Ground improvement using chemical methods: a review. Heliyon 7:e07678. https://doi.org/10.1016/j.heliyon.2021.e07678
4. Jones RM (2018) Mechanics of composite materials, 2e éd. CRC Press, Boca Raton. https://doi.org/10.1201/9781498711067
5. Kazemian S, Huat BBK, Prasad A, Barghchi M (2010) A review of stabilization of soft soils by injection of chemical grouting
6. Ling HI, Liu H, Kaliakin VN, Leshchinsky D (2004) Analyzing dynamic behavior of geosynthetic-reinforced soil retaining walls. J Eng Mech 130(8):911–920. https://doi.org/10.1061/(ASCE)0733-9399(2004)130:8(911)
7. Sajini KS, Niranjana K (2021) Effect of coir fiber on compressibility behavior of clayey soil. In: Patel S, Solanki CH, Reddy KR, Shukla SK (eds) Proceedings of the Indian geotechnical conference 2019. Springer, Singapore, pp 581–589. https://doi.org/10.1007/978-981-33-6444-8_52
8. Vincenzini A, Augarde CE, Gioffrè M (2021) Experimental characterization of natural fibre–soil interaction: lessons for earthen construction. Mater Struct 54(3):110. https://doi.org/10.1617/s11527-021-01703-z
9. Binici H, Aksogan O, Shah T (2005) Investigation of fibre reinforced mud brick as a building material. Constr Build Mater 19(4):313–318. https://doi.org/10.1016/j.conbuildmat.2004.07.013
10. Hejazi SM, Sheikhzadeh M, Abtahi SM, Zadhoush A (2012) A simple review of soil reinforcement by using natural and synthetic fibers. Constr Build Mater 30:100–116. https://doi.org/10.1016/j.conbuildmat.2011.11.045

11. Puppala AJ, Musenda C (2000) Effects of fiber reinforcement on strength and volume change in expansive soils. Transp Res Rec 1736(1):134–140. https://doi.org/10.3141/1736-17
12. Kozłowski RM, Mackiewicz-Talarczyk M (eds) Handbook of natural fibres, 2nd edn, in The textile institute book series. Woodhead Publishing, p ii. https://doi.org/10.1016/B978-0-12-818782-1.09001-5
13. Bordoloi S, Hussain R, Garg A, Sreedeep S, Zhou W-H (2017) Infiltration characteristics of natural fiber reinforced soil. Transp Geotech 12:37–44. https://doi.org/10.1016/j.trgeo.2017.08.007
14. Karthikeyan G, Karthic S, Kolathayar S (2018) Experimental studies on strength performance of subgrade soil mixed with bottom ash and coir fiber. In: Shi X, Liu Z, Liu J (eds) Proceedings of GeoShanghai 2018 international conference: transportation geotechnics and pavement engineering. Springer, Singapore, pp 289–298. https://doi.org/10.1007/978-981-13-0011-0_31
15. Menezes LCP, Sousa DB, Fucale S, Ferreira SRM (2018) Analysis of the physical-mechanical behavior of clayey sand soil improved with coir fiber. Soils Rocks, n° 1, déc. 2018, Consulté le: 7 août 2025. [En ligne]. Disponible sur. https://www.soilsandrocks.com/sr-421031
16. Yang H, Li P, Yang E, Jiang X, Chen J (2023) The strength law of coir fiber-reinforced soil based on modified Duncan-Chang model. Front Earth Sci 10. https://doi.org/10.3389/feart.2022.1078624
17. Abdi MR, Wild S (1993) Sulphate expansion of lime-stabilized Kaolinite: I. Physical characteristics. Clay Miner 28(4):555–567. https://doi.org/10.1180/claymin.1993.028.4.06
18. Wei L, Chai SX, Zhang HY, Shi Q (2018) Mechanical properties of soil reinforced with both lime and four kinds of fiber. Constr Build Mater 172:300–308. https://doi.org/10.1016/j.conbuildmat.2018.03.248
19. Habibi AA, Fallah Tafti M, Narani S, Abbaspour M (2021) Effects of waste tire textile fibres on geotechnical properties of compacted lime-stabilized low plastic clays. Int J Geotech Eng 15(9):1118–1134. https://doi.org/10.1080/19386362.2021.1907069
20. Shen Y, Tang Y, Yin J, Li M, Wen T (2021) An experimental investigation on strength characteristics of fiber-reinforced clayey soil treated with lime or cement. Constr Build Mater 294:123537. https://doi.org/10.1016/j.conbuildmat.2021.123537
21. Hojjati F, Sarkar A (2019) Mechanical properties of soil reinforced with polypropylene fibre. Proc Inst Civ Eng Constr Mater 175(2):72–81. https://doi.org/10.1680/jcoma.18.00057
22. Li G, Zhang J, Liu J (2019) Experimental study on the shear behaviors of polypropylene fiber-reinforced sand. KSCE J Civ Eng 23(12):4992–5001. https://doi.org/10.1007/s12205-019-0794-7
23. Tiwari N, Satyam N, Patva J (2020) Engineering characteristics and performance of polypropylene fibre and silica fume treated expansive soil subgrade. Int J Geosynth Ground Eng 6(2):18. https://doi.org/10.1007/s40891-020-00199-x
24. De-yin W, Chao-sheng T, Jian LI, Bao-sheng LIU, Wei T, Kun ZHU (2013) Shear strength characteristics of fiber-reinforced unsaturated cohesive soils. Chin J Geotech Eng 35(10):1933–1940
25. Gao L, Hu G, Chen J, Ren K, Zhou Q, Yu X (2018) Dynamic properties of carbon-fiber-reinforced clay soil. Soil Mech Found Eng 55(5):317–324. https://doi.org/10.1007/s11204-018-9543-z
26. Malekzadeh M, Bilsel H (2012) Effect of polypropylene fiber on mechanical behaviour of expansive soils. Electron J Geotech Eng 17:55–63
27. Navagire OP, Srinivasan V, Patel A, Gowrisankar D (2025) Stabilization of expansive soils using polypropylene fiber for enhanced engineering properties: a study on black cotton soils from India. Indian Geotech J. https://doi.org/10.1007/s40098-025-01225-3
28. Tiwari N, Satyam N (2019) Experimental study on the influence of polypropylene fiber on the swelling pressure expansion attributes of silica fume stabilized clayey soil. Geosciences 9(9):377. https://doi.org/10.3390/geosciences9090377
29. Asadollahi F, Dabiri R (2017) Effects of glass fiber reinforced polymer on geotechnical properties of clayey soil. J Struct Eng Geotech 7(2):73–83

30. Perera ST, Saberian M, Zhu J, Roychand R, Li J (2022) Effect of crushed glass on the mechanical and microstructural behavior of highly expansive clay subgrade. Case Stud Constr Mater 17:e01244. https://doi.org/10.1016/j.cscm.2022.e01244
31. Yazici MF, Keski N (2021) Zeminlerin Doğal ve Sentetik Lifler ile Güçlendirilmesi Üzerine Bir Derleme Çalışması, Erzincan Üniversitesi Fen Bilim. Enstitüsü Derg 14(2):631–663. https://doi.org/10.18185/erzifbed.874339
32. Gowthaman S, Nakashima K, Kawasaki S (2018) A state-of-the-art review on soil reinforcement technology using natural plant fiber materials: past findings, present trends and future directions. Materials 11(4):553. https://doi.org/10.3390/ma11040553
33. Das SK, Samui P, Sabat AK (2011) Application of artificial intelligence to maximum dry density and unconfined compressive strength of cement stabilized soil. Geotech Geol Eng 29(3):329–342 (2011). https://doi.org/10.1007/s10706-010-9379-4
34. Shaban WM, Elbaz K, Yang J, Shen S-L (2021) A multi-objective optimization algorithm for forecasting the compressive strength of RAC with pozzolanic materials. J Clean Prod 327:129355. https://doi.org/10.1016/j.jclepro.2021.129355
35. Elbaz K, Shen S-L, Sun W-J, Yin Z-Y, Zhou A (2020) Prediction model of shield performance during tunneling via incorporating improved particle swarm optimization into ANFIS. IEEE Access 8:39659–39671. https://doi.org/10.1109/ACCESS.2020.2974058
36. Tiwari N, Satyam N (2021) Coupling effect of pond ash and polypropylene fiber on strength and durability of expansive soil subgrades: an integrated experimental and machine learning approach. J Rock Mech Geotech Eng 13(5):1101–1112. https://doi.org/10.1016/j.jrmge.2021.03.010
37. Garg A, Bordoloi S, Mondal S, Ni J-J, Sreedeep S (2020) Investigation of mechanical factor of soil reinforced with four types of fibers: an integrated experimental and extreme learning machine approach. J Nat Fibers 17(5):650–664. https://doi.org/10.1080/15440478.2018.1521763
38. Ikeagwuani CC, Nwonu DC (2022) Application of fuzzy logic and grey based Taguchi approach for additives optimization in expansive soil treatment. Road Mater Pavement Des 23(4):849–873. https://doi.org/10.1080/14680629.2020.1847726
39. Wang S, Chen F, Xue Q, Zhang P (2020) Splitting tensile strength of cement soil reinforced with basalt fibers. Materials 13(14). https://doi.org/10.3390/ma13143110
40. Chou J-S, Yang K-H, Lin J-Y (2016) Peak Shear strength of discrete fiber-reinforced soils computed by machine learning and metaensemble methods. J Comput Civ Eng 30(6):04016036. https://doi.org/10.1061/(ASCE)CP.1943-5487.0000595
41. Chou J-S, Ngo N-T (2018) Engineering strength of fiber-reinforced soil estimated by swarm intelligence optimized regression system. Neural Comput Appl 30(7):2129–2144. https://doi.org/10.1007/s00521-016-2739-0
42. Yu Z et al (2021) Artificial intelligence model for studying unconfined compressive performance of fiber-reinforced cemented paste backfill. Trans Nonferrous Met Soc China 31(4):1087–1102. https://doi.org/10.1016/S1003-6326(21)65563-2
43. Kumar DD, Harinder D (2020) ANN and SVM based modeling of stabilized soils. In: 2020 fourth international conference on I-SMAC (IoT in social, mobile, analytics and cloud) (I-SMAC), pp 748–755. https://doi.org/10.1109/I-SMAC49090.2020.9243557
44. Elgaali E, Alshammari H, Joda A, Alnaqbi F, Rajab H (2024) Artificial intelligence to simulate engineering properties of soil. In: 2024 Advances in science and engineering technology international conferences (ASET), pp 1–4. https://doi.org/10.1109/ASET60340.2024.10708656
45. Ndepete CP et al (2022) Exploring the effect of basalt fibers on maximum deviator stress and failure deformation of silty soils using ANN, SVM and FL supported by experimental data. Adv Eng Softw 172:103211. https://doi.org/10.1016/j.advengsoft.2022.103211
46. Tinoco J, Correia AAS, Venda Oliveira PJ (2021) Soil-cement mixtures reinforced with fibers: a data-driven approach for mechanical properties prediction. Appl Sci 11(17). https://doi.org/10.3390/app11178099

47. Houcine B, Mohamed R, Samir K, Sarra B (2023) Artificial intelligence for the prediction of the physical and mechanical properties of a compressed earth reinforced by fibers. J Eng Exact Sci 9(4). https://doi.org/10.18540/jcecvl9iss4pp15910-01e
48. Owusu-Ansah D, Tinoco J, Correia AAS, Oliveira PJV (2022) Prediction of elastic modulus for fibre-reinforced soil-cement mixtures: a machine learning approach. Appl Sci 12(17). https://doi.org/10.3390/app12178540
49. Hasanzadeh A, Vatin NI, Hematibahar M, Kharun M, Shooshpasha I (2022) Prediction of the mechanical properties of basalt fiber reinforced high-performance concrete using machine learning techniques. Materials 15(20). https://doi.org/10.3390/ma15207165
50. Yang Z, Yu C-H, Buehler MJ (2021) Deep learning model to predict complex stress and strain fields in hierarchical composites. Sci Adv 7(15):eabd7416. https://doi.org/10.1126/sciadv.abd7416
51. Guo P, Meng W, Xu M, Li VC, Bao Y (2021) Predicting mechanical properties of high-performance fiber-reinforced cementitious composites by integrating micromechanics and machine learning. Materials 14(12). https://doi.org/10.3390/ma14123143
52. Sharma A, Jain A, Gupta P, Chowdary V (2021) Machine learning applications for precision agriculture: a comprehensive review. IEEE Access 9:4843–4873. https://doi.org/10.1109/ACCESS.2020.3048415
53. Wadoux AMJ-C, Minasny B, McBratney A (2020) Machine learning for digital soil mapping: applications, challenges and suggested solutions, févr, Consulté le: 5 août 2025. [En ligne]. Disponible sur. https://eartharxiv.org/repository/view/434/
54. Consoli NC, Prietto PDM, Ulbrich LA (1998) Influence of fiber and cement addition on behavior of sandy soil. J Geotech Geoenviron Eng 124(12):1211–1214. https://doi.org/10.1061/(ASCE)1090-0241(1998)124:12(1211)
55. Divya PV, Viswanadham BVS, Gourc JP (2014) Evaluation of tensile strength-strain characteristics of fiber-reinforced soil through laboratory tests. J Mater Civ Eng 26(1):14–23. https://doi.org/10.1061/(ASCE)MT.1943-5533.0000772
56. Ateş A (2016) Mechanical properties of sandy soils reinforced with cement and randomly distributed glass fibers (GRC). Compos Part B Eng 96:295–304. https://doi.org/10.1016/j.compositesb.2016.04.049
57. Al-Refeai TO (1991) Behavior of granular soils reinforced with discrete randomly oriented inclusions. Geotext Geomembr 10(4):319–333. https://doi.org/10.1016/0266-1144(91)90009-L
58. Ahmad F, Bateni F, Azmi M (2010) Performance evaluation of silty sand reinforced with fibres. Geotext Geomembr 28(1):93–99. https://doi.org/10.1016/j.geotexmem.2009.09.017
59. Evangelou ED, Markou IN, Verykaki SE, Bantralexis KE (2023) Mechanical behavior of fiber-reinforced soils under undrained triaxial loading conditions. Geotechnics 3(3):874–893. https://doi.org/10.3390/geotechnics3030047
60. Patel SK, Singh B (2017) Experimental investigation on the behaviour of glass fibre-reinforced cohesive soil for application as pavement subgrade material. Int J Geosynth Ground Eng 3(2):13. https://doi.org/10.1007/s40891-017-0090-x
61. Rawat V, Satyam N (2024) Effect of fibre-reinforced microbial-induced calcite precipitation on the mechanical properties of coastal soil. Soil Use Manag 40(2):e13078. https://doi.org/10.1111/sum.13078
62. Shukla SK (2017) In: Shukla SK (ed) Soil reinforcing mechanisms and models, in fundamentals of fibre-reinforced soil engineering. Springer, Singapore, pp 111–144. https://doi.org/10.1007/978-981-10-3063-5_4
63. Bouaricha L, Djafar Henni A, Lancelot L (2019) Glass fiber effect on the undrained static response of chlef sand (Northern Algeria). In: Kallel A, Erguler ZA, Cui Z-D, Al Karrech, Karakus M, Kulatilake P, Shukla SK (eds) Recent advances in geo-environmental engineering, geomechanics and geotechnics, and geohazards. Springer International Publishing, Cham, pp 273–276. https://doi.org/10.1007/978-3-030-01665-4_63
64. Bouaricha L, Djafar Henni A (2020) Combined effect of layers number and the glass fiber type on the shear strength characteristics of chlef sandy soil, Iran. J Sci Technol Trans Civ Eng 44(1):107–114. https://doi.org/10.1007/s40996-019-00335-2

65. Sujatha ER, Atchaya P, Darshan S, Subhashini S (2021) Mechanical properties of glass fibre reinforced soil and its application as subgrade reinforcement. Road Mater Pavement Des 22(10):2384–2395. https://doi.org/10.1080/14680629.2020.1746387
66. Diambra A, Ibraim E, Muir Wood D, Russell AR (2010) Fibre reinforced sands: experiments and modelling. Geotext Geomembr 28(3):238–250. https://doi.org/10.1016/j.geotexmem.2009.09.010
67. Wang Y-X et al (2017) Laboratory investigation on strength characteristics of expansive soil treated with jute fiber reinforcement. Int J Geomech 17(11):04017101. https://doi.org/10.1061/(ASCE)GM.1943-5622.0000998
68. Hegde A, Sitharam TG (2015) Experimental and analytical studies on soft clay beds reinforced with bamboo cells and geocells. Int J Geosynth Ground Eng 1(2):13. https://doi.org/10.1007/s40891-015-0015-5
69. Nguyen MD, Yang KH, Lee SH, Wu CS, Tsai MH (2013) Behavior of nonwoven-geotextile-reinforced sand and mobilization of reinforcement strain under triaxial compression. Geosynth Int 20(3):207–225. https://doi.org/10.1680/gein.13.00012
70. Consoli NC, Montardo JP, Donato M, Prietto PD (2004) Effect of material properties on the behaviour of sand—Cement—Fibre composites. Proc Inst Civ Eng Ground Improv 8(2):77–90. https://doi.org/10.1680/grim.2004.8.2.77
71. Maher M, Ho Y (1993) Behavior of fiber-reinforced cemented sand under static and cyclic loads. Geotech Test J 16(3):330–338. https://doi.org/10.1520/GTJ10054J
72. Michalowski RL, Zhao A (1996) Failure of fiber-reinforced granular soils. J Geotech Eng 122(3):226–234. https://doi.org/10.1061/(ASCE)0733-9410(1996)122:3(226)
73. Mirzababaei M, Mohamed M, Arulrajah A, Horpibulsuk S, Anggraini V (2017) Practical approach to predict the shear strength of fibre-reinforced clay. Geosynth Int 25(1):50–66. https://doi.org/10.1680/jgein.17.00033
74. Ranjan G, Vasan RM, Charan HD (1996) Probabilistic analysis of randomly distributed fiber-reinforced soil. J Geotech Eng 122(6):419–426. https://doi.org/10.1061/(ASCE)0733-9410(1996)122:6(419)
75. Ashmawy AK, Sukumaran B (1997) Discussion: probabilistic analysis of randomly distributed fiber-reinforced soil. J Geotech Geoenviron Eng 123(10):986–988. https://doi.org/10.1061/(ASCE)1090-0241(1997)123:10(986)
76. Zornberg JG (2002) Discrete framework for limit equilibrium analysis of fibre-reinforced soil. Géotechnique 52(8):593–604. https://doi.org/10.1680/geot.2002.52.8.593
77. Michalowski RL, Čermák J (2003) Triaxial compression of sand reinforced with fibers. J Geotech Geoenviron Eng 129(2):125–136. https://doi.org/10.1061/(ASCE)1090-0241(2003)129:2(125)
78. Waldron LJ (1977) The shear resistance of root-permeated homogeneous and stratified soil. Soil Sci Soc Am J 41(5):843–849. https://doi.org/10.2136/sssaj1977.03615995004100050005x
79. Gray DH, Ohashi H (1983) Mechanics of fiber reinforcement in sand. J Geotech Eng 109(3):335–353. https://doi.org/10.1061/(ASCE)0733-9410(1983)109:3(335)
80. Jewell RA, Wroth CP (1987) Direct shear tests on reinforced sand. Géotechnique 37(1):53–68. https://doi.org/10.1680/geot.1987.37.1.53
81. Maher MH, Gray DH (1990) Static response of sands reinforced with randomly distributed fibers. J Geotech Eng 116(11):1661–1677. https://doi.org/10.1061/(ASCE)0733-9410(1990)116:11(1661)
82. Ding D, Hargrove SK (2006) Nonlinear stress-strain relationship of soil reinforced with flexible geofibers. J Geotech Geoenviron Eng 132(6):791–794. https://doi.org/10.1061/(ASCE)1090-0241(2006)132:6(791)
83. Michalowski RL (2008) Limit analysis with anisotropic fibre-reinforced soil. Géotechnique 58(6):489–501. https://doi.org/10.1680/geot.2008.58.6.489
84. Najjar SS, Sadek S, Alcoverro A (2013) Quantification of model uncertainty in shear strength predictions for fiber-reinforced sand. J Geotech Geoenviron Eng 139(1):116–133. https://doi.org/10.1061/(ASCE)GT.1943-5606.0000742

85. Baghbani A, Choudhury T, Costa S, Reiner J (2022) Application of artificial intelligence in geotechnical engineering: a state-of-the-art review. Earth-Sci Rev 228:103991. https://doi.org/10.1016/j.earscirev.2022.103991
86. Liu H, Su H, Sun L, Dias-da-Costa D (2024) State-of-the-art review on the use of AI-enhanced computational mechanics in geotechnical engineering. Artif Intell Rev 57(8):196. https://doi.org/10.1007/s10462-024-10836-w
87. Shao W et al (2023) The application of machine learning techniques in geotechnical engineering: a review and comparison. Mathematics 11(18):3976. https://doi.org/10.3390/math11183976
88. Edincliler A, Cabalar AF, Cagatay A, Cevik A (2012) Triaxial compression behavior of sand and tire wastes using neural networks. Neural Comput Appl 21(3):441–452. https://doi.org/10.1007/s00521-010-0430-4.
89. Gul N, Mir BA (2022) Parametric study of glass fiber reinforced fine-grained soil with emphasis on microstructural analysis. Int J Geotech Eng 16(6):716–728. https://doi.org/10.1080/19386362.2022.2049524
90. Ghani S, Thapa I, Adhikari DK, Waris KA (2024) Soil categorization and liquefaction prediction using deep learning and ensemble learning algorithms. Transp Infrastruct Geotechnol 12(1):22. https://doi.org/10.1007/s40515-024-00474-8
91. Şehmusoğlu EH, Kurnaz TF, Erden C (2025) Estimation of soil liquefaction using artificial intelligence techniques: an extended comparison between machine and deep learning approaches. Environ Earth Sci 84(5):130. https://doi.org/10.1007/s12665-025-12116-4
92. Naik JK, Muduli PK, Behera GC (2025) Multigene genetic programming-based modeling for evaluation of liquefaction potential of soil using shear wave velocity. Transp Infrastruct Geotechnol 12(6):182. https://doi.org/10.1007/s40515-025-00642-4
93. Karna P, Muduli PK, Sultana P (2024) Evaluation of liquefaction potential of soil using soft computing techniques. Indian Geotech J 54(2):489–499. https://doi.org/10.1007/s40098-023-00786-5
94. Alqawasmeh H, Alzubi Y, Mahamied A (2023) State-of-the-art review: fiber-reinforced soil as a proactive approach for liquefaction mitigation and risk management. J Eng 2023(1):8737304. https://doi.org/10.1155/2023/8737304
95. Bharti JP, Samui P (2024) Predicting slope failure with intelligent hybrid modeling of ANFIS with GA and PSO. Multiscale Multidiscip Model Exp Des 7(4):4539–4555. https://doi.org/10.1007/s41939-024-00492-6
96. Mustafa R, Kumar A, Kumar S, Sah NK, Kumar A (2024) Application of soft computing techniques for slope stability analysis. Transp Infrastruct Geotechnol 11(6):3903–3940. https://doi.org/10.1007/s40515-024-00446-y
97. Tiwari S, Das SK, Mohanty M (2025) A critical performance assessment of the machine learning algorithms for slope stability analysis. Neural Comput Appl 37(16):9735–9757. https://doi.org/10.1007/s00521-024-10471-0
98. Kurnaz TF, Erden C, Dağdeviren U, Demir AS, Kökçam AH (2024) Comparison of machine learning algorithms for slope stability prediction using an automated machine learning approach. Nat Hazards 120(8):6991–7014. https://doi.org/10.1007/s11069-024-06490-8
99. Zhao F, Zheng Y (2022) Shear strength behavior of fiber-reinforced soil: experimental investigation and prediction model. Int J Geomech 22(9):04022146. https://doi.org/10.1061/(ASCE)GM.1943-5622.0002502
100. Yang S, Ling S (2022) An artificial intelligence-based motion trajectory prediction of fibrous matters. Adv Intell Syst 4(1):2100136. https://doi.org/10.1002/aisy.202100136
101. Kumar S, Singh D (2023) Prediction of UCS and CBR behavior of fiber-reinforced municipal solid waste incinerator bottom ash composites using experimental and machine learning methods. Constr Build Mater 367:130230. https://doi.org/10.1016/j.conbuildmat.2022.130230
102. Rezazadeh Eidgahee D, Shiri H (2024) Machine learning–enhanced modeling of stress–strain behavior of frozen sandy soil. Geotechnics 4(4):1228–1245. https://doi.org/10.3390/geotechnics4040062

103. Kannan G, Sujatha ER (2023) Prediction of strength parameters of fibre reinforced soil using machine learning algorithms. In: Muthukkumaran K, Ayothiraman R, Kolathayar S (ed) Soil dynamics, earthquake and computational geotechnical engineering. Springer Nature, Singapore, pp 43–54. https://doi.org/10.1007/978-981-19-6998-0_5
104. Chou J-S, Truong D-N, Le T-L, Thu Ha Truong T (2021) Bio-inspired optimization of weighted-feature machine learning for strength property prediction of fiber-reinforced soil. Expert Syst Appl 180(115042). https://doi.org/10.1016/j.eswa.2021.115042
105. Tien Bui D, Hoang N-D, Nhu V-H (2019) A swarm intelligence-based machine learning approach for predicting soil shear strength for road construction: a case study at Trung Luong National Expressway Project (Vietnam). Eng Comput 35(3):955–965. https://doi.org/10.1007/s00366-018-0643-1
106. Armaghani DJ, Mirzaei F, Shariati M, Trung NT, Shariati M, Trnavac D (2020) Hybrid ANN-based techniques in predicting cohesion of sandy-soil combined with fiber. Geomech Eng 20(3):191–205
107. Chumacero JMG, Torres PLA, Portilla CCCL, Perez SPM (2023) Influence of artificial intelligence on the optimization of the dosage of natural hydraulic lime, plastic and metallic fibers on the geological characteristics of a treated soil. Rev Ing Constr 38(3). https://doi.org/10.7764/RIC.00080.21
108. He S, Li J (2009) Modeling nonlinear elastic behavior of reinforced soil using artificial neural networks. Appl Soft Comput 9(3):954–961. https://doi.org/10.1016/j.asoc.2008.11.013
109. Sungur A, Yazıcı MF, Keskin N (2022) Prediction of the shear strength of glass fiber-reinforced clay soil by adaptive neuro-fuzzy inference system (ANFIS). El-Cezeri 9(4):1255–1264. https://doi.org/10.31202/ecjse.1133184
110. Arabet L, Hidjeb M, Belaabed F (2022) A comparative study of reinforced soil shear strength prediction by the analytical approach and artificial neural networks. Eng Technol Appl Sci Res 12(6). https://doi.org/10.48084/etasr.5394
111. Festugato L, Menger E, Benezra F, Kipper EA, Consoli NC (2017) Fibre-reinforced cemented soils compressive and tensile strength assessment as a function of filament length. Geotext Geomembr 45(1):77–82. https://doi.org/10.1016/j.geotexmem.2016.09.001
112. Tamassoki S, Daud NNN, Wang S, Roshan MJ (2023) CBR of stabilized and reinforced residual soils using experimental, numerical, and machine-learning approaches. Transp Geotech 42:101080. https://doi.org/10.1016/j.trgeo.2023.101080
113. Gholampour A, Mansouri I, Kisi O, Ozbakkaloglu T (2020) Evaluation of mechanical properties of concretes containing coarse recycled concrete aggregates using multivariate adaptive regression splines (MARS), M5 model tree (M5Tree), and least squares support vector regression (LSSVR) models. Neural Comput Appl 32(1):295–308. https://doi.org/10.1007/s00521-018-3630-y
114. Zeraatpisheh M, Ayoubi S, Jafari A, Tajik S, Finke P (2019) Digital mapping of soil properties using multiple machine learning in a semi-arid region, central Iran. Geoderma 338:445–452. https://doi.org/10.1016/j.geoderma.2018.09.006
115. Shiri F, Perumal T, Mustapha N, Mohamed R (2024) A comprehensive overview and comparative analysis on deep learning models. J Artif Intell 6:301–360. https://doi.org/10.32604/jai.2024.054314
116. Nguyen QH et al (2021) Influence of data splitting on performance of machine learning models in prediction of shear strength of soil. Math Probl Eng 2021(1):4832864. https://doi.org/10.1155/2021/4832864
117. Vijaya Ramnath B et al (2013) Evaluation of mechanical properties of abaca–jute–glass fibre reinforced epoxy composite. Mater Des 51:357–366. https://doi.org/10.1016/j.matdes.2013.03.102
118. Senthilkumar K et al (2018) Mechanical properties evaluation of sisal fibre reinforced polymer composites: a review. Constr Build Mater 174:713–729. https://doi.org/10.1016/j.conbuildmat.2018.04.143
119. Wu H, Lin X, Zhou A (2020) A review of mechanical properties of fibre reinforced concrete at elevated temperatures. Cem Concr Res 135(106117). https://doi.org/10.1016/j.cemconres.2020.106117

120. Yazici MF, Keskin SN (2024) Experimental investigation of the mechanical properties of polypropylene fiber-reinforced clay soil and development of predictive models: effects of fiber length and fiber content. Arab J Sci Eng 49(10):13593–13611. https://doi.org/10.1007/s13369-023-08688-7
121. Raja MNA, Shukla SK, Khan MUA (2022) An intelligent approach for predicting the strength of geosynthetic-reinforced subgrade soil. Int J Pavement Eng 23(10):3505–3521. https://doi.org/10.1080/10298436.2021.1904237
122. Daghistani F, Abuel-Naga H (2023) Evaluating the influence of sand particle morphology on shear strength: a comparison of experimental and machine learning approaches. Appl Sci 13(14). https://doi.org/10.3390/app13148160
123. Bouaricha L, Khatti J, Yazici MF, Chehat A, Henni AD, Lancelot L (2025) Effect of polypropylene fibre, moisture content, relative density, and vertical stresses on mechanical properties of poorly graded sandy soil: a geotechnical and hybrid XGBoost-based investigation. Constr Build Mater 493:143286. https://doi.org/10.1016/j.conbuildmat.2025.143286

An Interpretable Deep Neural Framework for Topic-Aware Analysis of Fandom Conflicts on Social Platforms

Anam Naz and Hikmat Ullah Khan

Abstract Social media platforms generate huge volumes of user-generated content on daily basis, which often includes toxic comments that highlight online conflicts in the form of social fandom war. These conflicts pose active challenges for digital content moderation. With the enhancements of natural language processing, researchers have developed advanced methods to detect and interpret toxic behavior in online communication. Deep learning has played a significant role in this progress, empowering models to capture complex and semantic patterns that traditional rule-based or shallow machine learning approaches are unable to detect them accurately. The main aim of this study is to integrate advanced large language models, Yi-34B characterized by bidirectional attention mechanisms and dual-stage training to enhance the classification of toxic social media comments. The approach combines both traditional textual features such as term frequency-inverse document frequency, parts-of-speech tagging, and bigrams with deep semantic features including advanced sentence embedding and topic modeling using Latent Dirichlet Allocation method. Empirical analysis demonstrates that the proposed model achieves the highest accuracy of 98%, compared to BERT and RoBERTa. Furthermore, explainable AI techniques are used to interpret model decisions and transparency based on global (SHAP) and Local (LIME) features explanation. In addition, statistical significance tests validate the model's accuracy and per-class performance, which ensure its state-of-the-art effectiveness in social media toxicity classification.

Keywords Artificial intelligence · Natural language processing · Deep learning · Feature engineering · Large language models · Toxic comments · Social media · Topic modelling · Hate speech detection · Fandom war

A. Naz · H. U. Khan
Department of Information Technology, University of Sargodha, Sargodha, Punjab, Pakistan
e-mail: anamtahir130@gmail.com

H. U. Khan (✉)
Department of Computer Science, University of Wah, Rawalpindi, Pakistan
e-mail: dr.hikmat.niazi@gmail.com

A. Azizi et al. (eds.), *Advanced Decision-Making Under Uncertainty*, Emerging Trends in Mechatronics, https://doi.org/10.1007/978-981-95-8696-7_7

1 Introduction

Social media platforms have become significant in digital environment creating massive amount of user-generated content (UGC) that highlight diverse opinions, personalities, and behaviors [1]. The online platforms such as Facebook, Instagram X, and Reddit enable users to communicate freely, sharing their thoughts and views through text, images, and videos [2]. However, interpreting this complex and dynamic content requires advanced tools. Thanks to artificial intelligence (AI), especially natural language processing (NLP), which emerged as a significant tool in this domain [3]. NLP approaches help in automated analysis of textual content to understand user behavior [4], sentiments [5], and interactions [6], making it crucial for various applications such as content moderation [7] and behavioral studies [8].

Among the various patterns of communication on social media, toxic comments frequently enhance as a form of fandom war which reflects the meaning of cyberbullying [9], trolling, abusive language [10], hate speech, arguments, and rudeness [11]. This digital language poses challenges to the healthy environment of online communities and the safety of users. Traditional ML and DL methods have been employed to detect and classify these toxic behaviors, typically focusing on individual categories separately [12]. However, such approaches often face limitations in capturing patterns and overlapping toxic behaviors across different classes, restricting their effectiveness in real-world social media environments [13]. Recent advances in LLMs can analyze context, semantics, and language patterns, highlight enhanced capabilities for detecting toxic comments [14]. This study aims to improve the classification of toxic fandom war content on social media, addressing the complexities and subtleties of online discourse with robust AI techniques, as different fandom patterns are shown in Table 1.

In this work, we use a publicly available dataset containing diverse toxic comments with five classes equally balanced categories from UGC. Using state-of-the-art LLM which further compared with traditional ML, DL, and transformer variants to classify toxic classes. Traditional textual features including TF-IDF, bigram, PoS tagging and deep features integrate with sentence embedding are also applied for understanding and detection of toxic content in social media fandom wars. The key contributions for this study are follows:

- integration of a powerful LLM Yi-34B to capture semantic and contextual patterns in social media toxic comments, achieving a high accuracy of 98%,
- comparison of the proposed LLM against traditional ML, DL, and transformer-based models such as BERT and RoBERTa to benchmark performance improvements,
- application of explainable AI techniques, including LIME and SHAP, for interpreting model decisions and enhancing transparency in toxic comment classification, and
- exploration of statistical tests to assess the significance of model performance across overall accuracy and per-class effectiveness, providing deeper insights into classification robustness.

Table 1 Fandom war dialogues nature types based on their characteristics and features

Type	Definition	Intent	Typical language style	Emotional tone
Trolling	Deliberate provocation to incite reactions	To provoke, disrupt, or annoy	Sarcastic, mocking, playful, inflammatory	Mischievous, provocative
Argument	Logical disagreement or debate	To persuade or counter opinions	Formal or informal reasoning, assertive	Defensive, assertive
Cyberbullying	Intentional online harassment or intimidation	To harm, intimidate, or isolate	Threatening, aggressive, repetitive	Hostile, aggressive
Rudeness	Discourteous or impolite behavior	To offend or disrespect	Blunt, curt, disrespectful	Impolite, insensitive
Abusive language	Use of offensive, vulgar, or hateful terms	To insult or demean	Profanity, slurs, derogatory terms	Angry, hateful

The remainder of this paper is organized as follows: Sect. 2 reviews related literature on social media toxicity detection and transformer models; Sect. 3 provides the description of problem statement with formulation. Section 4 describes the dataset and methodology; Sect. 5 presents experimental results and analyses; Sect. 6 discusses explainability and statistical findings; and Sect. 7 concludes with insights and directions for future work.

2 Related Work

Social media fandom wars are often waged with toxic language, misinformation, and hateful content. Recent works also show that deep transformer models combined with pre-trained word embeddings achieved significantly better classification performance on this kind of social media text than was obtained with older techniques using pre-trained static word embeddings or shallow models interpreters pretrained on neural tasks [15]. The study obtained a 93% performance result for multi-class tweet sentiment classification with BERT fine-tuning with a Bi-LSTM layer, outperforming a CNN baseline's 89%. Typically, the contextual embeddings of transformers have richer representations than static embeddings, and they have been widely used to achieve state-of-the-art performance in various NLP problems [16]. Transformer-based models for hate speech and offensive language detection in social media utilized fine-tuned BERT with Twitter hate speech datasets and incorporated a racial bias mitigation strategy to combat the racial prediction bias [17]. In a study [18] tackled multilingual hate speech detection via ensembles of several BERT variants. They obtained the best results in their experiments when ensemble models with parallel fusion model with 85% accuracy on an English hate dataset and 89% on a Korean corpus of hate, surpassing their single model baselines. Similarly, monolingual vs. multilingual transformers are employed for low resource languages [19] and presented that the language-specific BERT models marginally outperformed a multilingual model on hate speech corpora. However cross-lingual transfer was still beneficial.

FakeBERT combined BERT with parallel CNN layers to capture semantic and local cues in news posts [20]. This hybrid model achieved an impressive 88.9% accuracy on a fake news dataset, a substantial gain over earlier LSTM models. Another different approach by incorporating stance detection features [21] used BERT embeddings to analyze a news article's content and its stance toward a given claim. By interpreting how the article's tone agreed or disagreed with known facts, their model attained 85.3% accuracy in early fake news detection. More recently, a reinforced Transformer introduced [22] for multimodal fake news. They fine-tuned a transformer in three stages and applied a hybrid PSO-D optimization, enabling the model to capture inconsistencies between an image and caption. A comparison between a CNN–SVM ensemble and a BERT model for cyberbullying detection is introduced. The fine-tuned BERT achieved 97.3% accuracy on two Twitter cyberbullying datasets performing slightly better than an ensemble [23]. A peculiar complication in the dialogue between fans is sarcasm which is frequently used to ridicule rival fan groups [24], explored using a BERT-LSTM model for the purpose of Hindi-English code-mixed tweets. They utilized a pre-trained multilingual BERT to obtain embedding and passed these embeddings to an LSTM which identifies long-range context. In another study [25] proposed graph convolution networks combined with BERT for sarcasm detection in which the goal is to capture local as well as global relationships among words.

The study [26] introduced a model for event detection in crises composed of Mobile BERT embeddings of tweets and using an enhanced PSO algorithm to identify informative features, and their approach can achieve better precision for crisis events on social media versus LSTM based models. For rumor detection [27] present a comprehensive survey of transformer-based models. They find that BERT and its variations are effective on rumor classification on Twitter due to contextual capture and provide a taxonomy on approaches that include network propagation structure as well as vanilla BERT classifiers. One of the trends is the integration of network information [28]. They find that BERT and its variations are effective on rumor classification on Twitter due to contextual capture and provide a taxonomy on approaches that include network propagation structure as well as vanilla BERT classifiers. One of the trends is the integration of network information [29] on transformers in the topic of disaster and crisis social media analytics and found that BERT-based models are trending against the event classification. In general, Transformer-based fake news classifiers compose an important component in curbing the fandom wars, as they are very accurate at identifying deceptive posts, but that it must be carefully tuned to prevent false positives.

3 Problem Statement

Due to rapid increase in UGC, the social media platform becomes a war ground for users which poses a challenge to distinguish toxic comments like threats and insults. Such poisonous exchanges cause damage at a personal level to the quality of online interaction and perpetuate negative sentiments in the digital world. While bringing attention to the issue, differences in rate of change and sensitivity of offending speech is a complex problem that requires advanced computational techniques sensitive to patterns, linguistic tendencies and meaning. Formally, let be a dataset $D = \{(a_i, b_i\}_{i=1}^{N}$, where each sample a_i is a social media comment C which consists of unstructured text a_i, and $b_i \in \{1, 2, \ldots, C\}$ represents one of the predefined toxicity classes that can cover. The answer is to learn a function $f : A \rightarrow B$ which correctly predicts the class label b_i, for each input comment a_i, to classify toxic content, then make a timely and accurate response in online platforms.

4 Proposed Research Methodology

The proposed method is shown in Fig. 1, followed to identify toxic comments from social media wars. The approach starts with extensive data preprocessing to normalize the content which is followed by a complex feature extraction based on text features. Moreover, interpretable AI tools and statistical analyses also support the strength of the model, confirming that the proposed framework can achieve superior performance on a large scale, thus serving as a complete and truthful solution to toxicity detection in social media fandom discussion.

4.1 Data Collection and Preprocessing

The dataset used in this study is collected from a public repository GitHub from user-generated social media text[1] that was labeled in 5 balanced classes: abusive language, trolling, cyberbullying, rudeness, and arguments. To effectively model the data, a complete preprocessing pipeline was used. Raw text was pre-processed by cleaning processes (i.e., removal of stop words, digits, punctuation and special characters) to reduce noise and unnecessary information, defined in Eq. 1.

$$x_i^{(1)} = \text{Clean}(x_i) = x_i\{\text{stopwords, digits, punctuation, special characters}\} \quad (1)$$

After that, the text was normalized by transforming all the characters into lowercase shown in Eq. 2 and then tokenized to split the text into meaningful segments, defined in Eq. 3.

[1] https://github.com/VisionLangAI/Social-War \last accessed 1st July 2025.

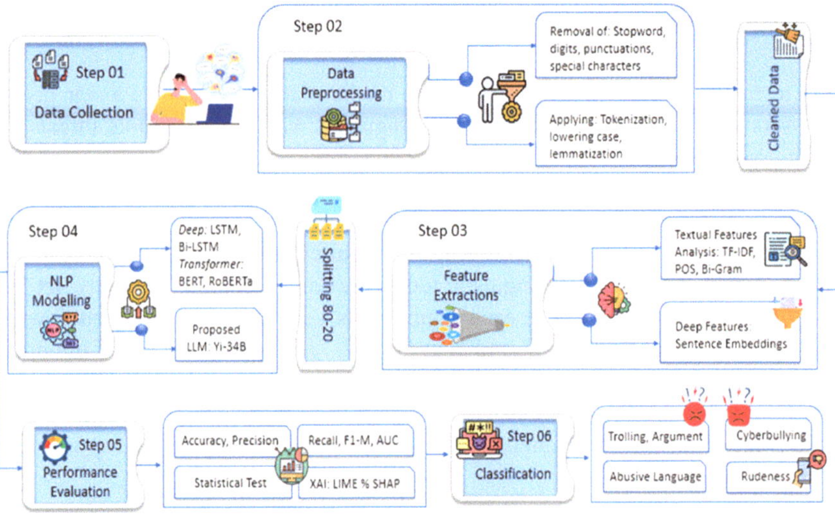

Fig. 1 Framework analysis of proposed research methodology

$$x_i^{(2)} = \text{Normalize}\left(x_i^{(1)}\right) = \text{Lowercase}\left(x_i^{(1)}\right) \tag{2}$$

$$x_i^{(3)} = \text{Tokenize}\left(x_i^{(2)}\right) = [t_1, t_2, \ldots, t_k] \tag{3}$$

This was followed by lemmatization, followed in Eq. 4, which converted the words to their naked base or dictionary form in parallel and established uniformity and comparison between texts, using in Eq. 5.

$$x_i' = \text{Lemmatize}\left(x_i^{(3)}\right) = [l_1, l_2, \ldots, l_k] \tag{4}$$

This preprocessing pipeline transformed the raw social media comments into an organized and clean form for processing in downstream natural language processing tasks [30], resulting in an improved and more robust toxic comment classification.

$$x_i' = \mathcal{P}(x_i) = \text{Lemmatize} \circ \text{Tokenize} \circ \text{Normalize} \circ \text{Clean}(x_i) \tag{5}$$

4.2 Feature Extraction

The study employs various state-of-the-art feature extraction techniques to capture both lexical and semantic information in the social media texts. We describe the TF-IDF representation t_j within a document d over the whole corpus denoted by dataset D.. The weight (TF-IDF) for this was proposed embeddings term j in document i that determined how many times termed using function *tfidf* (), computed on the base of weights $w_{i,j}$, using Eq. 6.

$$w_{i,j} = \frac{f_{i,j}}{\sum_k f_{i,k}} \times \log \frac{|\mathcal{D}|}{\left|\left\{d : t_j \in d\right\}\right|} \tag{6}$$

In addition we used POS tagging to perform a syntactic c features by lemmatizing of each token t_k with the tagged τ_1 (word, pos) pair to building feature vectors $p_i = [\tau_1, \tau_2, \ldots, \tau_m]$ along the structures of grammar that is an essential variant functionality to set for toxicity detection, calculated from Eq. 7. The combination model Bigram further emphasizes the relationship of token pairs t_k, t_{k+1} as shown, respectively and which are formulated to be

$$P(t_k, t_{k+1}) = \frac{\text{Count}(t_k, t_{k+1})}{\sum_{x,y} \text{Count}(x, y)} \tag{7}$$

Other than lexical features, the text also employs deep semantic representation through paragraph representation, counted by sentences $s = (w_1, w_2, ...w_n)$. It operates by inputting a sentence and outputting an embedding vector $s \in \mathbb{R}^d$ from a transformer-based encoder (with weights) e_{w_i} defined as in Eq. 8.

$$s = \Phi(S) = \frac{1}{n} \sum_{i=1}^{n} e_{w_i} + \text{AttentionWeights} \tag{8}$$

These techniques, when combined, allow a holistic representation of social media text incorporating surface-level and deep contextual features.

4.3 Proposed Model

The Yi-34B model is a state-of-the-art large language model which is built in a manner that it is sensitive to semantic and syntactic nuances of social media text, and a fine-grained aspect of toxic attitudes toward fandom wars. This is a two-stage scheme of a bidirectional transformer shown in Fig. 2 which jointly optimizes global multi-head self-attention with multi-level positional-encodings and residual learning to learn fine-grained long-range contextual dependencies.

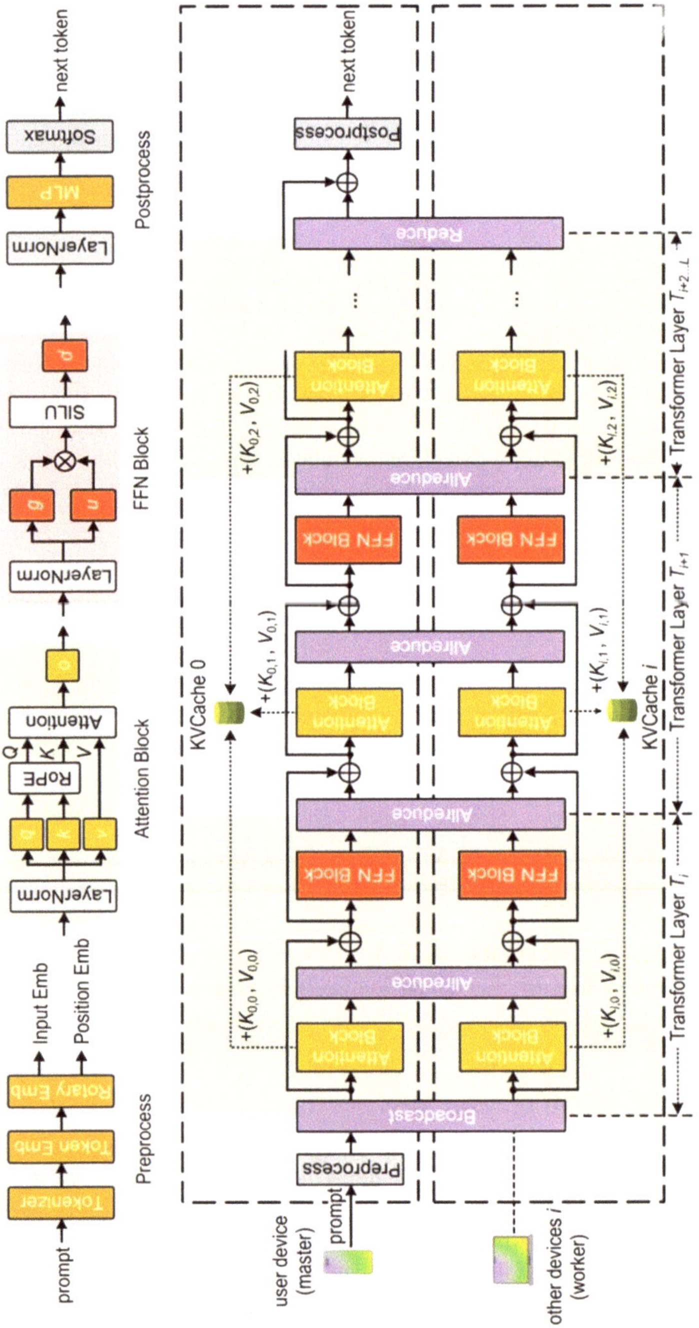

Fig. 2 Architecture of proposed model

The model takes an input token sequence as input using Eq. 9.

$$X = \{x_1, x_2, \ldots, x_n\} \tag{9}$$

Each token x_i is transformed into a high-dimensional embedding vector $e_i \in \mathbb{R}^d$ via an embedding function $\mathcal{E}(.)$, define in Eq. 10.

$$E = \mathcal{E}(X) = \{e_1, e_2, \ldots, e_n\} \tag{10}$$

The core of Yi-34B lies in its Multi-head self-attention mechanism, computed as in Eq. 11.

$$\text{Attention}(Q, K, V) = \text{softmax}\left(\frac{QK^{\mathrm{T}}}{\sqrt{d_k}} + B\right)V \tag{11}$$

where $Q, K, V \in \mathbb{R}^{n*dk}$ are the query, key, and value matrices respectively, and $B \in \mathbb{R}^{n*n}$ is a learnable relative positional bias matrix enhancing the attention scores by encoding token position relationships.

The Multi-head attention is extended over h parallel heads, concatenated and linearly projected in Eq. 12.

$$\text{MultiHead}(E) = \text{Concat}(\text{head}_1, \ldots, \text{head}_h)W^O \tag{12}$$

With each head, using Eq. 13.

$$\text{head}_i = \text{Attention}\left(EW_i^Q, EW_i^K, EW_i^V\right) \tag{13}$$

here $W_i^{Q,K,V} \in \mathbb{R}^{d*d_k} and\ W^O \in \mathbb{R}^{hd_k*d}$ are learned projection matrices.

Following attention, Yi-34B applies a position-wise Feed-Forward Network (FFN) per token, shown in Eq. 14.

$$\text{FFN}(z) = \sigma(zW_1 + b_1)W_2 + b_2 \tag{14}$$

where $\sigma(.)$ denotes the GELU activation, $W_1 \in \mathbb{R}^{d*d_{ff}}$, $W_2 \in \mathbb{R}^{d_{ff}*d}$ and b_1, b_2 are biases.

Each transformer block uses pre-layer normalization and residual connections to stabilize gradients, shown in Eq. 15.

$$z' = \text{LayerNorm}(z + \text{Sublayer}(z)) \tag{15}$$

where Sublayer(.) is either multi-head attention or FFN.

The dual-stage transformer stack involves two cascaded stages:

- Stage 1: Captures broad context and general semantic features using L_1 transformer layers.
- Stage 2: Refines toxic feature representation through L_2 deeper transformer layers specialized for toxicity nuances.

Finally, the classification output $\hat{y} \in \mathbb{R}^C$ (for $C = 5$ toxicity classes) is computed by using Eq. 16.

$$\hat{y} = \text{softmax}(hW_o + b_o) \tag{16}$$

where $h \in \mathbb{R}^d$ is the pooled representation from the final transformer layer, and $W_o \in \mathbb{R}^{d*C}$ and $b_o \in \mathbb{R}^C$ are classification parameters based on weight and bias values respectively.

The interaction of all layers with relative positional bias, dual-stage deep transformer blocks, and pre-normalization helps the model to trace and detect complex patterns.

5 Experimental Setup

The experimental environment involves the combination of the suggested and baselines models on a publicly available social media toxicity data. Experimental conditions were used to train and assess models to ensure that they were compared fairly.

5.1 Baseline Models

This study compares with traditional ML and DL approaches to ensure the robustness of proposed model. The choice of these models was based on their power and limitations as shown in Table 2. Deep learning models with the capability of capturing the temporal dependence between sequential data include LSTM and Bi-LSTM that process the up and down streams to learn the context better. In terms of the transformer, they exploit attention-focused processes that hold contextual relationships in text where RoBERTa utilizes superior training and performance improvement over BERT [31]. These models would serve as potent benchmarks to evaluate the effectiveness of the proposed model in hitting the target on toxic social media comments classification.

Table 2 Comparative summary of baseline and proposed model

Model	Type	Key characteristics	Strengths	Limitations
LSTM	Deep learning	Sequential model capturing long-term dependencies	Effective for sequential data	Limited context (unidirectional)
Bi-LSTM	Deep learning	Processes sequences bidirectionally for richer context	Improved context understanding	Higher computational cost
BERT	Transformer	Uses multi-head self-attention; pre-trained on large corpora	Strong contextual embeddings	Resource-intensive, large models
RoBERTa	Transformer	Optimized BERT with dynamic masking, more training data	Improved accuracy and robustness	Even more computationally heavy
Yi-34B	LLM	Multi-layer transformer architecture, dual-stage attention, bidirectional context modeling, large-scale pretrained on diverse corpora	Exceptional ability to capture complex linguistic patterns and subtle toxic behavior; achieves highest accuracy (98%)	High computational resource requirements; potential latency in real-time applications; needs large data for fine-tuning

5.2 *Machine Requirements*

Experiments were developed in a high-performance computing environment with GPUs to support the computational demands of deep learning and transformer-based models. All training and inferences smoothly deployed, using NVIDIA Tesla V100 GPU with VRAM, Intel Xeon Gold CPU, and 128 GB Ram system, as displayed in Table 3. The software stack included Python 3.9 and key libraries including TensorFlow and PyTorch, Hugging Face Transformers and Scikit-learn. In addition, higher level natural language preprocessing of texts was performed by using libraries NLTK or SpaCy and explainability analysis using SHAP and LIME.

5.3 *Performance Measures*

The commonly used evaluation metrics on social media toxic comments to have a holistic evaluation are displayed in Table 4. Accuracy is an aggregate measure for the correctness of the predictions, and precision, recall, and F1-score, which are measures that record the quality of predictions at the class label level. The Area Under the Receiver Operating Characteristic Curve (AUC-ROC) evaluates how well

Table 3 Analysis of machine and libraries requirements

Component	Specification and version	Purpose
GPU	NVIDIA Tesla V100 (128GB VRAM)	Accelerated deep learning training
CPU	Intel Xeon Gold 6248 (20 cores, 2.5GHz)	High-performance processing
RAM	128 GB	Large-scale data handling
Programming language	Python 3.9	Model development and scripting
Deep learning Libs	TensorFlow 2.8, PyTorch 1.13	Model training and experimentation
Transformer library	Hugging face transformers 4.25	Pre-trained transformer models
NLP libraries	NLTK 3.7, SpaCy 3.5	Text preprocessing and tokenization
Explainability tools	SHAP 0.40, LIME 0.2	Model interpretability analysis
Evaluation tools	Scikit-learn 1.0	Metrics and data splitting

Table 4 Analysis of performance measures

Metrics	Description	Formulation
Accuracy	Proportion of correctly predicted samples	$\frac{TP+TN}{TF+FN+FP+TP}$
Precision	Proportion of true positives among all predicted positives	$\frac{TP}{TP+FP}$
Recall	Proportion of true positives detected among all actual positives	$\frac{TP}{TP+FN}$
F1-score	Harmonic mean of precision and recall	$\frac{2(Precision*Recall)}{Precision+Recall}$
AUC-ROC	Area under ROC curve, measures separability of classes	$\sum_{i=1}^{n-1} \frac{(FP_{i+1}-FP_i).(TP_{i+1}-TP_i)}{2}$

the model differentiates between two classes for different threshold values. Together, these measurements provide a strong basis for evaluating the effectiveness of the model to deal with the imbalanced as well as with the nuanced text classification tasks.

5.4 *Statistical Analysis*

The performance measures of the model need to be verified for their significance and reliability using the statistical analysis. In this paper, the hypothesis test technique, p

Table 5 Description analysis of statistical test

Metric	Description	Equation
p-value	The probability of obtaining results at least as extreme as observed, assuming null hypothesis is true	$p = P(\text{Data}H_0)$
t-test	Compares means of two groups to assess if they differ significantly	$t = \frac{\overline{x_1} - \overline{x_2}}{s_p\sqrt{\frac{1}{n_1} + \frac{1}{n_2}}}$
ANOVA	Tests differences among three or more group means	$F = \frac{\text{Between -group variance}}{\text{Within -group variance}}$
Chi-square	Tests association between categorical variables	$\chi^2 = \sum \frac{(O_i - E_i)^2}{E_i}$

value, t test, ANOVA analysis, and Chi-square test are used to evaluate the differences of the model accuracy and per-class performance, shown in Table 5.

5.5 *Explainable AI*

Interpretability solutions are indispensable elements to interpret and understand the decision-making process of complex machine learning models, in sensitive applications such as social media toxicity detection. The work utilizes two state-of-the-art XAI methodologies including LIME and SHAP to generate local and global explanations of the model prediction making the model (more) transparent and trustworthy, describe in Table 6.

Table 6 Description analysis of explainable AI interpretation

Method	Description	Computation
LIME	Local interpretable model agnostic Explanations; explains individual predictions by approximating the model locally $g \in G$ with an interpretable model loss analysis $\mathcal{L}$, to compute complexity penalty $\Omega(g)$	$\mathcal{L}(f, g, \pi_x) = g \in G \cdot \mathcal{L}(f, g, \pi_x) + \Omega(g)$
SHAP	SHapley Additive exPlanations; provides unified measure of feature importance S based on cooperative game theory	$\phi_i = \sum_{S \subseteq N \setminus \{i\}} \frac{\lvert S\rvert!(\lvert N\rvert - \lvert S\rvert - 1)!}{\lvert N\rvert!} \left[f_{S \cup \{i\}}\left(x_{S \cup \{i\}}\right) - f_S(x_S) \right]$

6 Results and Discussion

These visualizations present crucial information about the structure and content of a dataset. Figure 3 pie chart indicates each category of toxic behaviors in dataset, exists at exactly 20%. This balance is very important for robust models according to our classification task since it avoids a bias to any class. Figure 4, regarding sentence length distributions across classes, presents distinct differences between how users convey each toxicity type. The distribution is concentrated around the shorter sentence lengths (3–6 words) with the highest number of comments. There also seem to be fewer long sentences in Abusive Language comments, possibly due to different writing styles. The difference in the length of sentences in different classes may also have an impact on feature extraction and model performance, which requires the model to handle variable inputs gracefully.

From the top 10-word frequency plot in Fig. 5, the key words frequently occurred, which reflects a high usage of casual and conversational language used in the comments. These are simply common words, indicative of a mass medium using the medium in a normal way via good cheer, humor, and avuncular slang. The TF-IDF terms visualization in Fig. 6, showing commonly occurring words that also have a certain contextuality throughout the dataset—such as owned, inconsiderate, and annoying. It is very likely that these words have little strength to discriminate between different class types of comments based on different kinds of comments even if they are common because there while the common words might overlap, the underlying more sophisticated TF-IDF words might be better at discriminating between comment classes for classification purposes. The bigram frequency plot in Fig. 7 further enhances this linguistic analysis by revealing popular two-word

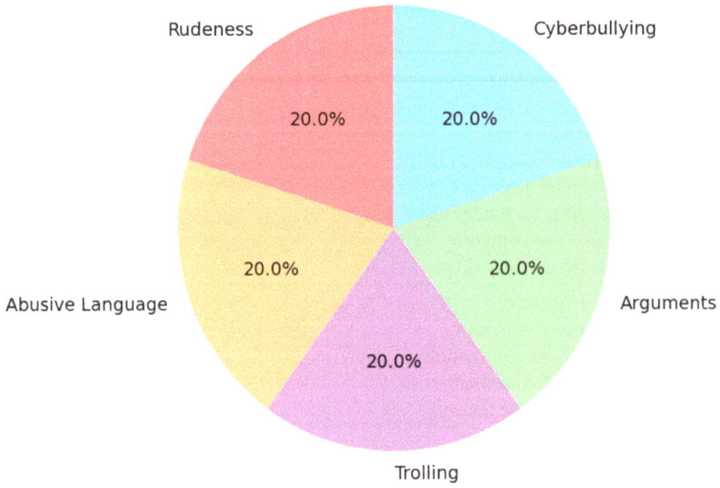

Fig. 3 Distribution of social media comments labels

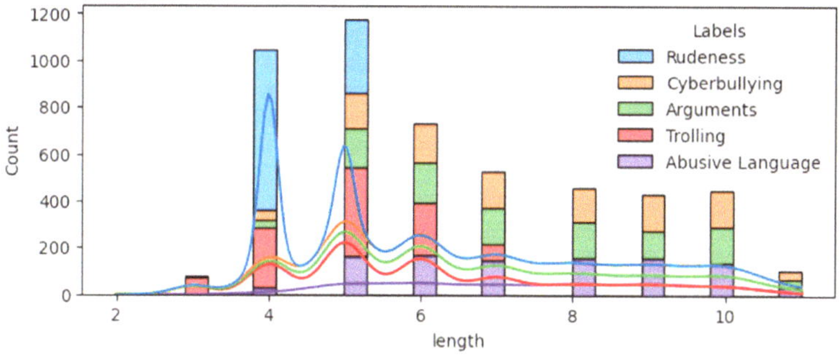

Fig. 4 Analysis of sentence length by class

patterns such as nice try, owned hard, potentially indicating sarcastic or bellicose tones of the very toxic online language. These kinds of bigrams can catch contextual meaning which single words may not and can improve performance of model for toxicity detection.

Moreover, the POS tag distribution plot in Fig. 8, nouns and adjectives appear to be the most prevalent form of words in the comments, which make up more than half of the tags, and proper nouns are also very common. The 5 theme-level topics obtained through the LDA analysis of the corpus of posts indicate the significant thematic groups of the social media comments, in regard to the expressed concerns and emotional tensions manifested by the users presented in Fig. 9. The dimension with the highest percentage of 25 is dissatisfaction with the behavior of others, which is typified by constant irritations and quarreling when online. The second one is Critique and Sarcasm (22%), which shows that a great deal of biting wit and negative feedback circulates in the form of a communication ethos on these sites. Self-Perception & Failure (20%) was that many comments assume that people have their own fears, insecurities, and poor self-images that are being projected and indicate the psychological nature of online communication. The message of

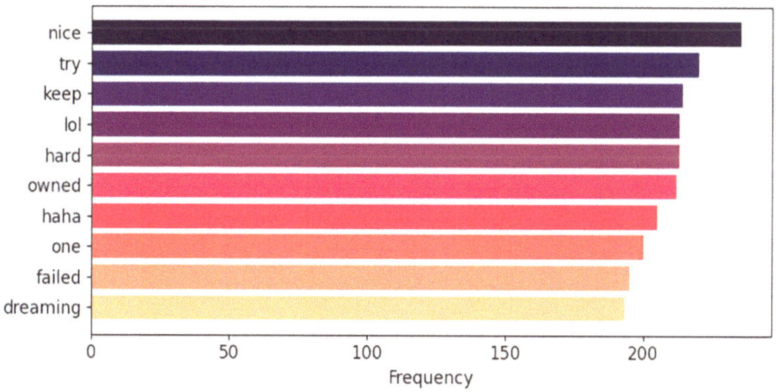

Fig. 5 Analysis of top 10 words by frequencies showing fandom definition

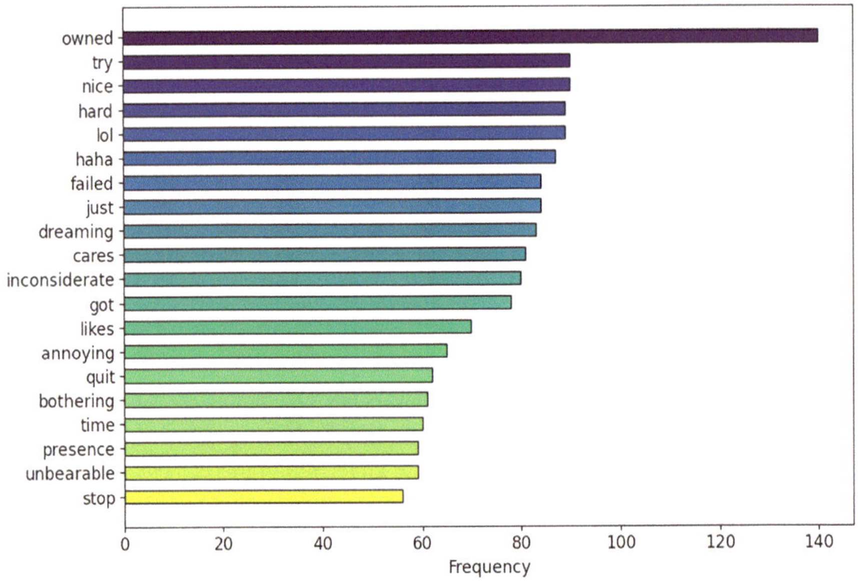

Fig. 6 Analysis of top terms by TF-IDF

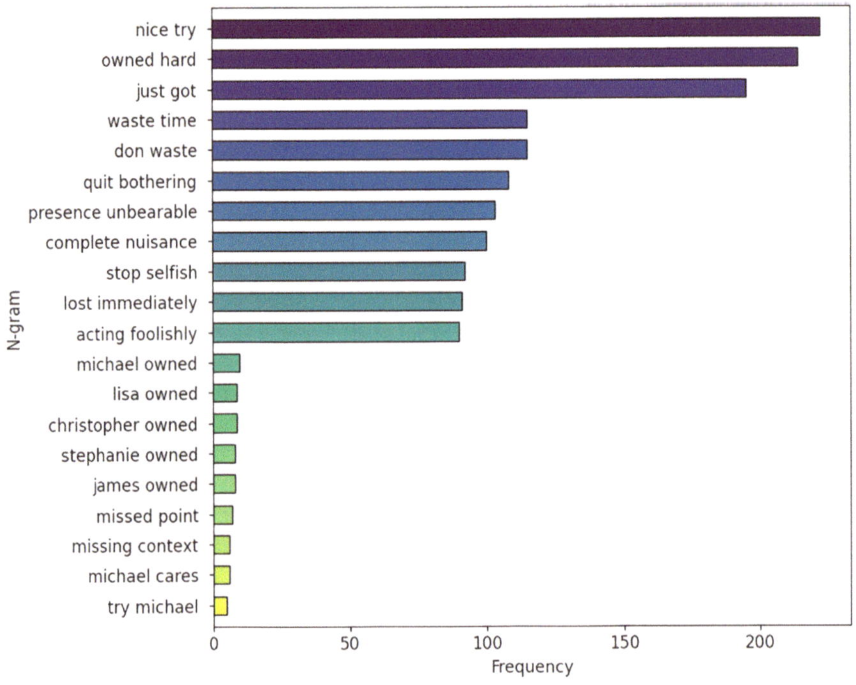

Fig. 7 Analysis of top terms by bigram frequency

Social Media Engagement & Complaints covers 18 percent of the gathered tweets and is associated with the use of social media and social media-related complaints. Finally, Time Management & Productivity Issues (15%) consists of concerns of being productive and the effect our online activity is having on real life.

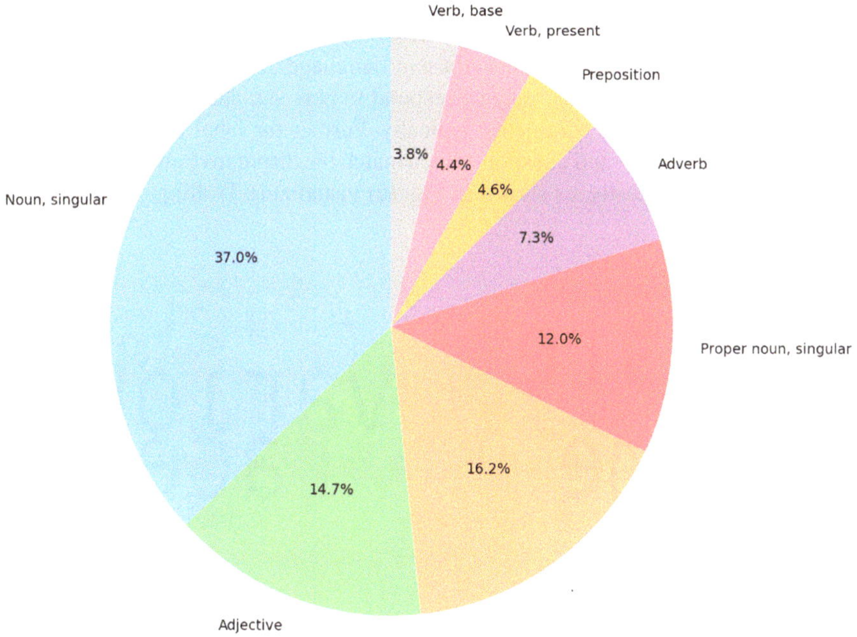

Fig. 8 Analysis of top terms distributions by PoS tagging

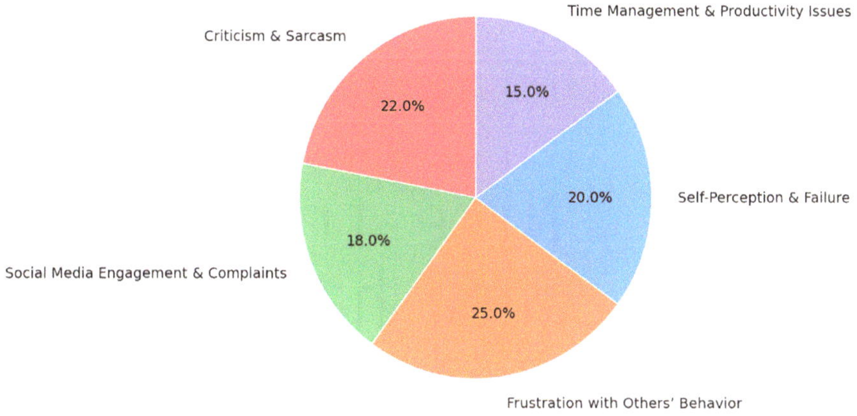

Fig. 9 Analysis of topic modeling using LDA

The collections of word clouds in Fig. 10 offer informative visual abstracts for the most prevalent and characteristic wordings in each social media comment class. As in (a) showing Rudeness label illustrates the words waste, time, inconsideration, or bad as the frustration and negative attitude typically observed in the rude feedback, conveying lots of irritation and frustration. Cyberbullying in (b) demonstrates the words such as various, attack, and nature are indicative of the diverse and aggressive nature of cyberbullying textual content, where users tend to mention harm and attack.

In the Fig. (c), which represents Abusive Language, the words like article, one, person seem conspicuous that may correspond to personal attacks or the criticisms as present in the abusive comments, typically. Further the label in (d) contains the words whether, answer, and question, which match the disputative and argumentative nature of a debate. Lastly, as shown in Fig. (e) visualizing Trolling, words such as

Fig. 10 Word cloud analysis of each label **a** Rudeness **b** Cyberbullying **c** Abusive language **d** Argument **e** Trolling

nice, owned, and failed are prevalent in the visualization. These words indicate the use of sarcastic language to bait or provoke others over the internet.

As the feature analysis of the dataset, shown in Fig. 11 conducted on the model of the LIWC demonstrates, the data reflects the definite stylistic and psychological differences in five categories of social media comments. Abusive language, arguments, and cyberbullying have a relatively greater use of long forms of words (approx. 29%) and high dictionary word content (approx. 87–89%) which indicates more structured language than Rudeness and Trolling which have shorter comments and less dictionary coverage. Rudeness is characterized by an extremely high pronoun (17%), first-person, and present tense verb (24 desired targets of aggression between individuals). Trolling demonstrates a rather curious effect of tiny word counts, lack of pronouns or articles, and high influx of negative terms of emotion (≈4.7%), and verbs in present or past tense which suggests brief yet emotionally heavy provocations. Abusive Language and Arguments resemble each other in moderate distribution of pronouns and balance in proportion of emotion words, but with more leaning of Arguments on prepositions, that: they are more likely to use complex phrases in disputing speech. The analysis reveals that Rudeness is based on immediacy and personal pronouns, Trolling is short but high-emotional, and the Abusive Language, Arguments, and Cyberbullying are more structured and long-term, and this difference can contribute to automatic classification of online hostility.

The development of architecture in models has been to ensure that there is constant improvement in performance as they become more sophisticated. Classical architectures such as LSTM with sentence embeddings can only recommend mid-level accuracy of 60.83% with low precision, recall and F1 as in Table 7. It means that

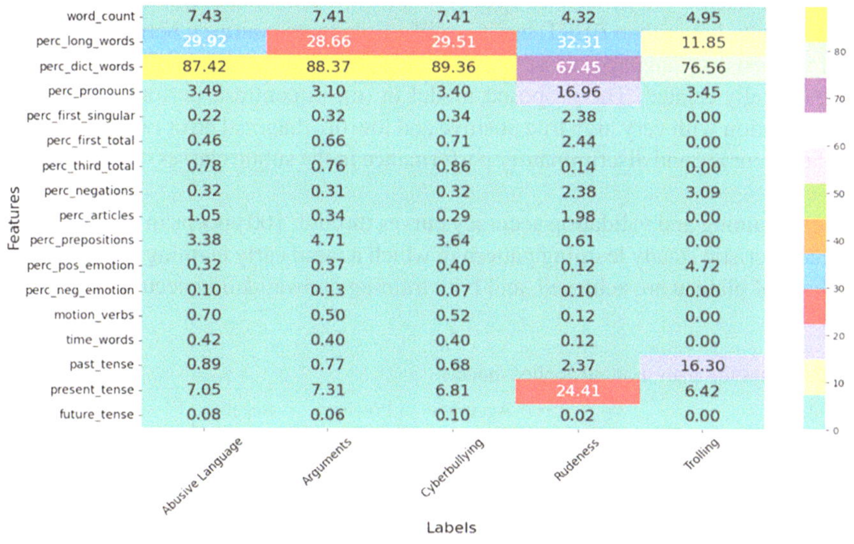

Fig. 11 Linguistic and psychological dimensions by label

the LSTM model can discover certain sequential relationships in text, but it fails in revealing the details and context of social media toxicity classification significantly. This is surpassed by the Bi-LSTM accuracy of 73.36 as it is seen that the ability to process text in both directions to consider both the past and future context allows us to better learn the patterns that are inherent in the data. Comparatively, the present technique of transformer-based models displays significant enhancements. It has high accuracy rate with high precision and recall given that it can model more compositional context using the self-attention mechanism. RoBERTa goes a step further to achieve a higher accuracy of 91.06 per cent because the pretraining and optimization strategies have been enhanced to identify more complex textual trends. The invented Yi-34B LLM is much more effective as the accuracy can be described as 98.03 percent and both the precision and recall as well as F1-score are rated at 98 percent. It proves that LLM is quite skilled at deciphering subtle and complicated features of harmful social media like sarcasm, implicit biases, and a broad scope of various linguistic expressions. The LSTM model has an AUC score of almost 100 (99.39) which confirms its strength and reliability in the detection of toxic and non-toxic comments.

As it can be seen from confusion matrices for the five models in Fig. 12, the classification performance is improved in a progression from the classic LSTM to the advanced Yi-34B LLM. LSTM matrix in (a) shows significant misclassifications of all the classes, notably poor discrimination between arguments, cyberbullying, and abusive language. The Bi-LSTM in (b) performs better, particularly in identifying arguments and trolling, but it still confuses some other classes, like abusive language and cyberbullying. BERT in (c) improves class separation even more with true positive but some misclassifications for arguments and rudeness. However, despite its general strength, we observe that RoBERTa shown in (d) is only slightly better than BERT, but it is still a little more confused between rudeness and the abusive language classes, which indicated that these two classes are quite linguistically related. The proposed model in (e), in contrast, performs near-ideal classification with very high true positive and low off-diagonal error rates, indicating higher accuracy and discriminatory performance in the subtle classes of social media toxicity.

The training and validation accuracy curves through 100 epochs in Fig. 13 shows a characteristic steady learning pattern in which a rapid early learning and thereafter a phase of plateau are achieved, and both training and validation accuracy approach

Table 7 Results analysis of all applied models

Model	Accuracy	Precision	Recall	F1-score	AUC
LSTM + sentence embeddings	60.83	59.28	57.39	58.32	61.28
Bi-LSTM + sentence embeddings	73.36	69.28	71.93	70.64	77.29
BERT	85.86	83.48	81.48	82.47	87.93
RoBERTa	91.06	89.84	87.12	88.46	92.48
Yi-34B LLM	**98.03**	**96.30**	**97.32**	**96.81**	**99.39**

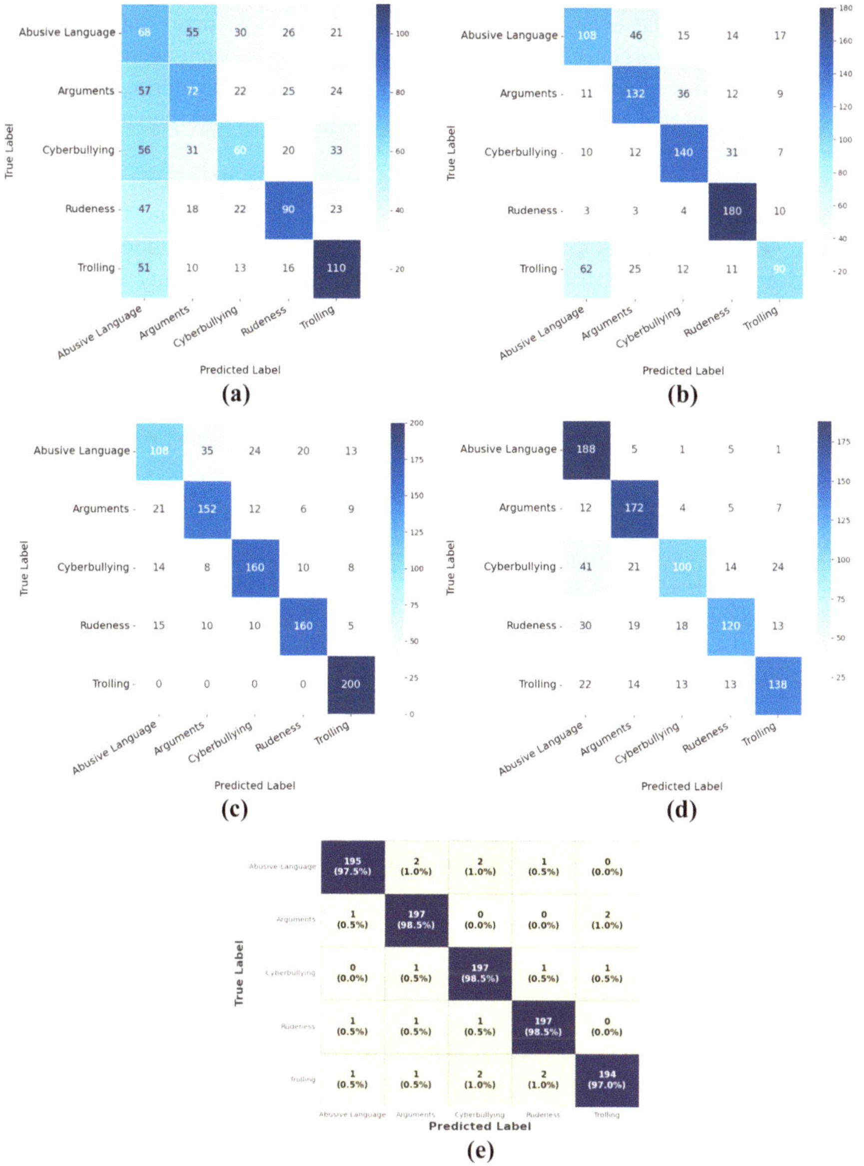

Fig. 12 Confusion matrix analysis of **a** LSTM **b** Bi-LSTM **c** BERT **d** RoBERTa **e** Proposed model

each other. It means that learning is done efficiently with small overfitting, the accuracy of validation is very near to the training cure. It is also observed that the accuracy is high and converging since higher number of epochs in the later part of training, showing the network is learning on its training data and generalizing on the testing data. The convergent behavior of training and validation accuracy supports confidence in the model's generalization and prevents the model from overfitting shown in Fig. 14. And the loss curves show that they decrease steadily during the training process, namely that the training loss falls smoothly down and the validation loss fluctuates but generally goes down as well.

All the class-wise confidence scores are very high, with the lowest of the social media toxicity classes publishing confidence values >75%, and most with scores >90%. The generalization ability of the model is positive as evidenced by high confidence scores in most classes, which demonstrates the certainty of the model

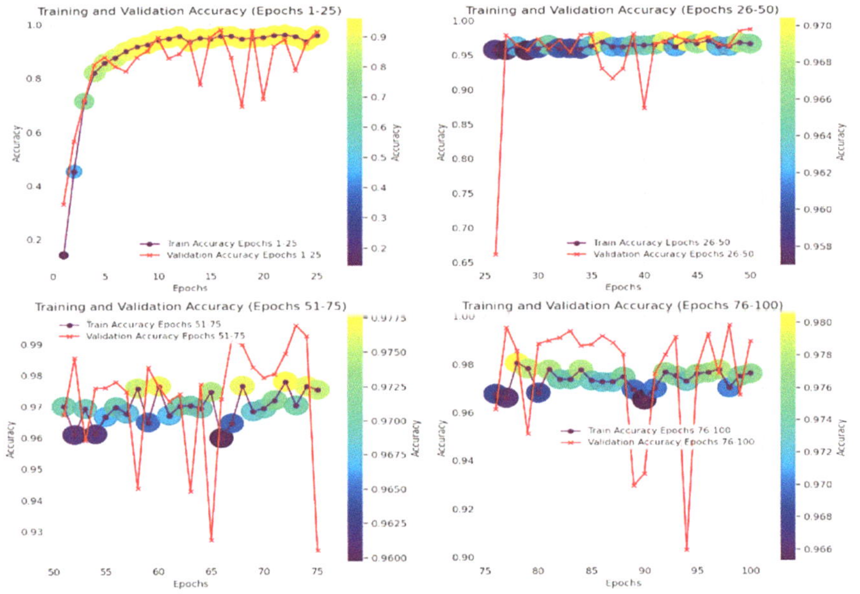

Fig. 13 Generalization of proposed model over epoch

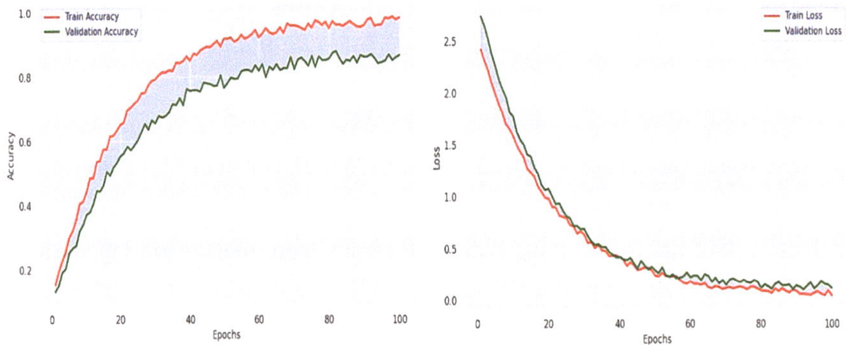

Fig. 14 Overall model performance accuracy and loss analysis

in its predictions and that it can discriminate the subtleties of the categories, shown in Fig. 15. Moreover, the training progress graph in Fig. 16 shows efficient use of the GPU and memory resources as the usage of the GPU is modest (~0.26GB) and the memory consumption is around 15.7%. Not only with the limitation of hardware requirements, but the model has also promised accuracy and convergence, demonstrating an optimized implementation and scalability.

The number depicts the word-level counterfactual embedding effect of the proposed Yi-34B LLM model. The bar plot in Fig. 17 shows how much the embeddings change when explicitly toxic words are substituted with neutral ones. The larger the cosine distance, the larger the semantic change that is reflected by the model that can reflect the toxicity-to-neutrality transition. But a closer examination will show that such words as an idiot, loser, and worthless provide the most significant embedding shifts in substituting them with the neutral words such as person, individual, and useless. This means that the Yi-34B considers these tokens as heavy handed in semantics of abuse or humiliation. That they are replaced with neutral descriptors causes a fundamental change in meaning, which creates a notable change in the embedding space. This demonstrates that the model is vulnerable to high intensity toxic cues. In contrast to them, hate and ugly are showing moderate variation when they are replaced with other words that are not harsh such as dislike and plain. Yi-34B however feels that there was a reduction of negativity although the mood is somewhat critical. At the lower end of the scale, the words stupid or dumb undergo rather low embedding when replaced with silly or uninformed. Such names are rather inaccurate and Yi-34B places them close to their poisonous property in the semantic space. It implies that these replacements may render negative connotation less conspicuous, though it is not necessary that it is entirely softened.

In general, the Fig. highlights the fact that Yi-34B is a strong semantic respondent to the toxic language. High embedding shifts indicate their ability to differentiate overtly abusive expression and more neutral, whereas moderate to low shifts indicate less obvious differentiation in which there is a softening of negativity but not its elimination. This counterfactual experiment supports the robustness of Yi-34B in addressing the toxicity detection and mitigation tasks, providing some insight to what lexical items contribute to the toxicity classification tasks most effectively.

The CLS attention distribution to the Arguments class (as Fig. 18a shows) focuses on the tokens like today (0.25), business (0.22), month (0.20), and open (0.18). These weights indicate that Yi-34B LLM model emphasizes on contextual and reasoning related terms in the classification of argumentative sentences. The fact that the model is interested in such tokens is indicative of the model being able to identify the rational discourse markers that differentiate argumentation and other types of toxic or casual communication. Figure 18b, Trolling class, gives the greatest weight to dreaming (0.40), LOL (0.35). These tokens represent the sarcastic and ridiculing character of trolling content. The attention weights of the class of Abusive Language can be seen in Fig. 18c where such tokens as somebody (0.27), growth (0.25), and notes (0.20) are highlighted. This shows that Yi-34B can pick up on even the subtleties of abuse where the toxicity is not necessarily caused by literal insults but also by the framing to be demeaning in its context. Furthermore, Fig. 18d, the weights of the attention related to

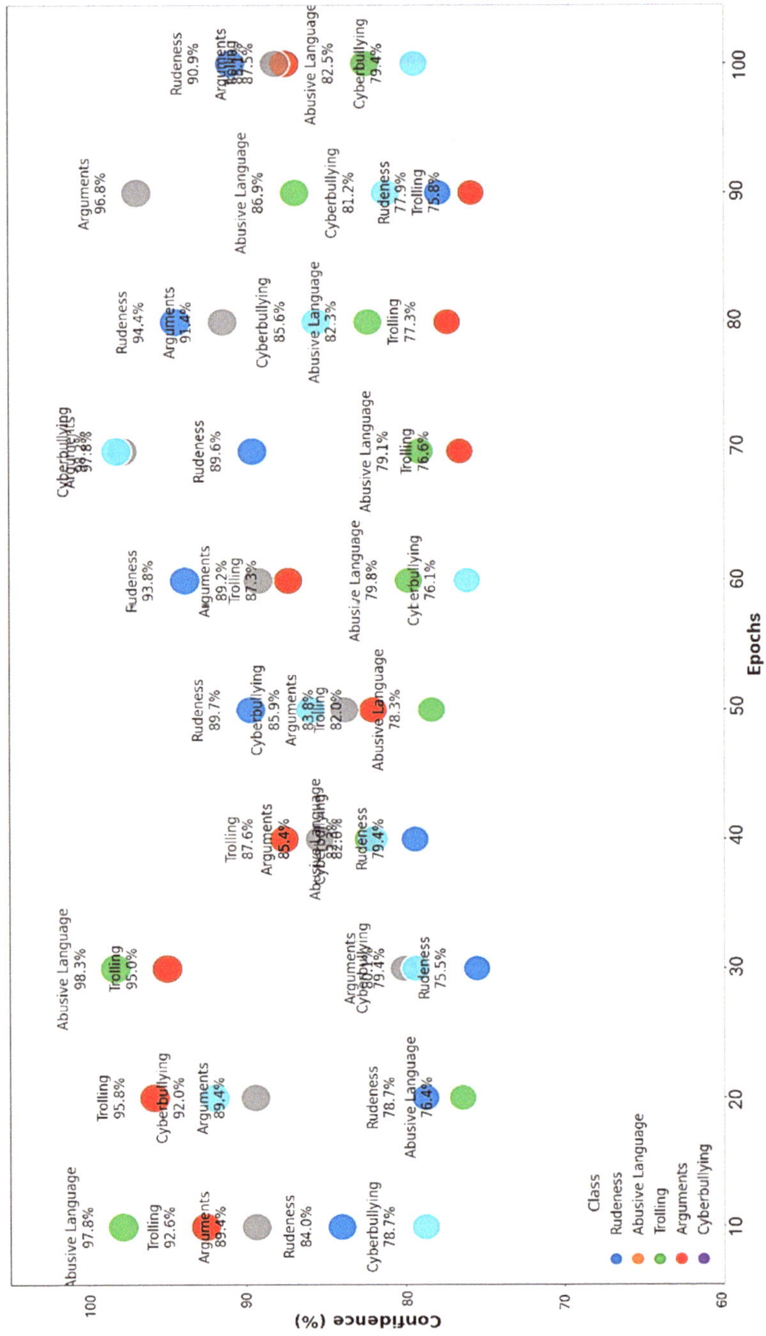

Fig. 15 Confidence score analysis for each class based on unseen data predictions

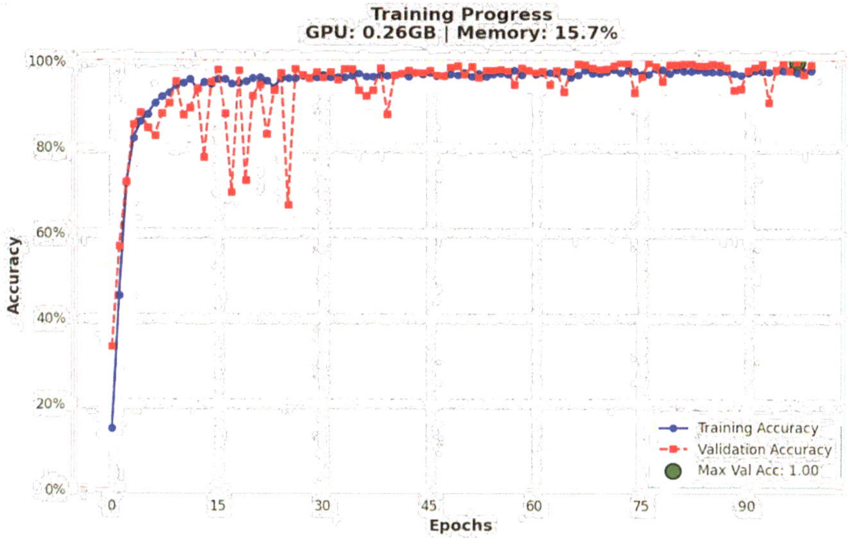

Fig. 16 Training analysis and model with resource utilization

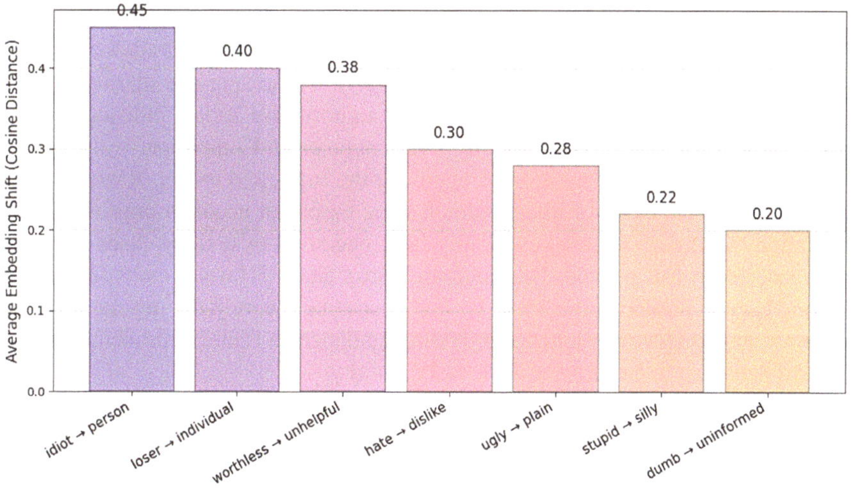

Fig. 17 Word level counterfactual impact on embedding space using proposed model embeddings layer

the Rudeness class show the tokens quit (0.30), bothering (0.35) and everyone (0.25). The model is very clear that directive imperatives and dismissive verbs go hand in hand with rudeness. The high value on bothering specifically indicates that the model is sensitive to the words that convey the emotion of annoyance and lack of politeness that is typical of rudeness. Lastly, Fig. 18e shows attention of the Cyberbullying class, where it focuses on tokens like fight (0.28), scientist (0.25), decide (0.22), and party (0.20).

These weights imply that the model identifies the expression which is oriented toward aggression or a targeted expression such as fight and targeting the context-setting nouns the model in question moderately. The extent of the dissemination of weights demonstrates that cyberbullying can be identified by a combination of abusive verbs and specific mentioning, which corresponds to the contextual explanation capacity of the model at large. The interpretability of the model is emphasized in this analysis, where the attention weights can exhibit class-specific linguistic attributes to lead to classification decisions.

The qualitative error analysis presented in Fig. 19 gives a detailed picture of sentence-level predictions made with the help of the proposed model on unknown data. The number represents eight representative samples, whereby each bar is proportional by the length of the token, and the color is of the category of prediction outcome. The entire sentence is presented with the labels of the real and predicted ones, with the pinpointed instances of the model being right or wrong in the classification. The instances that are classified properly indicate that the model is strong enough to separate between the toxic and the non-toxic expressions. These illustrations affirm that the model is effective in extracting rudeness using directive language and identification of rational discourse markers as being part of non-toxic argumentative speech. These mistakes indicate that the model occasionally has difficulties with overt personal attacks in contexts where they are contained in generally constructed sentences, not reflecting their poisonous severity. These spurious alarms indicate that the model is sensitive to some formal or authoritative lexical indicators which can trigger it to falsely predict toxicity where none exist. Comprehensively, Fig. 19 highlights the weaknesses and the strengths of the suggested model. It works effectively on clear-cut cases of toxic and non-toxic cases but needs to improve disambiguation of borderline cases especially false negatives of toxicity cases that have been incorrectly categorized. This qualitative error analysis layer is essential because it completes quantitative measures by demonstrating the way the misclassification occurs at the sentence level, hence providing guidance on future improvements, e.g., context-sensitive fine-tuning or adversarial training.

Yi-34B LLM model is created with a specially designed set of hyperparameters in terms of effective learning process and generalization shown in Table 8. Deep transformer architecture in form of multiple encoder layers and attention heads is efficient to represent complicated associations in text data. We use the adaptive optimizer, as well as regularization techniques to improve during training to prevent over-fitting and achieve a stable convergence. These precautions along with sequence length and gradient clipping enable the training process by being more robust and model can process long context. The same set of hyperparameter choices is partly what makes the model so effective at the task of toxic comment classification, which is an ideal balance between computational complexity and high accuracy and generalization.

The results of the statistical significance test confirm entirely that Yi-34B LLM model performs well compared to others when applying traditional methods as it is demonstrated in Fig. 20. The p-values of the paired t-test and Wilcoxon test of

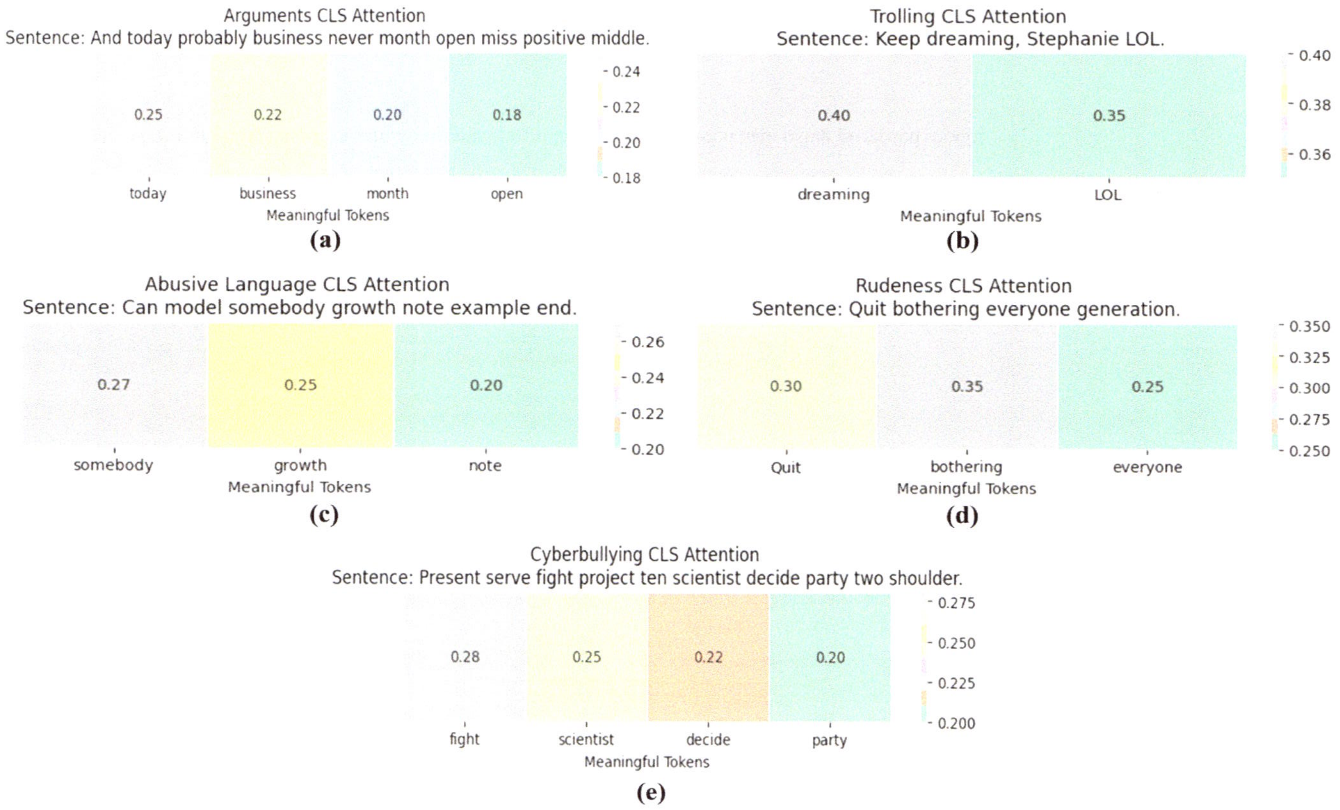

Fig. 18 CLS attention mechanism assigning weights to words strength based on sentences for each class: **a** argument **b** trolling **c** abusive language **d** rudeness **e** cyberbullying

Fig. 19 Qualitative error analysis based on sample dataset for prediction from unseen data using proposed model

Table 8 Analysis of hyperparameter settings

Hyperparameter	Value
Epochs	100
Batch size	32
Learning rate	3e-5
Optimizer	AdamW
Weight decay	0.01
Transformer layers	48
Attention heads	32
Attention head dimension	32
Feed-forward dimension	4096
Dropout rate	0.1
Max sequence length	512
Activation function	GELU
Positional encoding	Learned positional embeddings
Gradient clipping	1.0

both accuracy and loss are very low, which is why it is possible to conclude that the gains obtained, including the impressive accuracy of 98.03% and the significantly lower loss by the Yi-34B LLM are statistically significant and do not appear because of chance. The heatmap of statistical significance demonstrates that the observed differences in the model performance are statistically significant. In the Pearson correlation p-value we see a little more significant value of the accuracy (0.0602) which is slight but still at the same time shows that there is a significant correlation and it is further evidence of the same that the model is consistently learned. The class-level significance test shows high statistical significance in each of the tests across all toxic categories, which shows how strong the model has performed across each group, in Fig. 21. The p-values of these latter two are significantly lower than 0.05 on the t-test, ANOVA, and Chi-square tests, and this shows the excellent ability of the model to discriminate against the four classes. The p-values of the arguments, though still significant, are comparatively greater than the others of the usages indicating that slight variations or difficulty in modeling of this type are possible but also are quite understandable.

The interpretation of the results by means of SHAP analysis of Yi-34B LLM model determines certain informative tokens behind the brilliant toxic comment classifiers. The bar plot of Fig. 22 depicts absolute SHAP values of feature significance across classes, with words in the chart like owned, cares, and failed feature the most features importance especially to classes like trolling and rudeness. This appears to suggest that this model is being trained to learn to pick up fine, class-specific expressions occasioned during the classification decision. Moreover, the radar plot in Fig. 23 demonstrates the contribution of words to classes using LIME, and it exhibits certain linguistic tendencies of abusive language, cyberbullying, and

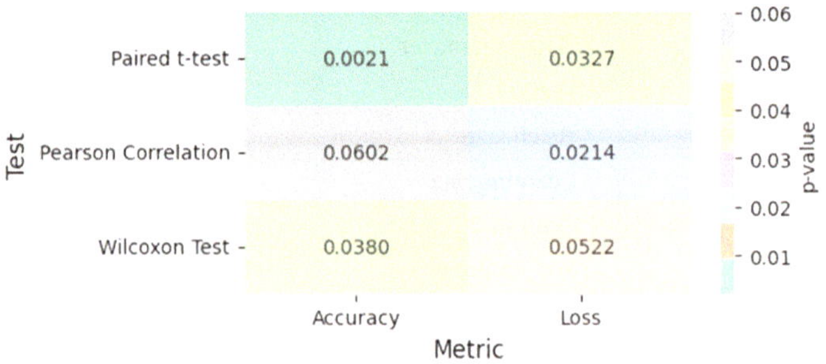

Fig. 20 Statistical test analysis based on proposed model accuracy and loss using p-values

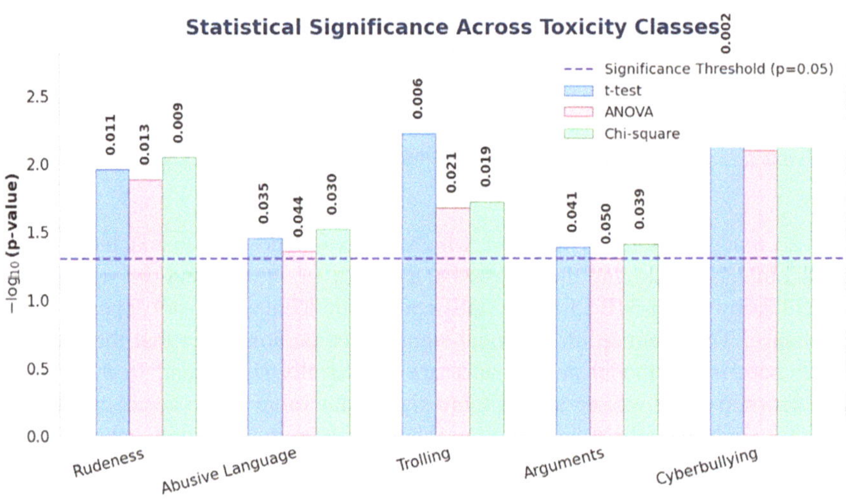

Fig. 21 Statistical analysis using various tests across each label

arguments, which means that the model has the potential to acquire deep semantic knowledge. method to be applied in the context of detecting toxicity in social media.

Figure 24 shows the comparison of shared LLM training on the validation accuracy across epochs, which clearly demonstrates the superiority of Yi-34B LLM model that consistently outperforms the other shared models with the peak accuracy close to 98%, and that significantly outperforms the traditional LSTM and Bi-LSTM. The Yi-34B LLM leads in the entire training process as well, possessing performance advantages of 7.2% and 16.2% over previous models. It is particularly salient in both the early learning phase (epochs 40–50) and the stabilization phase (40–100 epochs) that Yi-34B learns rapidly and develops a steady baseline, and more simplistic models experience both slower growth and a lower ceiling. This large performance gap highlights the utility of proposed LLM in modeling complex patterns of toxicity in social media comments and outperforms popular deep learning models as well as

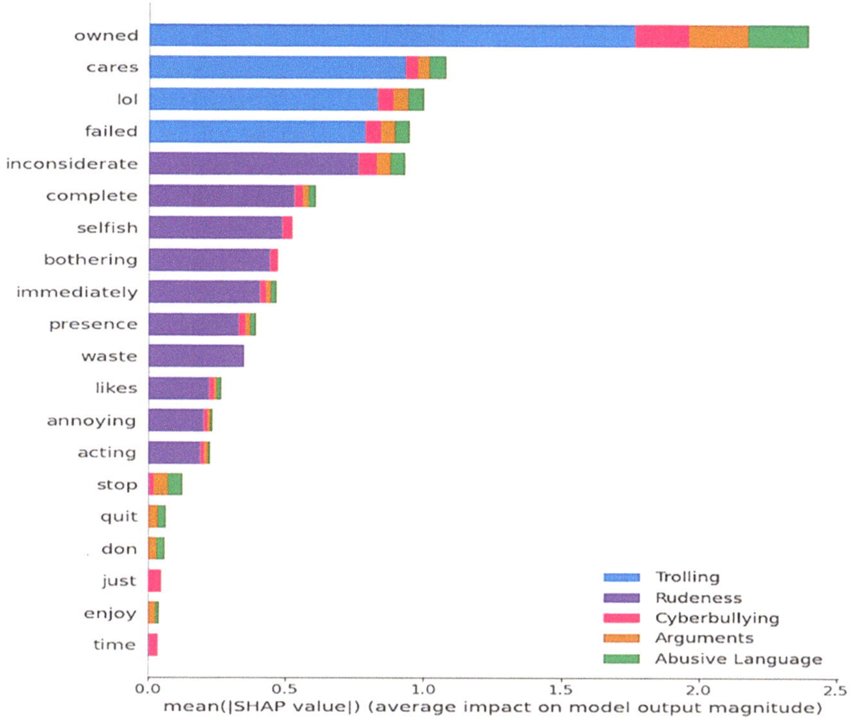

Fig. 22 SHAP features importance visualization depicting the average impact of key tokens on the LLM model's toxic comment classification across different classes

representations based on transformers in terms of learning rate and final prediction power.

The comparison between Yi-34B LLM with existing models indicates considerable improvements in the classification of toxic comments on social media during fandom war, shown in Table 9. Previous models have been mainly task-specific ones, and have been fine-tuned with BERT, BERTLSTM, and others, however, the Yi-34B LLM outperforms all of these, with a remarkable 98% accuracy in five toxicity categories. This is significantly better than the traditional transformer-based models that typically achieve accuracy in the range of 78% to 95%. The proposed model shows its strength for dealing with multiple toxicity categories in a balanced data set shows its potential capturing capability which are further extended applications in content moderation and analyses.

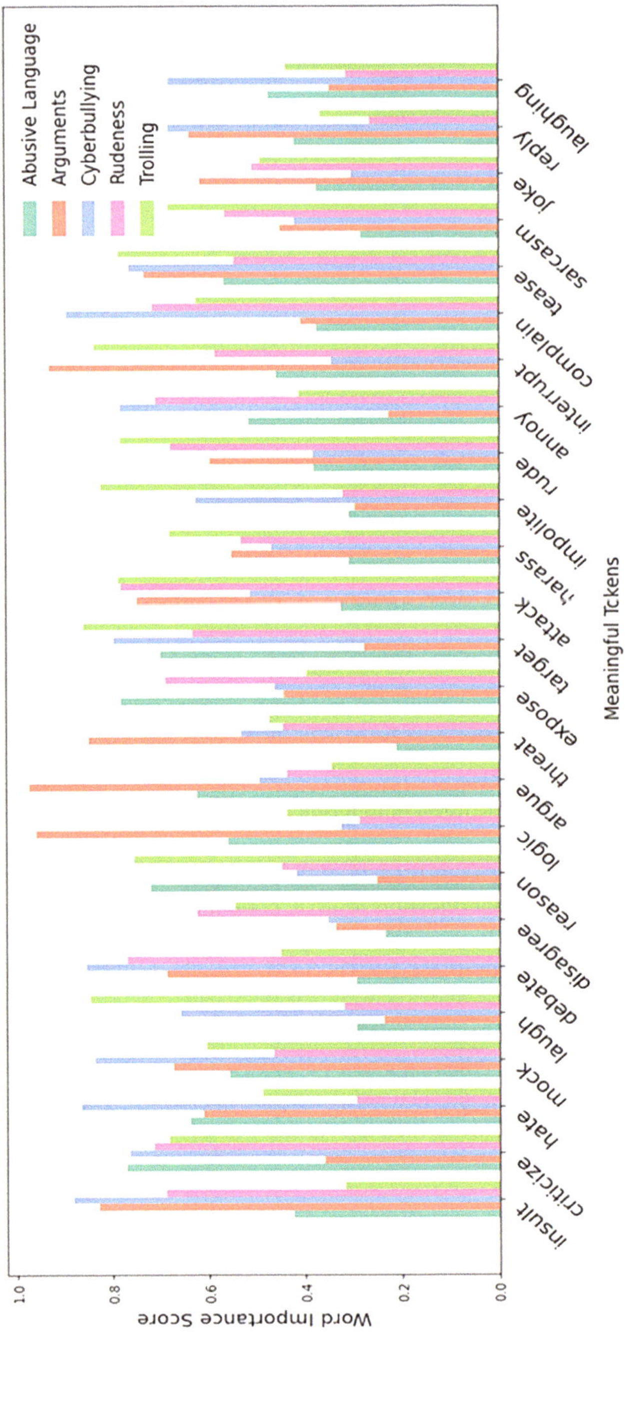

Fig. 23 LIME explanation plot illustrating the contribution of individual words toward the model's prediction for each class

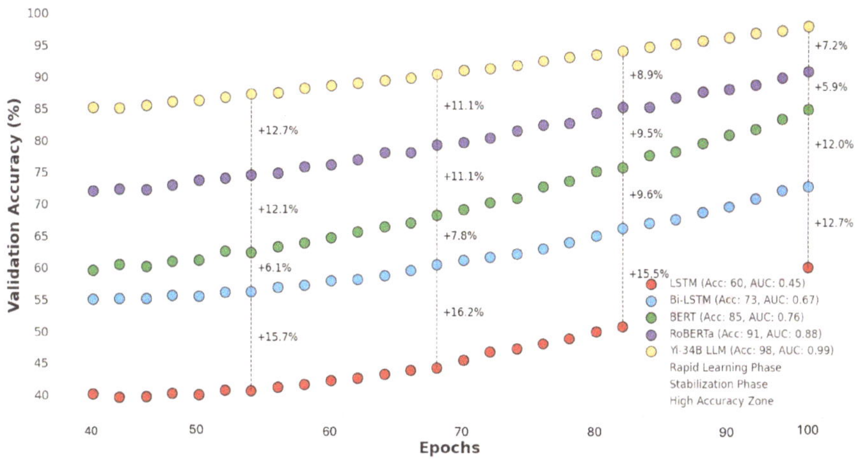

Fig. 24 Model comparison analysis illustrating performance gap

Table 9 Comparison with state-of-the-art models

Refs	Year	Approach	Dataset	Features	Result (%)	Research area
[17]	2020	BERT	Twitter hate speech	BERT embeddings; n-gram	F1: 85	Hate speech detection
[20]	2021	BERT + CNN	Facebook fake news	BERT semantic vectors	Acc: 88	Fake news classification
[24]	2022	BERT-LSTM	Twitter sarcasm	Word + textual embeddings	F1: 81	Sarcasm detection
[16]	2023	BERT + Bi-LSTM	Twitter sentiment	Word embeddings	Acc: 93 (3-class)	Sentiment classification
[13]	2024	XLNeT	Stress detection	Word embeddings	Acc: 95	Mental health
[18]	2025	Ensemble BERT	Hate speech data	BERT models via meta-learner	Acc: 89	Hate speech detection
Proposed		**Yi-34B LLM**	**Social media comments**	**Pre-trained Embeddings**	**Acc: 98% (5-classes)**	**Fandom war**

7 Conclusion

The proposed study significantly contributes to the identification of toxic remarks on social media platforms because it combines the strength of the LLM powered by both the DL and NLP technologies. An excellent accuracy of 98% with the help of the proposed Yi-34B model that compressed both classic ML models and transformer models. Such high-performing performance is a vivid example of the great capacity of the model to retrieve complex linguistic patterns and toxic behaviors which depict a new standard in classifying toxicity according to CLS embeddings tokens and weight settings. This also supported the results performance using explainable AI techniques inclusion of statistical validations based on log of p-values. In the next direction, future studies are needed on applying domain-related contextual information to enhance model relevance, the abilities should be extended to multilingual data sets to have a wider application, and the architecture must be made more efficient to determine which settings can reduce computational needs. Moreover, it will be necessary to increase the diversity of training data sets and create interpretability tools to enhance the resilience and transparency of models and make their implementation more credible and feasible. Overall, this paper provides a solid groundwork of the next generation toxic comment detection systems that would result in safer and more inclusive Internet spaces.

Authors Contributions The authors Anam Naz and Hikmat Ullah Khan equally contributed to conducting this study from planning to writing.

AI Use Disclosure Statement The authors used generative AI technologies only for proofreading. This includes minor corrections to grammar, spelling, and syntax. No AI was used for ideation, data analysis, or writing original content.

Funding Statement No external funds received for this study.

Data Availability The dataset is freely available at: https://github.com/VisionLangAI/Social-War

Ethics Concerns Authors declare that they have no conflict of interest.

References

1. Boer M, Stevens GWJM, Finkenauer C, de Looze ME, van den Eijnden RJJM (2021) Social media use intensity, social media use problems, and mental health among adolescents: investigating directionality and mediating processes. Comput Human Behav 116:106645. https://doi.org/10.1016/j.chb.2020.106645
2. Saura JR, Dwivedi YK, Palacios-Marqués D (2022) Editorial: online user behavior and user-generated content. Front Psychol 13:895467. https://doi.org/10.3389/FPSYG.2022.895467
3. Naz A, Khan HU, Bukhari A, Alshemaimri B, Daud A, Ramzan M (2025) Machine and deep learning for personality traits detection: a comprehensive survey and open research challenges. Artif Intell Rev 58(8):239. https://doi.org/10.1007/s10462-025-11245-3
4. Butler R (2020) A systematic literature review of the factors affecting smartphone user threat avoidance behaviour. Emerald Group Holdings Ltd. https://doi.org/10.1108/ICS-01-2020-0016
5. Ishfaq U, Khan HU, Shabbir D (2025) Exploring the role of sentiment analysis with network and temporal features for finding influential users in social media platforms. Soc Netw Anal Min 14(1):241. https://doi.org/10.1007/s13278-024-01396-6
6. Kumar Sharma H, Kshitiz K, Shailendra (2018) NLP and machine learning techniques for detecting insulting comments on social networking platforms. In: 2018 International conference on advances in computing and communication engineering (ICACCE), pp 265–272. https://doi.org/10.1109/ICACCE.2018.8441728

7. Lee LW, Dabirian A, McCarthy IP, Kietzmann J (2020) Making sense of text: artificial intelligence-enabled content analysis. Eur J Mark 54(3):615–644. https://doi.org/10.1108/EJM-02-2019-0219
8. Naz A, Khan HU, Alsahfi T, Alhajlah M, Alshemaimri B, Daud A (2025) Using transformers and Bi-LSTM with sentence embeddings for prediction of openness human personality trait. PeerJ Comput Sci 11:e2781. https://doi.org/10.7717/peerj-cs.2781
9. Chia ZL, Ptaszynski M, Masui F, Leliwa G, Wroczynski M (2021) Machine Learning and feature engineering-based study into sarcasm and irony classification with application to cyberbullying detection. Inf Process Manag 58(4). https://doi.org/10.1016/j.ipm.2021.102600
10. Ibrohim MO, Budi I (2023) Hate speech and abusive language detection in Indonesian social media: progress and challenges. Elsevier Ltd. https://doi.org/10.1016/j.heliyon.2023.e18647
11. Mahmood A, Khan HU, Ramzan M (2020) On modelling for bias-aware sentiment analysis and its impact in twitter. J Web Eng 19(1):1–28. https://doi.org/10.13052/jwe1540-9589.1911
12. Pillow DR, Kohler J, Bowers C, Mills S, Crabtree MA (2025) Expressing and negotiating identities in social media ecosystems: a typology of users and their associated personality profiles. Pers Individ Dif 232:112824. https://doi.org/10.1016/j.paid.2024.112824
13. Kasanneni Y, Duggal A, Sathyaraj R, Raja SP (2025) Effective analysis of machine and deep learning methods for diagnosing mental health using social media conversations. IEEE Trans Comput Soc Syst 12(1):274–294. https://doi.org/10.1109/TCSS.2024.3487168
14. Kjell ONE, Sikström S, Kjell K, Schwart HA (2022) Natural language analyzed with AI-based transformers predict traditional subjective well-being measures approaching the theoretical upper limits in accuracy. Sci Rep 12(1). https://doi.org/10.1038/s41598-022-07520-w
15. Ramos G et al (2024) A comprehensive review on automatic hate speech detection in the age of the transformer. Social Netw Anal Min 14(1):1–25. https://doi.org/10.1007/S13278-024-01361-3
16. Bello A, Ng SC, Leung MF (2023) A BERT framework to sentiment analysis of tweets. Sensors 23(1):506. https://doi.org/10.3390/S23010506
17. Mozafari M, Farahbakhsh R, Crespi N (2020) Hate speech detection and racial bias mitigation in social media based on BERT model. PLoS ONE 15(8):e0237861. https://doi.org/10.1371/JOURNAL.PONE.0237861
18. Yoo S, Jeon E, Hyeon J, Cho J (2025) Adaptive ensemble techniques leveraging BERT based models for multilingual hate speech detection in Korean and english. Sci Rep 15(1):1–20. https://doi.org/10.1038/S41598-025-88960-Y
19. Mohamed O, Kassem AM, Ashraf A, Jamal S, Mohamed EH (2023) An ensemble transformer-based model for Arabic sentiment analysis. Soc Netw Anal Min 13(1):1–14. https://doi.org/10.1007/S13278-022-01009-0
20. Kaliyar RK, Goswami A, Narang P (2021) FakeBERT: fake news detection in social media with a BERT-based deep learning approach. Multimed Tools Appl 80(8):11765–11788. https://doi.org/10.1007/S11042-020-10183-2
21. Karande H, Walambe R, Benjamin V, Kotecha K, Raghu TS (2021) Stance detection with BERT embeddings for credibility analysis of information on social media. PeerJ Comput Sci 7:1–20. https://doi.org/10.7717/PEERJ-CS.467
22. Faizz Ahmad KS, Pamidimukkala SG, Sathe AP, GNVG S, Ch K (2025) Hybrid optimization driven fake news detection using reinforced transformer models. Sci Rep 15(1):1–16. https://doi.org/10.1038/S41598-025-99936-3
23. Saini H, Mehra H, Rani R, Jaiswal G, Sharma A, Dev A (2024) Enhancing cyberbullying detection: a comparative study of ensemble CNN–SVM and BERT models. Soc Netw Anal Min 14(1):1–18. https://doi.org/10.1007/S13278-023-01158-W
24. Pandey R, Singh JP (2023) BERT-LSTM model for sarcasm detection in code-mixed social media post. J Intell Inf Syst 60(1):235–254. https://doi.org/10.1007/S10844-022-00755-Z
25. Mohan A, Nair AM, Jayakumar B, Muraleedharan S (2023) Sarcasm detection using bidirectional encoder representations from transformers and graph convolutional networks. Procedia Comput Sci 218:93–102. https://doi.org/10.1016/J.PROCS.2022.12.405

26. Dahou A, Mabrouk A, Ewees AA, Gaheen MA, Abd Elaziz M (2023) A social media event detection framework based on transformers and swarm optimization for public notification of crises and emergency management. Technol Forecast Soc Change 192:122546. https://doi.org/10.1016/J.TECHFORE.2023.122546
27. Anggrainingsih R, Hassan GM, Datta A (2024) Transformer-based models for combating rumours on microblogging platforms: a review. Artif Intell Rev 57(8):1–69. https://doi.org/10.1007/S10462-024-10837-9
28. Karpagalingam T, Britto JJ, Haripriya R, Shreenee N (2024) Rumor detection using BERT-based social circle and interaction network model. Soc Netw Anal Min 14(1):1–14. https://doi.org/10.1007/S13278-024-01362-2
29. Sufi F (2024) A sustainable way forward: systematic review of transformer technology in social-media-based disaster analytics. Sustainability 16(7):2742. https://doi.org/10.3390/SU16072742
30. Naz A, Khan HU, Alesawi S, Abouola OI, Daud A, Ramzan M (2024) AI knows you: deep learning model for prediction of extroversion personality trait. IEEE Access:1. https://doi.org/10.1109/ACCESS.2024.3486578
31. Muqadas A, Khan HU, Ramzan M, Naz A, Alsahfi T, Daud A (2025) Deep learning and sentence embeddings for detection of clickbait news from online content. Sci Rep 15(1). https://doi.org/10.1038/s41598-025-97576-1

Optimal Site Selection for E-Waste Recycling and Upcycling Centre: A Complex Linear Diophantine Fuzzy MCGDM Approach

Dinanath Choudhary, Sudipa Choudhury, and Apu Kumar Saha

Abstract There are serious environmental and socioeconomic issues due to the explosive increase of electronic waste, especially in the ecologically fragile North-East region of India. Selecting optimal sites for recycling and up-cycling centres requires addressing multiple conflicting criteria under uncertainty. This study presents a novel Complex Linear Diophantine Fuzzy Set (CLDFS) based Multi-Criteria Group Decision-Making (MCGDM) framework that integrates Step-wise Weight Assessment Ratio Analysis (SWARA) for deriving criteria weights and Combinative Distance-based Assessment (CODAS) for ranking alternatives. CLDFS, a recent extension of fuzzy set theory, extends membership representation in the complex plane, effectively handling ambiguity, uncertainty, and hesitancy in expert evaluations. The analysis incorporates judgments from three experts, evaluating ten criteria across nine alternative sites. Results identify Agartala city of Tripura, NE India as the most suitable location. Comparative analysis confirms the strength and superiority of the projected methodology, providing a mathematically rigorous, versatile, and practical decision-support tool for sustainable e-waste management.

Keywords Complex Linear Diophantine Fuzzy set · SWARA · CODAS · E-Waste recycling and up-cycling centres

---

D. Choudhary · S. Choudhury · A. K. Saha (✉)
Department of Mathematics, National Institute of Technology Agartala, Agartala, Tripura, India
e-mail: apusaha_nita@yahoo.co.in

D. Choudhary
e-mail: dnchoudharymbi@gmail.com

S. Choudhury
e-mail: sudipachoudhury032@gmail.com

A. Azizi et al. (eds.), *Advanced Decision-Making Under Uncertainty*, Emerging Trends in Mechatronics, https://doi.org/10.1007/978-981-95-8696-7_8

1 Introduction

Global electronic waste (e-waste) generation has touched alarming levels, with over 62 million metric tons produced in 2023 alone yet nearly 20% is formally recycled [1]. This exponential rise, driven by the rapid expansion of the digital economy and increased consumer demand for electronic devices, presents a growing environmental and public health challenge worldwide [2]. India, ranked among the top five e-waste producers, faces a dual dilemma: managing the benefits of technological advancement while addressing the eco-friendly significances of improper e-waste disposal [3].

Within India, the North-Eastern Region (NER) comprising eight states is particularly vulnerable [4]. This region, characterized by its ecological sensitivity, rugged terrain, infrastructural limitations, and geographic isolation, poses unique challenges for sustainable e-waste administration [5]. Currently, the absence of formal collection and recycling infrastructure in the NER has led to informal and unsafe practices and manual dismantling by unregulated sectors [6]. These operations put workers' health at serious danger and contaminate the land, water, and air [7], while also resulting in the loss of valuable recoverable materials [8]. This not only hampers environmental safety but also undermines efforts to transition toward a circular economy [9].

Establishing formal e-waste recycling and upcycling centres in the NER is essential for mitigating these impacts [10]. However, their success hinges not merely on infrastructure development but on strategic, context-sensitive site selection that accounts for the region's complex environmental, social, technical, and economic conditions [11]. Experts must consider variables such as ecological vulnerability, population density, existing transport and digital infrastructure, land availability, and proximity to e-waste sources [12]. These often conflicting and interdependent criteria make the selection process inherently complex [13].

These information measurements are important components of research and play a critical role in making decisions. Taking decisions is a real-world situation where the goal is to accomplish an objective that aligns with certain restrictions is commonly referred to as multi-criteria decision-making (MCDM), and Multi-criteria group decision-making (MCGDM) is the term used when multiple decision-makers participate in the process [14]. MCDM is a group of techniques that, through examining decision-makers' assessments and viewpoints regarding the predetermined objective, analyses the benefits and drawbacks of options based on several relevant restrictions [15].

Conventional decision models typically linear or deterministic struggle to manage the dynamic, uncertain nature of such environments, particularly when expert input is qualitative or expressed in vague, linguistic terms [16]. To address such problems, a variety of strategies, such as Fuzzy Set (FS) [17], Interval-Valued Fuzzy Set (IVFS) [18], Intuitionistic Fuzzy Set (IFS) [19], Pythagorean Fuzzy Set (PyFS) [20], q-Rung Orthopair Fuzzy Set (q-ROFS) [21], are used. Fuzzy MCDM methods, through the use of membership and non-membership degrees, allow for a more precise evaluation of uncertain criteria like environmental impact, cost, and feasibility in site selection, offering a mathematical advantage over traditional methods [22].

Membership degree (MD) ranges in [0, 1] are used to specify a stepwise evaluation of objects in FSs [23], however, in various situations, just MD is unable to decipher the unclear data [24], so the non-membership degree (NMD) is a prerequisite, and they might or might not be connected [25]. When different experts assign MD and NMD to different things, it becomes difficult to comply with such standards [26]. Then ultimately the concept of IFS [27], PyFS [28], q-ROFS [29] are introduced and has been applied in different real life application effectively. But, it has come to notice that PyFS and q-ROFS occasionally failed to completely represent data uncertainty [30]. In order to overcome this limitation, Riaz and Hashmi [31] created a special FS variation, known as the Linear Diophantine Fuzzy Set (LDFS), which offers a new viewpoint for decision analysis research and incorporates reference parameters (RPs) along with the MD and NMD. When compared to other FSs, the LD-FS concept is more intriguing and useful with the advantage of the inclusion of RPs.

Complex Fuzzy Sets (CFS) [32, 33] are useful in a range of contexts because of their ability to simultaneously record uncertainty and complex-valued features. The domain of the CFS extends to the the complex plane, as opposed to being restricted to the interval [0, 1] [32]. When both amplitude and phase terms are included, a complex-valued membership stands out from the others [34]. While "amplitude" still includes the concept of fuzziness, "phase" denotes the affirmation of a complicated fuzzy set, which calls for the study of the second dimension of membership [35]. Based on this complex value feature, LDFS has been extended in to Complex Linear Diophantine Fuzzy Set (CLDFS) [36].

The concept of aggregation operator (AO) has grown much attention lately and is now one of the popular subjects in MCDM challenges [37]. In decision analysis, triangular norm (TN) and triangular co-norm (TCN) are necessary for completing FSs, which are vital systems that play a significant part in information unification [38]. On the basis of various TNs and TCNs, a large number of AOs have been established for various extensions of FSs and have been applied in different fields. Riaz et al. [39] created a novel MCDM technique using LDF Weighted Average (LDFWA) and LDF Weighted Geometric (LDFWG) operators. Zia et al. [40] proposed a novel Fuzzy MCDM approach based on the CLDF Aggregation Operator (CLDFAO). This methodology combines the flexibility of fuzzy logic with the structural rigor of linear Diophantine equations to aggregate linguistic expert judgments more effectively [41]. The CLDFAO enhances decision robustness by accommodating uncertainty, integrating subjective knowledge, and ensuring logical consistency through mathematically grounded relationships [39].

This study closes a significant gap by creating a context-sensitive, uncertainty-aware, and technically rigorous decision-support framework for e-waste recycling site selection in India's NER [42]. The proposed CLDFAO-based MCGDM approach not only addresses the limitations of existing models but also offers a transferable methodology for other ecologically fragile and logistically complex regions.

1.1 Research Gap and Motivation

After a thorough literature review, several critical gaps in existing methodologies were identified:

- Current fuzzy set frameworks face significant limitations in representing complex-valued MD and NMD, especially in decision-making scenarios involving linguistic ambiguity and high uncertainty. These frameworks are mostly restricted to the real unit interval, limiting their ability to capture subtle expert perceptions. The CLDFS framework overcomes this by allowing MD and NMD values to be freely defined over the complex domain. Additionally, reference parameters improve interpretability, enabling more precise modelling of uncertain information.
- Aggregation of expert opinions is vital in fuzzy MCDM, yet most existing AOs are not suited for complex-valued fuzzy data. To fill this gap, this study utilized two operators CLDFWA (Complex Linear Diophantine Fuzzy Weighted Average) and CLDFWG (Complex Linear Diophantine Fuzzy Weighted Geometric), designed specifically for the CLDFS framework. These operators robustly aggregate uncertain and imprecise expert evaluations.
- The aggregation operators improve adaptability and consistency in aggregating expert judgments for conflicting and interdependent criteria. Unlike conventional operators limited to real-valued fuzzy data, these are designed within the CLDFS framework to handle complex-valued uncertainty. To enhance robustness, the framework integrates SWARA for flexible criteria weighting and CODAS for reliable alternative ranking. This combined approach offers greater generalizability, transparency, and effectiveness, making it well-suited for complex, real-world decision problems involving linguistic ambiguity and multifaceted expert evaluations.
- A key application of this approach is in selecting sites for e-waste recycling and upcycling centres in India's NER an area that remains underexplored. The NER's ecological sensitivity, infrastructural challenges, and geographic isolation create unique issues that traditional models struggle to address. This study proposes a CLDFS-based MCGDM framework that integrates environmental and contextual complexities, supporting more informed, adaptable, and sustainable site selection.

1.2 Main Objective

The primary contributions of this study are listed below, taking into account the aforementioned research gaps and goals:

- Exploration of CLDFSs: This study utilizes the CLDFS framework by examining its fundamental operations like addition, multiplication, and power highlighting their essential properties. This advances the mathematical modelling of uncertainty in decision-making involving complex fuzzy information.

- Application of Existing Aggregation Operators: The research employs established aggregation operators within the CLDFS framework, CLDFWA and CLDFWG operators. Their capacity to effectively aggregate complex-valued expert judgments is investigated.
- Integration with SWARA: The proposed methodology integrates the CLDFS based aggregation operators into a comprehensive MCDM framework. As part of this integration, the Stepwise Weight Assessment Ratio Analysis (SWARA) method is employed to evaluate the relative importance of the selected criteria. This enhances the objectivity and relevance of the decision-making process by capturing expert judgment systematically.
- Integration with CODAS: Within the same MCDM framework, the Combinative Distance-based Assessment (CODAS) algorithm is employed for ranking the alternatives. The use of CODAS enables effective differentiation among alternatives based on their distances from an ideal solution, making the evaluation more robust, interpretable, and suitable for complex fuzzy environments.
- Case Study on E-Waste Site Selection in North-Eastern India: The proposed CLDFS based MCGDM methodology is applied to determine optimal sites for e-waste recycling and upcycling centres in India's North-Eastern Region. The case study considers the region's ecological sensitivity, infrastructural constraints, and social factors, delivering a sustainable and context-sensitive decision-making solution. Comparative analyses are performed to evaluate consistency and robustness.

2 Preliminaries

In this section, some of the fundamental theories have been discussed which will be utilized in the subsequent segments.

2.1 Fuzzy Set Theory

Fuzzy set theory [17] advances classical set concepts by assigning each element in a universe (U) a MD ranging from 0 to 1. This graded membership captures uncertainty and partial truth, providing a nuanced framework that underpins complex decision-making, artificial intelligence, and control systems where binary classification falls short.

Let U denote a non–empty universe of discourse. The following fundamental concepts are recalled.

Definition 2.1.1 [17]: FS (F) on U is defined as $F = \{(x, \mu_F(x))|x \in \mathrm{U}\}$, where $\mu_F : \mathrm{U} \rightarrow [0, 1]$ signifies the MD of x such that $0 \leq \mu_F(x) \leq 1$ for all $x \in \mathrm{U}$.

Definition 2.1.2 [18]: CFS ($\mathbb{F}$) on U is defined as $\mathbb{F} = \left\{\left(x, \left\langle \mu_{\mathbb{F}}(x)e^{2\pi i(\theta_{\mathbb{F}}(x))}\right\rangle\right)|x \in \text{U}\right\}$ where $\mu_{\mathbb{F}}(x)e^{2\pi i(\theta_{\mathbb{F}}(x))}$ signifies the complex valued MD of x such that $0 \leq \mu_{\mathbb{F}}(x), \theta_{\mathbb{F}}(x) \leq 1$ for all $x \in \text{U}$.

Definition 2.1.3 [19]: IFS (I) on U is defined as $I = \{(x, \langle \mu_I(x), \nu_I(x)\rangle)|x \in \text{U}\}\}$, where $\mu_I : \text{U} \to [0, 1]$ and $\nu_I : \text{U} \to [0, 1]$ signify the MD and NMD respectively of x, such that $0 \leq \mu_I(x) + \nu_I(x) \leq 1$ for all $x \in \text{U}$.

Definition 2.1.4 [43]: CIFS ($\mathbb{I}$) on U is defined as $\mathbb{I} = \left\{\left(x, \left\langle \mu_{\mathbb{I}}(x)e^{2\pi i(\theta_{\mathbb{I}}(x))}, \nu_{\mathbb{I}}(x)e^{2\pi i(\emptyset_{\mathbb{I}}(x))}\right\rangle\right)|x \in \text{U}\right\}$, where $\mu_{\mathbb{I}}(x)e^{2\pi i(\theta_{\mathbb{I}}(x))} and \nu_{\mathbb{I}}(x)e^{2\pi i(\emptyset_{\mathbb{I}}(x))}$ signify the complex valued MD and NMD respectively of x, such that $0 \leq \mu_{\mathbb{I}}(x) + \nu_{\mathbb{I}}(x) \leq 1$ and $0 \leq \theta_{\mathbb{I}}(x) + \emptyset_{\mathbb{I}}(x) \leq 1$ for all $x \in \text{U}$.

Definition 2.1.5 [44] **:** PyFS (P) on U is defined as $P = \{(x, \langle \mu_P(x), \nu_P(x)\rangle)|x \in \text{U}\}$, where $\mu_P(x) and \nu_P(x)$ signify the MD and NMD respectively of x, such that $0 \leq (\mu_P(x))^2 + (\nu_P(x))^2 \leq 1$ for all $x \in \text{U}$.

Definition 2.1.6 [20]: CPyFS ($\mathbb{P}$) on U is defined as $\mathbb{P} = \left\{\left(x, \left\langle \mu_{\mathbb{P}}(x)e^{2\pi i(\theta_{\mathbb{P}}(x))}, \nu_{\mathbb{P}}(x)e^{2\pi i(\emptyset_{\mathbb{P}}(x))}\right\rangle\right)|x \in \text{U}\right\}$, where $\mu_{\mathbb{P}}(x)e^{2\pi i(\theta_{\mathbb{P}}(x))}$ and $\nu_{\mathbb{P}}(x)e^{2\pi i(\emptyset_{\mathbb{P}}(x))}$ signify the complex valued MD and NMD respectively of x, such that $0 \leq (\mu_{\mathbb{P}}(x))^2 + (\nu_{\mathbb{P}}(x))^2 \leq 1$ and $0 \leq (\theta_{\mathbb{P}}(x))^2 + (\emptyset_{\mathbb{P}}(x))^2 \leq 1$ for all $x \in \text{U}$.

Definition 2.1.7 [21]: q-ROFS (Q) on U is defined as $Q = \left\{\left(x, \left\langle \mu_Q(x), \nu_Q(x)\right\rangle\right)|x \in \text{U}\right\}$, where $\mu_Q(x) and \nu_Q(x)$ signify the MD and NMD respectively of x, such that for $q \geq 1$, we have, $0 \leq \left(\mu_Q(x)\right)^q + \left(\nu_Q(x)\right)^q \leq 1$ for all $x \in \text{U}$.

Definition 2.1.8 [45]: Cq-ROFS ($\mathbb{Q}$) on U is defined as $\mathbb{Q} = \left\{\left(x, \left\langle \mu_{\mathbb{Q}}(x)e^{2\pi i(\theta_{\mathbb{Q}}(x))}, \nu_{\mathbb{Q}}(x)e^{2\pi i(\emptyset_{\mathbb{Q}}(x))}\right\rangle\right)|x \in \text{U}\right\}$, where $\mu_{\mathbb{Q}}(x)e^{2\pi i(\theta_{\mathbb{Q}}(x))}$ and $\nu_{\mathbb{Q}}(x)e^{2\pi i(\emptyset_{\mathbb{Q}}(x))}$ signify the complex valued MD and NMD respectively of x, such that for $q \geq 1$, we have, $0 \leq \left(\mu_{\mathbb{Q}}(x)\right)^q + \left(\nu_{\mathbb{Q}}(x)\right)^q \leq 1$ and $0 \leq \left(\theta_{\mathbb{Q}}(x)\right)^q + \left(\emptyset_{\mathbb{Q}}(x)\right)^q \leq 1$, for all $x \in \text{U}$.

Definition 2.1.9 [36]: LDFS (L) on U is defined as $L = \{(x, \langle \mu_L(x), \nu_L(x)\rangle, \langle \alpha_L, \beta_L\rangle)|x \in \text{U}\}$, Where $\mu_L(x) and \nu_L(x)$ signify the MD and NMD respectively of x, and α_L, β_L are reference parameters such that, $\forall x \in \text{U}$, we have,

$$0 \leq \mu_L(x), \nu_L(x) \leq 1, \quad 0 \leq \alpha_L + \beta_L \leq 1, \quad 0 \leq \alpha_L \mu_L(x) + \beta_L \nu_L(x) \leq 1$$

Definition 2.1.10 [36]: CLDFS ($\mathbb{L}$) on U is defined as $\mathbb{L} = \left\{\left(x, \left\langle \mu_{\mathbb{L}}(x)e^{2\pi i(\theta_{\mathbb{L}}(x))}, \nu_{\mathbb{L}}(x)e^{2\pi i(\emptyset_{\mathbb{L}}(x))}\right\rangle, \langle \alpha_{\mathbb{L}}, \beta_{\mathbb{L}}\rangle\right)|x \in \text{U}\right\}\}$, where $\mu_{\mathbb{L}}(x)e^{2\pi i(\theta_{\mathbb{L}}(x))}$ and $\nu_{\mathbb{L}}(x)e^{2\pi i(\emptyset_{\mathbb{L}}(x))}$ signify the complex valued MD and NMD respectively of x, and $\alpha_{\mathbb{L}}, \beta_{\mathbb{L}}$ are reference parameters such that $\forall x \in \text{U}$, $\mu_{\mathbb{L}}(x), \nu_{\mathbb{L}}(x), \theta_{\mathbb{L}}(x), \emptyset_{\mathbb{L}}(x), \alpha_{\mathbb{L}}, \beta_{\mathbb{L}} \in [0, 1]$,

$$0 \le \alpha_{\mathbb{L}} + \beta_{\mathbb{L}} \le 1, \quad 0 \le \alpha_{\mathbb{L}}\mu_{\mathbb{L}}(x) + \beta_{\mathbb{L}}\nu_{\mathbb{L}}(x) \le 1, 0 \le \alpha_{\mathbb{L}}\theta_{\mathbb{L}}(x) + \beta_{\mathbb{L}}\emptyset_{\mathbb{L}}(x) \le 1.$$

Definition 2.1.11 [36]: Let $\mathbb{A}_j = \left(\left\langle \mu_{\mathbb{A}_j} e^{2\pi i\left(\theta_{\mathbb{A}_j}\right)}, \nu_{\mathbb{A}_j} e^{2\pi i\left(\emptyset_{\mathbb{A}_j}\right)}\right\rangle, \left\langle \alpha_{\mathbb{A}_j}, \beta_{\mathbb{A}_j}\right\rangle\right), j = 1, 2$ be CLDFNs with $r > 0$. Here are some fundamental operations as:

a. $\mathbb{A}_1 \oplus \mathbb{A}_2 = \begin{pmatrix} \left\langle \left(\mu_{\mathbb{A}_1} + \mu_{\mathbb{A}_2} - \mu_{\mathbb{A}_1}\mu_{\mathbb{A}_2}\right) e^{2\pi i\left(\theta_{\mathbb{A}_1} + \theta_{\mathbb{A}_2} - \theta_{\mathbb{A}_1}\theta_{\mathbb{A}_2}\right)}, \nu_{\mathbb{A}_1}\nu_{\mathbb{A}_2} e^{2\pi i\left(\emptyset_{\mathbb{A}_1}\emptyset_{\mathbb{A}_2}\right)} \right\rangle, \\ \left\langle \alpha_{\mathbb{A}_1} + \alpha_{\mathbb{A}_2} - \alpha_{\mathbb{A}_1}\alpha_{\mathbb{A}_2}, \beta_{\mathbb{A}_1}\beta_{\mathbb{A}_2} \right\rangle \end{pmatrix}$

b. $\mathbb{A}_1 \otimes \mathbb{A}_2 = \begin{pmatrix} \left\langle \mu_{\mathbb{A}_1}\mu_{\mathbb{A}_2} e^{2\pi i\left(\theta_{\mathbb{A}_1}\theta_{\mathbb{A}_2}\right)}, \left(\nu_{\mathbb{A}_1} + \nu_{\mathbb{A}_2} - \nu_{\mathbb{A}_1}\nu_{\mathbb{A}_2}\right) e^{2\pi i\left(\emptyset_{\mathbb{A}_1} + \emptyset_{\mathbb{A}_2} - \emptyset_{\mathbb{A}_1}\emptyset_{\mathbb{A}_2}\right)} \right\rangle, \\ \left\langle \alpha_{\mathbb{A}_1}\alpha_{\mathbb{A}_2}, \beta_{\mathbb{A}_1} + \beta_{\mathbb{A}_2} - \beta_{\mathbb{A}_1}\beta_{\mathbb{A}_2} \right\rangle \end{pmatrix}$

c. $\mathbb{A}_1 \vee \mathbb{A}_2 = \begin{pmatrix} \left\langle max\left(\mu_{\mathbb{A}_1}, \mu_{\mathbb{A}_2}\right) e^{2\pi i\left(max\left(\theta_{\mathbb{A}_1}, \theta_{\mathbb{A}_2}\right)\right)}, min\left(\nu_{\mathbb{A}_1}, \nu_{\mathbb{A}_2}\right) e^{2\pi i\left(min\left(\emptyset_{\mathbb{A}_1}, \emptyset_{\mathbb{A}_2}\right)\right)} \right\rangle, \\ \left\langle max\left(\alpha_{\mathbb{A}_1}, \alpha_{\mathbb{A}_2}\right), min\left(\beta_{\mathbb{A}_1}, \beta_{\mathbb{A}_2}\right) \right\rangle \end{pmatrix}$

d. $\mathbb{A}_1 \wedge \mathbb{A}_2 = \begin{pmatrix} \left\langle min\left(\mu_{\mathbb{A}_1}, \mu_{\mathbb{A}_2}\right) e^{2\pi i\left(min\left(\theta_{\mathbb{A}_1}, \theta_{\mathbb{A}_2}\right)\right)}, max\left(\nu_{\mathbb{A}_1}, \nu_{\mathbb{A}_2}\right) e^{2\pi i\left(max\left(\emptyset_{\mathbb{A}_1}, \emptyset_{\mathbb{A}_2}\right)\right)} \right\rangle, \\ \left\langle min\left(\alpha_{\mathbb{A}_1}, \alpha_{\mathbb{A}_2}\right), max\left(\beta_{\mathbb{A}_1}, \beta_{\mathbb{A}_2}\right) \right\rangle \end{pmatrix}$

e. $\mathbb{A}^c = \left(\left\langle \nu_{\mathbb{A}} e^{2\pi i(\emptyset_{\mathbb{A}})}, \mu_{\mathbb{A}} e^{2\pi i(\theta_{\mathbb{A}})}, \beta_{\mathbb{A}}, \alpha_{\mathbb{A}} \right\rangle\right)$

f. $r\mathbb{A} = \left(\left\langle (1 - (1 - \mu_{\mathbb{A}})^r) e^{2\pi i(1-(1-\theta_{\mathbb{A}})^r)}, \nu_{\mathbb{A}}^r e^{2\pi i(\emptyset_{\mathbb{A}}^r)} \right\rangle, \left\langle (1 - (1 - \alpha_{\mathbb{A}})^r), \beta_{\mathbb{A}}^r \right\rangle\right)$

g. $\mathbb{A}^r = \left(\left\langle \mu_{\mathbb{A}}^r e^{2\pi i(\theta_{\mathbb{A}}^r)}, (1 - (1 - \nu_{\mathbb{A}})^r) e^{2\pi i(1-(1-\emptyset_{\mathbb{A}})^r)} \right\rangle, \left\langle \alpha_{\mathbb{A}}^r, (1 - (1 - \beta_{\mathbb{A}})^r) \right\rangle\right)$

h. $\mathbb{A}_1 \le \mathbb{A}_2$, if $\mu_{\mathbb{A}_1} \le \mu_{\mathbb{A}_2}, \theta_{\mathbb{A}_1} \le \theta_{\mathbb{A}_2}, \alpha_{\mathbb{A}_1} \le \alpha_{\mathbb{A}_2}, \nu_{\mathbb{A}_1} \ge \nu_{\mathbb{A}_2}, \emptyset_{\mathbb{A}_1} \ge \emptyset_{\mathbb{A}_2}$ and $\beta_{\mathbb{A}_1} \ge \beta_{\mathbb{A}_2}$

Definition 2.1.12 [36]: Let $\mathbb{A} = \left(\left\langle \mu_{\mathbb{A}} e^{2\pi i(\theta_{\mathbb{A}})}, \nu_{\mathbb{A}} e^{2\pi i(\emptyset_{\mathbb{A}})} \right\rangle, \langle \alpha_{\mathbb{A}}, \beta_{\mathbb{A}} \rangle\right)$ be CLDFNs, then the score function (SF) of $\mathbb{A}$ is a mapping $s : CLDFN(\mathrm{U}) \to [-1, 1]$, and is given by:

$$s(\mathbb{A}) = \frac{(\mu_{\mathbb{A}} + \theta_{\mathbb{A}} + \alpha_{\mathbb{A}}) - (\nu_{\mathbb{A}} + \emptyset_{\mathbb{A}} + \beta_{\mathbb{A}})}{3} \quad (1)$$

Definition 2.1.13 [36]: Let $\mathbb{A} = \left(\left\langle \mu_{\mathbb{A}} e^{2\pi i(\theta_{\mathbb{A}})}, \nu_{\mathbb{A}} e^{2\pi i(\emptyset_{\mathbb{A}})} \right\rangle, \langle \alpha_{\mathbb{A}}, \beta_{\mathbb{A}} \rangle\right)$ be CLDFNs, the accuracy function (AF) of $\mathbb{A}$ is a mapping $a : CLDFN(\mathrm{U}) \to [0, 1]$, and is given by:

$$a(\mathbb{A}) = \frac{(\mu_{\mathbb{A}} + \theta_{\mathbb{A}} + \alpha_{\mathbb{A}}) + (\nu_{\mathbb{A}} + \emptyset_{\mathbb{A}} + \beta_{\mathbb{A}})}{5} \quad (2)$$

Property: Let $\mathbb{A}_k = \left(\left\langle \mu_{\mathbb{A}_k} e^{2\pi i(\theta_{\mathbb{A}_k})}, \nu_{\mathbb{A}_k} e^{2\pi i(\emptyset_{\mathbb{A}_k})} \right\rangle, \left\langle \alpha_{\mathbb{A}_k}, \beta_{\mathbb{A}_k} \right\rangle\right)$, (k = 1, 2), then the comparison between $\mathbb{A}_1 and \mathbb{A}_2$ is given as follows-

a. If $s(\mathbb{A}_1) < s(\mathbb{A}_2)$, then $\mathbb{A}_1 < \mathbb{A}_2$
b. If $s(\mathbb{A}_1) = s(\mathbb{A}_2)$, then

- $a(\mathbb{A}_1) < a(\mathbb{A}_2)$, then $\mathbb{A}_1 < \mathbb{A}_2$
- $a(\mathbb{A}_1) = a(\mathbb{A}_2)$, then $\mathbb{A}_1 = \mathbb{A}_2$

Remark: Let $\mathbb{A}$ be a set of all CLDFNs and $s : \mathbb{A} \rightarrow [-1, 1]$ be a score function, then

a. $a(\mathbb{A})$ is increasing based on $\mu_{\mathbb{A}} e^{2\pi i(\theta_{\mathbb{A}})}$ and $\alpha_{\mathbb{A}}$
b. $a(\mathbb{A})$ is decreasing based on $\nu_{\mathbb{A}} e^{2\pi i(\emptyset_{\mathbb{A}})}$ and $\beta_{\mathbb{A}}$

2.2 *Triangular Norm and Triangular Co-Norm*

In fuzzy set theory, TN and TCN serve as essential binary operations on the interval [0,1], representing the fuzzy logic equivalents of conjunction (AND) and disjunction (OR), respectively [46]. These operators form the backbone of handling imprecise and uncertain information of real-world scenarios. By adhering to important algebraic properties such as commutativity, associativity, monotonicity, and possessing a neutral element they ensure reliable and coherent fuzzy inference. Their application is particularly vital in multi-criteria decision-making, where various fuzzy criteria must be combined in a logically consistent and interpretable way. This exploration focuses on enhancing the understanding of TN and TCNs and highlighting their pivotal role in decision processes under uncertainty.

Definition 2.2.1 [46] **:** A function $\mathrm{T} : [0, 1]^2 \rightarrow [0, 1]$ is said to be a TN if for all $x_1, x_2, x_3 \in [0, 1]$, the following axioms hold:

a. $\mathrm{T}(x_1, x_2) = \mathrm{T}(x_2, x_1)$
b. $\mathrm{T}(x_1, \mathrm{T}(x_2, x_3)) = \mathrm{T}(\mathrm{T}(x_1, x_2), x_3)$
c. $\mathrm{T}(x_1, x_2) \leq \mathrm{T}(x_1, x_3) if x_2 \leq x_3$
d. $\mathrm{T}(x_1, 1) = x_1$

Some important examples of TNs:

a. Minimum TN:$\mathrm{T}_{\wedge}(x_1, x_2) = \min(x_1, x_2)$
b. Product TN:$\mathrm{T}_{\mathrm{P}}(x_1, x_2) = x_1 x_2$
c. Lukasiewicz TN:$\mathrm{T}_{\mathrm{L}}(x_1, x_2) = \max(x_1 + x_2 - 1, 0)$
d. Drastic TN:$\mathrm{T}_{\mathrm{D}}(x_1, x_2) = \begin{cases} x_1, & \text{if } x_2 = 1 \\ x_2, & \text{if } x_1 = 1 \\ 0, & \text{else} \end{cases}$

Definition 2.2.2 [46] **:** For $x_1, x_2, x_3 \in [0, 1]$, TCN is a map $\perp : [0, 1]^2 \rightarrow [0, 1]$ satisfying the subsequent features:

a. $\perp(x_1, x_2) = \perp(x_2, x_1)$
b. $\perp(x_1, \perp(x_2, x_3)) = \perp(\perp(x_1, x_2), x_3)$
c. $\perp(x_1, x_2) \leq \perp(x_1, x_3), if x_2 \leq x_3$
d. $\perp(x_1, 0) = x_1$

Some important examples of TCNs:

e. Maximum TCN: $\perp_{\vee}(x_1, x_2) = \max(x_1, x_2)$
f. Probabilistic TCN: $\perp_{P}(x_1, x_2) = x_1 + x_2 - x_1 x_2$
g. Lukasiewicz TCN: $\perp_{L}(x_1, x_2) = \min(x_1 + x_2, 1)$

$$\text{Drastic TCN:}\perp_{D}(x_1, x_2) = \begin{cases} x_1, & \text{if } x_2 = 0 \\ x_2, & \text{if } x_1 = 0 \\ 1, & \text{else}. \end{cases}$$

2.3 Complex Linear Diophantine Fuzzy Aggregation Operators

This segment reviews the AOs used to combine CLDF information for MCDM problems. These involve weighted and ordered weighted operators of the averaging and geometric types.

Definition 2.3.1 [36] **:** Let us consider a family of CLDFNs over U such as $\mathbb{A}_j = \left(\left\langle \mu_j e^{2\pi i(\theta_j)}, \nu_j e^{2\pi i(\emptyset_j)} \right\rangle, \left\langle \alpha_j, \beta_j \right\rangle\right)$, and $W = (w_1, w_2, \ldots, w_n)^{\mathrm{T}}$ a weight vector (WV) such that $w_j > 0$ and $\sum_{j=1}^{n} w_j = 1$. Then, the map $CLDFWA(\mathrm{U})^n \rightarrow (\mathrm{U})$ is termed as CLDFWA operator and is defined as:

$$CLDFWA(\mathbb{A}_1, \mathbb{A}_2, \ldots, \mathbb{A}_n) = \oplus(w_j \mathbb{A}_j) for (j = 1, 2, 3, \ldots n)$$

$$= \left(\begin{array}{c} \left\langle \left(1 - \prod_{j=1}^{n} (1 - \mu_j)^{w_j}\right) e^{2\pi i \left(1 - \prod_{j=1}^{n} (1 - \theta_j)^{w_j}\right)}, \prod_{j=1}^{n} v_j^{w_j} e^{2\pi i \prod_{j=1}^{n} \phi_j^{w_j}} \right\rangle, \\ \left\langle \left(1 - \prod_{j=1}^{n} (1 - \alpha_j)^{w_j}\right), \prod_{j=1}^{n} \beta_j^{w_j} \right\rangle \end{array} \right) \quad (3)$$

Particular Case: When $\mathrm{W} = \left(\frac{1}{\mathrm{n}}, \frac{1}{\mathrm{n}}, \ldots, \frac{1}{\mathrm{n}}\right)^{\mathrm{T}}$, then the CLDFWA operator becomes *CLDFA* operator as $\mathrm{CLDFA}(\mathbb{A}_1, \mathbb{A}_2, \ldots, \mathbb{A}_\mathrm{n}) = \frac{1}{\mathrm{n}}\left(\oplus \mathbb{A}_\mathrm{j}\right)$ for $(\mathrm{j} = 1, 2, 3, \ldots \mathrm{n})$.

Definition 2.3.2 [36]: Let us consider a family of CLDFNs over U such as $\mathbb{A}_j = \left(\left\langle \mu_j e^{2\pi i(\theta_j)}, \nu_j e^{2\pi i(\emptyset_j)} \right\rangle, \left\langle \alpha_j, \beta_j \right\rangle\right)$, and $W = (w_1, w_2, \ldots, w_n)^{\mathrm{T}}$ a WV, $w_j > 0$ and $\sum_{j=1}^{n} w_j = 1$. Then, the map $CLDFWG(\mathrm{U})^n \rightarrow (\mathrm{U})$ is termed as CLDFWG operator and is defined as:

$$CLDFWG(\mathbb{A}_1, \mathbb{A}_2, \ldots, \mathbb{A}_n) = \otimes \mathbb{A}_j^{w_j} for (j = 1, 2, 3, \ldots n)$$

$$= \left(\begin{array}{c} \left\langle \prod_{j=1}^{n} \mu_j^{w_j} e^{2\pi i \prod_{j=1}^{n} \theta_j^{w_j}}, \left(1 - \prod_{j=1}^{n} (1-\nu_j)^{w_j}\right) e^{2\pi i \left(1 - \prod_{j=1}^{n} (1-\emptyset_j)^{w_j}\right)} \right\rangle, \\ \left\langle \prod_{j=1}^{n} \alpha_j^{w_j}, \left(1 - \prod_{j=1}^{n} (1-\beta_j)^{w_j}\right) \right\rangle \end{array} \right) \quad (4)$$

Particular Case: When $W = \left(\frac{1}{n}, \frac{1}{n}, \ldots, \frac{1}{n}\right)^T$, then the CLDFWG operator becomes *CLDFG* operator as $CLDFG(\mathbb{A}_1, \mathbb{A}_2, \ldots, \mathbb{A}_n) = \left(\otimes \mathbb{A}_j\right)^{\frac{1}{n}}$ for (j = 1, 2, 3, ... n).

Definition 2.3.3 [36]: Let us consider a family of CLDFNs over U such as $\mathbb{A}_j = \left(\mu_j e^{2\pi i(\theta_j)}, \nu_j e^{2\pi i(\emptyset_j)}, \alpha_j, \beta_j\right)$, and $W = (w_1, w_2, \ldots, w_n)^T$ a WV, $w_j > 0$ and $\sum_{j=1}^{n} w_j = 1$. Then, the map $CLDFOWA(U)^n \rightarrow (U)$ is termed as CLDFOWA operator and is defined as:

$$CLDFOWA(\mathbb{A}_1, \mathbb{A}_2, \ldots, \mathbb{A}_n) = \oplus\left(w_j \mathbb{A}_{\xi(j)}\right) for (j = 1, 2, 3, \ldots n)$$

$$= \left(\begin{array}{c} \left\langle \left(1 - \prod_{j=1}^{n} \left(1 - \mu_{\xi(j)}\right)^{w_j}\right) e^{2\pi i \left(1 - \prod_{j=1}^{n} \left(1-\theta_{\xi(j)}\right)^{w_j}\right)}, \prod_{j=1}^{n} \nu_{\xi(j)}^{w_j} e^{2\pi i \prod_{j=1}^{n} \emptyset_{\xi(j)}^{w_j}} \right\rangle, \\ \left\langle \left(1 - \prod_{j=1}^{n} \left(1 - \alpha_{\xi(j)}\right)^{w_j}\right), \prod_{j=1}^{n} \beta_{\xi(j)}^{w_j} \right\rangle \end{array} \right) \quad (5)$$

where $(\xi(1), \xi(2), \xi(3), \ldots, \xi(n))$ is a permutation of $(1, 2, 3, \ldots, n)$ such that $\mathbb{A}_{\xi(j-1)} \geq \mathbb{A}_{\xi(j)}$

Particular Case: When $W = \left(\frac{1}{n}, \frac{1}{n}, \ldots, \frac{1}{n}\right)^T$, then the CLDFOWA operator becomes *CLDFOA* operator as $CLDFOA(\mathbb{A}_1, \mathbb{A}_2, \ldots, \mathbb{A}_n) = \frac{1}{n}\left(\oplus \mathbb{A}_{\xi(j)}\right)$ for (j = 1, 2, 3, ... n).

Definition 2.3.4 [36]: Let us consider a family of CLDFNs over U such as $\mathbb{A}_j = \left(\left\langle \mu_j e^{2\pi i(\theta_j)}, \nu_j e^{2\pi i(\emptyset_j)} \right\rangle, \left\langle \alpha_j, \beta_j \right\rangle\right)$, and $W = (w_1, w_2, \ldots, w_n)^T$ a WV, $w_j > 0$ and $\sum_{j=1}^{n} w_j = 1$. Then, the map $CLDFOWG(U)^n \rightarrow (U)$ is termed as CLDFOWG operator and is defined as:

$$CLDFOWG(\mathbb{A}_1, \mathbb{A}_2, \ldots, \mathbb{A}_n) = \otimes \mathbb{A}_{\xi(j)}^{w_j} for (j = 1, 2, 3, \ldots n)$$

$$= \left(\begin{array}{c} \left\langle \prod_{j=1}^{n} \mu_{\xi(j)}^{w_j} e^{2\pi i \prod_{j=1}^{n} \theta_{\xi(j)}^{w_j}}, \left(1 - \prod_{j=1}^{n} \left(1 - \nu_{\xi(j)}\right)^{w_j}\right) e^{2\pi i \left(1 - \prod_{j=1}^{n} \left(1-\emptyset_{\xi(j)}\right)^{w_j}\right)} \right\rangle, \\ \left\langle \prod_{j=1}^{n} \alpha_{\xi(j)}^{w_j}, \left(1 - \prod_{j=1}^{n} \left(1 - \beta_{\xi(j)}\right)^{w_j}\right) \right\rangle \end{array} \right) \quad (6)$$

where $(\xi(1), \xi(2), \xi(3), \ldots, \xi(n))$ is a permutation of $(1, 2, 3, \ldots, n)$ such that $\mathbb{A}_{\xi(j-1)} \geq \mathbb{A}_{\xi(j)}$

Particular Case: When $W = \left(\frac{1}{n}, \frac{1}{n}, \ldots, \frac{1}{n}\right)^T$, then the CLDFOWG operator becomes *CLDF*OG operator as $\mathrm{CLDFOG}(\mathbb{A}_j, \mathbb{A}_j, \ldots, \mathbb{A}_n) = \left(\otimes \mathbb{A}_{\xi(j)}\right)^{\frac{1}{n}}$ for $(j = 1, 2, 3, \ldots n)$.

2.4 Stepwise Weight Assessment Ratio Analysis (SWARA)

The SWARA method enables decision-makers to articulate their judgments freely, without being restricted by predetermined scales or metrics [47]. By incorporating expert evaluations, this approach supports effective problem-solving and promotes balanced decision-making through the determination of weight values. As a flexible and robust decision support tool, SWARA is applicable across a wide range of contexts, offering solutions to both practical and scientific conflicts involving multiple competing objectives. The method proceeds through the following steps [48]:

Step 1 Each Decision Maker (DM) evaluates and ranks the criteria based on their relative significance, assigning the highest priority to the most significant criterion. The remaining criteria are then ordered in descending order of importance. In this context, let m represent the number of decision makers and n denote the number of criteria under consideration.

Step 2 The evaluations given by each Decision Maker (DM) are aggregated using the geometric mean, as shown in Eq. (7). The combined score of criteria is represented by $\overline{p}_j$ and is defined as:

$$\overline{p}_j = \left(\prod_{k=1}^{m} p_j^k\right)^{\frac{1}{m}}, p_j^k \text{ is the individual rating of criterion j by the } k^{th} \text{ DM} \tag{7}$$

Step 3 Initially, the criteria are organized in descending order based on their respective scores of relative importance. Then, starting from the second criterion, the comparative importance score s_j of each criterion j with respect to the preceding criterion $(j-1)$ is calculated. The comparative score is determined using the Eq. (8).

$$s_j = p_{j-1} - p_j, j = 2, \ldots, n. \tag{8}$$

Step 4 For each criterion, the coefficients C_j are derived through binary comparisons. These coefficients signify the position of criterion $(j + 1)$ relative to the criterion j. The C_j values are calculated by using the Eq. (9).

$$C_j = \begin{cases} 1, j = 1; \\ s_j + 1, j = 2, \ldots, n \end{cases} \tag{9}$$

Step 5 For all the criteria, the corrected weights $(s_j\prime)$ are calculated as:

$$S_j\prime = \begin{cases} 1, j = 1 \\ \frac{s_{j-1}\prime}{C_j}, j = 2, \ldots, n. \end{cases} \tag{10}$$

Step 6 The final weight of the criteria $(\mathrm{w_j})$ is evaluated by the equation:

$$w_j = \frac{s_j\prime}{\sum_{j=1}^{n} s_j\prime}, j = 1, 2, \ldots, n. \tag{11}$$

2.5 *Algorithm of Combinative Distance-Based Assessment (CODAS)*

The CODAS method [49], is a mathematically elegant MCDM approach that ranks alternatives by measuring their distance from the negative-ideal solution using both Euclidean (L2) and Taxicab (L1) metrics. By blending these distance measures, CODAS captures both magnitude and distribution of deviations, offering a precise, balanced, and computationally efficient evaluation. Its versatility has made it valuable in engineering, finance, healthcare, environmental management, and supply chain decisions, turning complex criteria into clear, quantitative rankings through rigorous mathematical reasoning.

The algorithms of the CODAS approach are given as follows:

Step 1 Prepare a fuzzy decision matrix with m alternatives and n criteria, with the intersection of each alternative and criteria x_{ij}, having matrix $\left(x_{ij}\right)_{m \times n}$. FDM is constructed as

$$D = x_{ij} = \begin{bmatrix} x_{11} & x_{12} & \cdots & x_{1n} \\ x_{21} & x_{22} & \ldots & x_{2n} \\ \vdots & \vdots & \vdots & \vdots \\ x_{i1} & x_{i2} & \cdots & x_{in} \\ \vdots & \vdots & \vdots & \vdots \\ x_{m1} & x_{m2} & \cdots & x_{mn} \end{bmatrix} \tag{12}$$

Step 2 Transform the fuzzy weighted normalized decision matrix by Eq. (13).

$$\mathrm{F} = f_{ij} = x_{ij}.w_j, i = 1, 2, \ldots m, andj = 1, 2, \ldots n \tag{13}$$

where w_j denotes the weights of the j th criterion.

Step 3 Determine the fuzzy negative ideal solution. The smallest value is chosen to determine the anti-ideal solution for each criterion since it reflects the lowest potential performance. This decision makes it easier to distinguish between the options and allows for a thorough analysis based on the lowest performance level that meets the requirement.
The equation for calculating the NIS(negative ideal solution) $= [NIJ]_{1\times n}$ as follows:

$$NIS_j = \min_i f_{ij} \tag{14}$$

where $i = 1, 2, .., m and j = 1, 2, .., n$

Step 4 Compute the Euclidean (E_i) and Taxicab distances (T_i) of alternatives from the fuzzy negative-ideal solution for each alternative using the following equation.

$$E_i = \sqrt{\sum_{j=1}^{n} \left(f_{ij} - NIS_j\right)^2} \tag{15}$$

$$T_i = \sum_{j=1}^{n} \left|f_{ij} - NIS_j\right| \tag{16}$$

Step 5 Formulate the assessment matrix ($\mathfrak{K}_{ik}$).

$$[\mathfrak{K}_{ik}]_{m\times n} = (E_i - E_k) + \varsigma(E_i - E_k) \times (H_i - H_k) \tag{17}$$

where $k \in \{1, 2, \ldots, m\}$ and $\varsigma(0.01 \leq \varsigma \leq 0.05)$ is a threshold parameter.

Step 6 Obtain the overall assessment score (Ψ). An overall score, or utility value, is obtained by adding up the weighted distances for each alternative, usually through the use of a summation algorithm. Higher scores indicate greater performance, and the alternatives are sorted in descending order based on these scores.

$$\Psi_i = \sum_{k=1}^{m} \mathfrak{K}_{ik} \tag{18}$$

3 Proposed Methodology

This study proposes a novel MCGDM strategy in CLDFS considering a set of criteria and a set of alternatives and a panel of experts from relevant fields. SWARA- CODAS approach in CLDFS has been proposed in this study that makes it possible to evaluate alternatives by methodically eliminating each criterion one at a time and allocating relative weights to them according to how little they affect the alternatives. This proposed methodology includes two stages of evaluation and the steps involved in these stages are given below.

First stage: CLDF based SWARA

Step 1 Initially, a decision matrix for the criteria is prepared based on each DMs opinion.

Step 2 These matrices are then aggregated by using CLDFWAAO, provided in Eq. (3). After that Eq. (1) is used to determine the score value of each criterion.

Step 3 Arranging the criteria in descending order following their respective score value, the comparative score is evaluated by using Eq. (8).

Step 4 The coefficient value of each criterion is determined by using Eq. (9). Then by the Eq. (10), the corrected weights are calculated.

Step 5 Finally, the weightages of the criteria are evaluated by using Eq. (11).

Second stage: CLDF based CODAS

Step 6 After obtaining the weights of the criteria, the DMs provided the linguistic preferences of each alternative with respect to the criteria.

Step 7 The set of criteria includes beneficiary and cost criteria both so it is essential to normalize the decision matrices. Equation (19) is utilized to normalize the CLDFNs, assigned for the alternatives based on the criteria.

$$x_{ij} = \begin{cases} x_{ij}, \textit{forbenefitcriteria} \\ \overline{x}_{ij}, \quad \textit{forcostcriteria} \end{cases} \tag{19}$$

Step 8 The decision matrices are then aggregated by using Eq. (3) where the weights of the criteria are utilized obtained by SWARA.

Step 9 The aggregated values of the alternatives are then de-fuzzified by using Eq. (1).

Step 10 The negative ideal solution is then identified for all the alternatives by using Eq. (14).

Step 11 Considering the NIS obtained in the last step, the Euclidian and Taxicab distances are calculated as per Eq. (15, 16).

Step 12 Equation (17) is then used to evaluate the assessment value.

Step 13 Finally, the overall assessment score is evaluated by using Eq. (18) and based on the result the alternatives are ranked.

4 Case Study

In this case study, a comprehensive exploration of the ideal site selection process is considered for establishing E-waste Recycling and Upcycling Centres in North-East India, employing a robust MCGDM framework. Recognizing the complex interplay of various factors influencing site suitability, the study incorporates ten well-defined criteria encompassing economic viability, environmental sustainability, technical feasibility, social acceptance, logistical convenience, and regulatory compliance. To thoroughly evaluate potential locations, nine distinct alternatives representing key urban and semi-urban centres across the North-Eastern states were identified and rigorously assessed. Each alternative was systematically analysed against the criteria, with expert input and quantitative techniques integrated to ensure an objective and balanced decision-making process. The goal of this study is to provide a scientifically grounded recommendation for selecting sites that maximize operational efficiency, environmental protection, and community benefits, thereby contributing to the maintainable organization of e-waste in this geographically and culturally unique region. Table 1 presents the selected criteria for site evaluation, while Table 2 lists the potential alternatives considered in the study.

Table 1 List of the selected criteria

Criteria name	Symbol	Criteria name	Symbol
Proximity to e-waste sources	C1	Regulatory feasibility and policy support	C6
Transport and infrastructure access	C2	Cost of establishment and operation	C7
Environmental risk and impact	C3	Scalability and tech flexibility	C8
Social acceptance and job potential	C4	Access to reuse/recycling markets	C9
Availability of land and utilities	C5	Potential for resource recovery	C10

Table 2 Description of the selected alternatives

Location	State	Symbol	Location	State	Symbol
Guwahati	Assam	A1	Dibrugarh	Assam	A5
Shillong	Meghalaya	A2	Itanagar	Arunachal Pradesh	A6
Imphal	Manipur	A3	Kohima	Nagaland	A7
Agartala	Tripura	A4	Gangtok	Sikkim	A8
			Aizawl	Mizoram	A9

4.1 Utilization of Proposed MCDM Method

CLDFS based MCGDM approach is utilized in the field of optimal site selection process for establishing E-Waste Recycling and Upcycling Centres in North-East India, employing a robust MCDM framework here.

Step 1: Selection of the experts:

First of all, a panel of three experts was considered. The first expert, an environmental engineer (weight 0.35), focused on environmental and technical aspects. The second, a municipality employee (weight 0.40), contributed insights on planning, regulations, and logistics. The third, a sustainability researcher (weight 0.25), provided input on economic viability and social acceptance. Their combined weighted opinions ensured a balanced site evaluation.

Step 2: Define criteria and alternatives:

The criteria and site alternatives for this study are identified from literature review, field surveys, and expert opinions. The selected criteria and alternatives are detailed in Tables 1 and 2, respectively.

Step 3: Assigning ratings to the criteria:

The DMs provided ratings for the criteria using CLDFN-based linguistic terms. These ratings for the criteria are displayed in Table 3.

Step 4: Weights of the criteria:

The ratings presented in Table 3 are aggregated using the proposed AO as defined in Eq. (3). The weights assigned to the DMs are incorporated to calculate the aggregated values for each criterion. Following aggregation, Eq. (1) is applied to determine the scores of the criteria. The aggregated ratings along with their corresponding score values are summarized in Table 4.

Step 5: Rating, Normalization, and Aggregation of Alternatives:

The selected DMs provided their linguistic ratings for the alternatives based on the chosen criteria using the CLDFN scale and the decision matrices are given in Tables 9, 10, and 11 in Appendix. Since the criteria include both benefit and cost types, the decision matrices were normalized accordingly before further analysis. Specifically, cost criteria were normalized using the appropriate normalization method, while benefit criteria remained unchanged. The transformed decision matrices are provided in Tables 12, 13, and 14 in Appendix. After normalization, the decision matrices from all DMs were aggregated by incorporating their respective weights, using Eq. (3) resulting in a final aggregated normalized decision matrix for the alternatives, which is given in Table 5.

Table 3 Ratings of the criteria

Criteria	DM1	DM2	DM3
C1	$\begin{pmatrix} \langle 0.80e^{2\pi i(0.29)}, 0.31e^{2\pi i(0.31)} \rangle, \\ \langle 0.30, 0.23 \rangle \end{pmatrix}$	$\begin{pmatrix} \langle 0.91e^{2\pi i(0.71)}, 0.72e^{2\pi i(0.91)} \rangle, \\ \langle 0.78, 0.21 \rangle \end{pmatrix}$	$\begin{pmatrix} \langle 0.80e^{2\pi i(0.29)}, 0.31e^{2\pi i(0.31)} \rangle. \\ \langle 0.30, 0.23 \rangle \end{pmatrix}$
C2	$\begin{pmatrix} \langle 0.87e^{2\pi i(0.94)}, 0.76e^{2\pi i(0.52)} \rangle, \\ \langle 0.75, 0.22 \rangle \end{pmatrix}$	$\begin{pmatrix} \langle 0.94e^{2\pi i(0.90)}, 0.72e^{2\pi i(0.65)} \rangle, \\ \langle 0.84, 0.11 \rangle \end{pmatrix}$	$\begin{pmatrix} \langle 0.68e^{2\pi i(0.45)}, 0.80e^{2\pi i(0.77)} \rangle. \\ \langle 0.16, 0.23 \rangle \end{pmatrix}$
C3	$\begin{pmatrix} \langle 0.94e^{2\pi i(0.90)}, 0.72e^{2\pi i(0.65)} \rangle, \\ \langle 0.84, 0.11 \rangle \end{pmatrix}$	$\begin{pmatrix} \langle 0.74e^{2\pi i(0.98)}, 0.64e^{2\pi i(0.96)} \rangle, \\ \langle 0.47, 0.35 \rangle \end{pmatrix}$	$\begin{pmatrix} \langle 0.94e^{2\pi i(0.90)}, 0.72e^{2\pi i(0.65)} \rangle. \\ \langle 0.84, 0.11 \rangle \end{pmatrix}$
C4	$\begin{pmatrix} \langle 0.65e^{2\pi i(0.15)}, 0.30e^{2\pi i(0.63)} \rangle, \\ \langle 0.70, 0.25 \rangle \end{pmatrix}$	$\begin{pmatrix} \langle 0.87e^{2\pi i(0.94)}, 0.76e^{2\pi i(0.52)} \rangle, \\ \langle 0.75, 0.22 \rangle \end{pmatrix}$	$\begin{pmatrix} \langle 0.87e^{2\pi i(0.94)}, 0.76e^{2\pi i(0.52)} \rangle, \\ \langle 0.75, 0.22 \rangle \end{pmatrix}$

(continued)

Table 3 (continued)

Criteria	DM1	DM2	DM3
C5	$\left(\begin{array}{c}\langle 0.74e^{2\pi i(0.98)}, 0.64e^{2\pi i(0.96)}\rangle,\\ \langle 0.47, 0.35\rangle\end{array}\right)$	$\left(\begin{array}{c}\langle 0.80e^{2\pi i(0.29)}, 0.31e^{2\pi i(0.31)}\rangle,\\ \langle 0.30, 0.23\end{array}\right)$	$\left(\begin{array}{c}\langle 0.65e^{2\pi i(0.15)}, 0.30e^{2\pi i(0.63)}\rangle,\\ \langle 0.70, 0.25\rangle\end{array}\right)$
C6	$\left(\begin{array}{c}\langle 0.91e^{2\pi i(0.71)}, 0.72e^{2\pi i(0.91)}\rangle,\\ \langle 0.78, 0.21\rangle\end{array}\right)$	$\left(\begin{array}{c}\langle 0.68e^{2\pi i(0.45)}, 0.80e^{2\pi i(0.77)}\rangle,\\ \langle 0.16, 0.23\end{array}\right)$	$\left(\begin{array}{c}\langle 0.91e^{2\pi i(0.71)}, 0.72e^{2\pi i(0.91)}\rangle,\\ \langle 0.78, 0.21\rangle\end{array}\right)$
C7	$\left(\begin{array}{c}\langle 0.65e^{2\pi i(0.15)}, 0.30e^{2\pi i(0.63)}\rangle,\\ \langle 0.70, 0.25\rangle\end{array}\right)$	$\left(\begin{array}{c}\langle 0.65e^{2\pi i(0.15)}, 0.30e^{2\pi i(0.63)}\rangle,\\ \langle 0.70, 0.25\end{array}\right)$	$\left(\begin{array}{c}\langle 0.50e^{2\pi i(0.45)}, 0.75e^{2\pi i(0.30)}\rangle,\\ \langle 0.35, 0.70\rangle\end{array}\right)$
C8	$\left(\begin{array}{c}\langle 0.43e^{2\pi i(0.43)}, 0.15e^{2\pi i(0.14)}\rangle,\\ \langle 0.45, 0.54\rangle\end{array}\right)$	$\left(\begin{array}{c}\langle 0.43e^{2\pi i(0.43)}, 0.15e^{2\pi i(0.14)}\rangle,\\ \langle 0.45, 0.54\end{array}\right)$	$\left(\begin{array}{c}\langle 0.74e^{2\pi i(0.98)}, 0.64e^{2\pi i(0.96)}\rangle,\\ \langle 0.47, 0.35\rangle\end{array}\right)$
C9	$\left(\begin{array}{c}\langle 0.57e^{2\pi i(0.75)}, 0.33e^{2\pi i(0.64)}\rangle,\\ \langle 0.38, 0.54\rangle\end{array}\right)$	$\left(\begin{array}{c}\langle 0.50e^{2\pi i(0.45)}, 0.75e^{2\pi i(0.30)}\rangle,\\ \langle 0.35, 0.70\end{array}\right)$	$\left(\begin{array}{c}\langle 0.57e^{2\pi i(0.75)}, 0.33e^{2\pi i(0.64)}\rangle,\\ \langle 0.38, 0.54\rangle\end{array}\right)$
C10	$\left(\begin{array}{c}\langle 0.68e^{2\pi i(0.45)}, 0.80e^{2\pi i(0.77)}\rangle,\\ \langle 0.16, 0.23\rangle\end{array}\right)$	$\left(\begin{array}{c}\langle 0.57e^{2\pi i(0.75)}, 0.33e^{2\pi i(0.64)}\rangle,\\ \langle 0.38, 0.54\end{array}\right)$	$\left(\begin{array}{c}\langle 0.43e^{2\pi i(0.43)}, 0.15e^{2\pi i(0.14)}\rangle,\\ \langle 0.45, 0.54\rangle\end{array}\right)$

Table 4 Final weights of the criteria

	Score	s_j	k_j	p_j	w_j
C3	0.356		1	1	0.115
C4	0.3543	0.0017	1.0017	0.998303	0.114
C8	0.2893	0.065	1.065	0.937374	0.108
C5	0.2578	0.0315	1.0315	0.908748	0.105
C1	0.2437	0.0141	1.0141	0.896113	0.103
C2	0.2403	0.0034	1.0034	0.893076	0.102
C6	0.1333	0.107	1.107	0.806754	0.093
C10	0.1167	0.0166	1.0166	0.79358	0.092
C7	0.0536	0.0631	1.0631	0.746478	0.086
C9	0.0369	0.0167	1.0167	0.734216	0.085

Step 6: Defuzzification of the Aggregated scores:

The aggregated normalized decision matrix obtained in Step 5 consists of fuzzy values expressed in the CLDFN linguistic scale. To facilitate precise ranking and further analysis, Eq. (1) is used to de-fuzzify these fuzzy values, which is presented in Table 6.

Step 7: Ranking of Alternatives:

After defuzzification, Eq. (14) is used to find the NIS and then using Eqs. (15) and (16), the Euclidian and Taxicab distances are evaluated. After that Eq. (17) used to find out the assessment score. Following this assessment score, the ranking of the alternatives is done accordingly and the result is presented in Table 7.

4.2 Result and Discussion

This study utilizes a CLDFAO-based MCGDM framework to systematically evaluate and prioritize potential sites for establishing E-Waste Recycling and Upcycling Centres in North-East India. Ten well-defined criteria were used to assess nine urban and semi-urban alternatives across the region (Tables 1 and 2). Each site was evaluated using expert input and fuzzy-based aggregation methods to ensure an objective, transparent, and robust decision-making process. The CLDFAO framework effectively handles uncertainty and enables a comprehensive comparison across all criteria. The results identify the most suitable sites that optimize operational efficiency, environmental protection, and social impact. These findings provide strategic guidance for stakeholders aiming to implement a sustainable e-waste management infrastructure in the region. Detailed rankings are provided below, followed by sensitivity analysis and comparative evaluations in the subsequent sub-sections.

Table 5 Aggregated decision matrix

	A1	A2	A3	A4	A5
C1	$\left(\left\langle 0.82e^{2\pi i(0.63)}, 0.45e^{2\pi i(0.33)}\right\rangle, \langle 0.54, 0.19\rangle\right)$	$\left(\left\langle 0.73e^{2\pi i(0.50)}, 0.66e^{2\pi i(0.20)}\right\rangle, \langle 0.19, 0.38\rangle\right)$	$\left(\left\langle 0.74e^{2\pi i(0.70)}, 0.35e^{2\pi i(0.42)}\right\rangle, \langle 0.55, 0.30\rangle\right)$	$\left(\left\langle 0.78e^{2\pi i(0.61)}, 0.37e^{2\pi i(0.36)}\right\rangle, \langle 0.63, 0.27\rangle\right)$	$\left(\left\langle 0.78e^{2\pi i(0.61)}, 0.37e^{2\pi i(0.36)}\right\rangle, \langle 0.63, 0.27\rangle\right)$
C2	$\left(\left\langle 0.85e^{2\pi i(0.65)}, 0.52e^{2\pi i(0.71)}\right\rangle, \langle 0.79, 0.19\rangle\right)$	$\left(\left\langle 0.79e^{2\pi i(0.53)}, 0.44e^{2\pi i(0.41)}\right\rangle, \langle 0.43, 0.38\rangle\right)$	$\left(\left\langle 0.83e^{2\pi i(0.63)}, 0.62e^{2\pi i(0.77)}\right\rangle, \langle 0.64, 0.32\rangle\right)$	$\left(\left\langle 0.85e^{2\pi i(0.72)}, 0.62e^{2\pi i(0.71)}\right\rangle, \langle 0.67, 0.27\rangle\right)$	$\left(\left\langle 0.86e^{2\pi i(0.73)}, 0.76e^{2\pi i(0.45)}\right\rangle, \langle 0.66, 0.24\rangle\right)$
C3	$\left(\left\langle 0.95e^{2\pi i(0.88)}, 0.47e^{2\pi i(0.43)}\right\rangle, \langle 0.48, 0.22\rangle\right)$	$\left(\left\langle 0.95e^{2\pi i(0.72)}, 0.49e^{2\pi i(0.41)}\right\rangle, \langle 0.42, 0.20\rangle\right)$	$\left(\left\langle 0.94e^{2\pi i(0.67)}, 0.47e^{2\pi i(0.28)}\right\rangle, \langle 0.26, 0.24\rangle\right)$	$\left(\left\langle 0.93e^{2\pi i(0.86)}, 0.57e^{2\pi i(0.37)}\right\rangle, \langle 0.31, 0.26\rangle\right)$	$\left(\left\langle 0.90e^{2\pi i(085)}, 0.65e^{2\pi i(0.41)}\right\rangle, \langle 0.45, 0.21\rangle\right)$
C4	$\left(\left\langle 0.81e^{2\pi i(0.61)}, 0.41e^{2\pi i(0.51)}\right\rangle, \langle 0.64, 0.16\rangle\right)$	$\left(\left\langle 0.92e^{2\pi i(0.84)}, 0.72e^{2\pi i(0.52)}\right\rangle, \langle 0.67, 0.20\rangle\right)$	$\left(\left\langle 0.85e^{2\pi i(0.63)}, 0.56e^{2\pi i(0.31)}\right\rangle, \langle 0.47, 0.23\rangle\right)$	$\left(\left\langle 0.83e^{2\pi i(0.51)}, 0.56e^{2\pi i(0.34)}\right\rangle, \langle 0.42, 0.27\rangle\right)$	$\left(\left\langle 0.85e^{2\pi i(0.41)}, 0.49e^{2\pi i(0.44)}\right\rangle, \langle 0.46, 0.21\rangle\right)$
C5	$\left(\left\langle 0.73e^{2\pi i(0.36)}, 0.46e^{2\pi i(0.41)}\right\rangle, \langle 0.26, 0.33\rangle\right)$	$\left(\left\langle 0.68e^{2\pi i(0.73)}, 0.65e^{2\pi i(0.41)}\right\rangle, \langle 0.30, 0.41\rangle\right)$	$\left(\left\langle 0.66e^{2\pi i(0.38)}, 0.69e^{2\pi i(0.39)}\right\rangle, \langle 0.22, 0.37\rangle\right)$	$\left(\left\langle 0.66e^{2\pi i(0.38)}, 0.69e^{2\pi i(0.39)}\right\rangle, \langle 0.22, 0.37\rangle\right)$	$\left(\left\langle 0.70e^{2\pi i(0.47)}, 0.54e^{2\pi i(0.32)}\right\rangle, \langle 0.21, 0.43\rangle\right)$
C6	$\left(\left\langle 0.90e^{2\pi i(0.85)}, 0.65e^{2\pi i(0.41)}\right\rangle, \langle 0.45, 0.21\rangle\right)$	$\left(\left\langle 0.87e^{2\pi i(0.67)}, 0.37e^{2\pi i(0.42)}\right\rangle, \langle 0.58, 0.19\rangle\right)$	$\left(\left\langle 0.84e^{2\pi i(0.77)}, 0.45e^{2\pi i(0.43)}\right\rangle, \langle 0.65, 0.15\rangle\right)$	$\left(\left\langle 0.84e^{2\pi i(0.77)}, 0.45e^{2\pi i(0.43)}\right\rangle, \langle 0.65, 0.15\rangle\right)$	$\left(\left\langle 0.89e^{2\pi i(0.81)}, 0.50e^{2\pi i(0.41)}\right\rangle, \langle 0.44, 0.18\rangle\right)$
C7	$\left(\left\langle 0.48e^{2\pi i(0.51)}, 0.52e^{2\pi i(0.40)}\right\rangle, \langle 0.50, 0.34\rangle\right)$	$\left(\left\langle 0.48e^{2\pi i(0.51)}, 0.50e^{2\pi i(0.61)}\right\rangle, \langle 0.56, 0.29\rangle\right)$	$\left(\left\langle 0.48e^{2\pi i(0.51)}, 0.50e^{2\pi i(0.61)}\right\rangle, \langle 0.56, 0.29\rangle\right)$	$\left(\left\langle 0.48e^{2\pi i(0.51)}, 0.50e^{2\pi i(0.61)}\right\rangle, \langle 0.56, 0.29\rangle\right)$	$\left(\left\langle 0.38e^{2\pi i(0.59)}, 0.55e^{2\pi i(0.41)}\right\rangle, \langle 0.36, 0.30\rangle\right)$
C8	$\left(\left\langle 0.57e^{2\pi i(0.61)}, 0.33e^{2\pi i(0.26)}\right\rangle, \langle 0.29, 0.28\rangle\right)$	$\left(\left\langle 0.59e^{2\pi i(0.68)}, 0.27e^{2\pi i(0.34)}\right\rangle, \langle 0.44, 0.16\rangle\right)$	$\left(\left\langle 0.57e^{2\pi i(0.61)}, 0.33e^{2\pi i(0.26)}\right\rangle, \langle 0.29, 0.28\rangle\right)$	$\left(\left\langle 0.57e^{2\pi i(0.61)}, 0.33e^{2\pi i(0.26)}\right\rangle, \langle 0.29, 0.28\rangle\right)$	$\left(\left\langle 0.49e^{2\pi i(0.44)}, 0.49e^{2\pi i(0.26)}\right\rangle, \langle 0.26, 0.72\rangle\right)$
C9	$\left(\left\langle 0.68e^{2\pi i(0.43)}, 0.54e^{2\pi i(0.47)}\right\rangle, \langle 0.70, 0.19\rangle\right)$	$\left(\left\langle 0.53e^{2\pi i(0.55)}, 0.60e^{2\pi i(0.27)}\right\rangle, \langle 0.51, 0.34\rangle\right)$	$\left(\left\langle 0.50e^{2\pi i(0.41)}, 0.50e^{2\pi i(0.46)}\right\rangle, \langle 0.58, 0.28\rangle\right)$	$\left(\left\langle 0.50e^{2\pi i(0.41)}, 0.50e^{2\pi i(0.46)}\right\rangle, \langle 0.58, 0.28\rangle\right)$	$\left(\left\langle 0.29e^{2\pi i(0.59)}, 0.53e^{2\pi i(0.42)}\right\rangle, \langle 0.39, 0.52\rangle\right)$
C10	$\left(\left\langle 0.69e^{2\pi i(0.54)}, 0.54e^{2\pi i(0.34)}\right\rangle, \langle 0.54, 0.33\rangle\right)$	$\left(\left\langle 0.67e^{2\pi i(0.65)}, 0.34e^{2\pi i(0.44)}\right\rangle, \langle 0.68, 0.23\rangle\right)$	$\left(\left\langle 0.89e^{2\pi i(0.83)}, 0.47e^{2\pi i(0.54)}\right\rangle, \langle 0.72, 0.16\rangle\right)$	$\left(\left\langle 0.87e^{2\pi i(0.68)}, 0.37e^{2\pi i(0.48)}\right\rangle, \langle 0.64, 0.16\rangle\right)$	$\left(\left\langle 0.82e^{2\pi i(0.80)}, 0.38e^{2\pi i(0.43)}\right\rangle, \langle 0.46, 0.21\rangle\right)$

(continued)

Table 5 (continued)

	A6	A7	A8	A9
C1	$\left(\langle 0.74e^{2\pi i(0.75)}, 0.69e^{2\pi i(0.35)}\rangle, \langle 0.29, 0.33\rangle\right)$	$\left(\langle 0.69e^{2\pi i(0.53)}, 0.84e^{2\pi i(0.25)}\rangle, \langle 0.15, 0.38\rangle\right)$	$\left(\langle 0.65e^{2\pi i(0.17)}, 0.32e^{2\pi i(0.67)}\rangle, \langle 0.75, 0.18\rangle\right)$	$\left(\langle 0.72e^{2\pi i(0.43)}, 0.73e^{2\pi i(0.33)}\rangle, \langle 0.21, 0.30\rangle\right)$
C2	$\left(\langle 0.75e^{2\pi i(0.58)}, 0.66e^{2\pi i(0.42)}\rangle, \langle 0.40, 0.42\rangle\right)$	$\left(\langle 0.84e^{2\pi i(0.54)}, 0.52e^{2\pi i(0.78)}\rangle, \langle 0.78, 0.22\rangle\right)$	$\left(\langle 0.76e^{2\pi i(0.65)}, 0.58e^{2\pi i(0.47)}\rangle, \langle 0.43, 0.33\rangle\right)$	$\left(\langle 0.78e^{2\pi i(0.61)}, 0.37e^{2\pi i(0.36)}\rangle, \langle 0.63, 0.27\rangle\right)$
C3	$\left(\langle 0.94e^{2\pi i(0.93)}, 0.53e^{2\pi i(0.43)}\rangle, \langle 0.69, 0.18\rangle\right)$	$\left(\langle 0.94e^{2\pi i(0.93)}, 0.53e^{2\pi i(0.43)}\rangle, \langle 0.69, 0.18\rangle\right)$	$\left(\langle 0.96e^{2\pi i(0.76)}, 0.48e^{2\pi i(0.43)}\rangle, \langle 0.59, 0.19\rangle\right)$	$\left(\langle 0.95e^{2\pi i(0.89)}, 0.52e^{2\pi i(0.50)}\rangle, \langle 0.70, 0.18\rangle\right)$
C4	$\left(\langle 0.90e^{2\pi i(0.80)}, 0.51e^{2\pi i(0.28)}\rangle, \langle 0.47, 0.24\rangle\right)$	$\left(\langle 0.89e^{2\pi i(0.78)}, 0.67e^{2\pi i(0.37)}\rangle, \langle 0.49, 0.21\rangle\right)$	$\left(\langle 0.92e^{2\pi i(0.84)}, 0.72e^{2\pi i(0.52)}\rangle, \langle 0.67, 0.20\rangle\right)$	$\left(\langle 0.92e^{2\pi i(0.84)}, 0.72e^{2\pi i(0.52)}\rangle, \langle 0.67, 0.20\rangle\right)$
C5	$\left(\langle 0.77e^{2\pi i(0.53)}, 0.59e^{2\pi i(0.44)}\rangle, \langle 0.52, 0.36\rangle\right)$	$\left(\langle 0.72e^{2\pi i(0.40)}, 0.42e^{2\pi i(0.24)}\rangle, \langle 0.23, 0.43\rangle\right)$	$\left(\langle 0.63e^{2\pi i(0.49)}, 0.80e^{2\pi i(0.30)}\rangle, \langle 0.16, 0.48\rangle\right)$	$\left(\langle 0.64e^{2\pi i(0.78)}, 0.76e^{2\pi i(0.32)}\rangle, \langle 0.25, 0.53\rangle\right)$
C6	$\left(\langle 0.91e^{2\pi i(0.36)}, 0.43e^{2\pi i(0.38)}\rangle, \langle 0.25, 0.20\rangle\right)$	$\left(\langle 0.91e^{2\pi i(0.72)}, 0.41e^{2\pi i(0.41)}\rangle, \langle 0.33, 0.23\rangle\right)$	$\left(\langle 0.89e^{2\pi i(0.68)}, 0.37e^{2\pi i(0.36)}\rangle, \langle 0.48, 0.18\rangle\right)$	$\left(\langle 0.92e^{2\pi i(0.32)}, 0.34e^{2\pi i(0.31)}\rangle, \langle 0.28, 0.20\rangle\right)$
C7	$\left(\langle 0.31e^{2\pi i(0.67)}, 0.57e^{2\pi i(0.33)}\rangle, \langle 0.16, 0.55\rangle\right)$	$\left(\langle 0.38e^{2\pi i(0.59)}, 0.55e^{2\pi i(0.41)}\rangle, \langle 0.36, 0.30\rangle\right)$	$\left(\langle 0.39e^{2\pi i(0.60)}, 0.54e^{2\pi i(0.61)}\rangle, \langle 0.43, 0.26\rangle\right)$	$\left(\langle 0.38e^{2\pi i(0.59)}, 0.55e^{2\pi i(0.41)}\rangle, \langle 0.36, 0.30\rangle\right)$
C8	$\left(\langle 0.64e^{2\pi i(0.58)}, 0.38e^{2\pi i(0.34)}\rangle, \langle 0.40, 0.29\rangle\right)$	$\left(\langle 0.59e^{2\pi i(0.68)}, 0.41e^{2\pi i(0.27)}\rangle, \langle 0.41, 0.16\rangle\right)$	$\left(\langle 0.59e^{2\pi i(0.68)}, 0.27e^{2\pi i(0.34)}\rangle, \langle 0.44, 0.16\rangle\right)$	$\left(\langle 0.60e^{2\pi i(0.57)}, 0.43e^{2\pi i(0.28)}\rangle, \langle 0.41, 0.33\rangle\right)$
C9	$\left(\langle 0.58e^{2\pi i(0.29)}, 0.47e^{2\pi i(0.45)}\rangle, \langle 0.68, 0.31\rangle\right)$	$\left(\langle 0.29e^{2\pi i(0.59)}, 0.53e^{2\pi i(0.42)}\rangle, \langle 0.39, 0.52\rangle\right)$	$\left(\langle 0.37e^{2\pi i(0.62)}, 0.61e^{2\pi i(0.54)}\rangle, \langle 0.40, 0.35\rangle\right)$	$\left(\langle 0.29e^{2\pi i(0.59)}, 0.53e^{2\pi i(0.42)}\rangle, \langle 0.39, 0.52\rangle\right)$
C10	$\left(\langle 0.57e^{2\pi i(0.59)}, 0.27e^{2\pi i(0.62)}\rangle, \langle 0.55, 0.22\rangle\right)$	$\left(\langle 0.55e^{2\pi i(0.46)}, 0.62e^{2\pi i(0.41)}\rangle, \langle 0.24, 0.69\rangle\right)$	$\left(\langle 0.69e^{2\pi i(043)}, 0.72e^{2\pi i(0.26)}\rangle, \langle 0.26, 0.39\rangle\right)$	$\left(\langle 0.54e^{2\pi i(0.59)}, 0.39e^{2\pi i(0.50)}\rangle, \langle 0.28, 0.34\rangle\right)$

Table 6 De-fuzzified values of the alternatives

	A1	A2	A3	A4	A5	A6	A7	A8	A9
C1	0.3386	0.0543	0.3052	0.3392	0.3392	0.1396	0.1386	0.1336	0.0030
C2	0.2924	0.1738	0.1344	0.2142	0.2702	0.0760	0.2139	0.1516	0.3392
C3	0.3969	0.3289	0.2924	0.3024	0.3108	0.4743	0.4743	0.4024	0.4487
C4	0.3223	0.3292	0.2804	0.2001	0.1976	0.380	0.3055	0.3292	0.3292
C5	0.0441	0.0787	0.0251	0.0251	0.0275	0.1434	0.0881	0.0806	0.0221
C6	0.3108	0.3820	0.4143	0.4143	0.3526	0.1661	0.3079	0.3828	0.2231
C7	0.0780	0.0546	0.0546	0.0546	0.0208	0.0208	0.0208	0.0015	0.0208
C8	0.2030	0.3151	0.2030	0.2030	0.1640	0.1966	0.2809	0.3151	0.1818
C9	0.2024	0.1287	0.0839	0.0839	0.0569	0.1052	0.0569	0.1147	0.0569
C10	0.1865	0.3286	0.4212	0.3934	0.3504	0.2032	0.1459	0.0035	0.0599

The criteria of Table 1 are then aggregated by the decision-makers (DMs) using CLDFN-based linguistic terms and the ratings of the criteria are presented in Table 3. These ratings were then aggregated using CLDFWAAO of Eq. (3) and the weights of the DMs were incorporated to calculate the aggregated values for the criterion. After aggregation, the SF, defined in Eq. (1), is applied to determine the score values of the criteria. The aggregated scores are then then arranged in descending order and SWARA has been incorporated to determine the weightages of the criteria by employing Eqs. (9), (10) and (11) and the calculated outputs are provided in Table 4.

Furthermore, the selected DMs provide their opinions in CLDFNs for the alternatives with respect to the identified criteria and the ratings are given in Appendix. the decision matrices are then normalized based on benefit and cost criteria. Equation (19) is applied here for normalization as criteria C7 and C9 are identified as cost criteria while the remaining criteria classified as benefit-type. These matrices are also included in Appendix as Tables 9, 10, 11.

The normalized decision matrices provided by the three DMs were then aggregated using the proposed CLDFWAAO, as presented in Eq. (3), incorporating the respective weights of the DMs to ensure a balanced and representative evaluation and the resulting aggregated values are included in Table 5. To facilitate precise ranking and further analysis, these fuzzy values were de-fuzzified using the score function defined in Eq. (1), and the resulting scores are presented in Table 6.

Subsequently, for CODAS approach, the NIS is then synthesized from the score values by using Eq. (14). Then Eqs. (15) and (16) is utilized to find out the distances and using these distances the assessment scores (Eq. (17) for all the alternatives are calculated. Equation (18) is used to evaluate the weights of the alternatives and the ranks of the alternatives are obtained accordingly.

The results, summarized in Table 7, indicate the ranking order of the alternatives as:

$$A4 \succ A3 \succ A5 \succ A1 \succ A2 \succ A7 \succ A6 \succ A9 \succ A8.$$

Table 7 Relative assessment score and ranking of the alternatives

	A1	A2	A3	A4	A5	A6	A7	A8	A9	Assessment	Rank
A1	0	0.02845	−0.1022	−0..109	−0.0825	0.10097	0.08601	0.11986	0.11585	0.157291866	4
A2	−0..0284	0	−0.1306	−0.1374	−0.1109	0.07243	0.0576	0.09133	0.08723	−0.098746717	5
A3	0.10275	0.13128	0	−0.0068	0.01976	0.2043	0.18885	0.22318	0.21947	1.082742805	2
A4	0.10965	0.13819	0.00685	0	0.02661	0.21125	0.19576	0.23013	0.22644	1.144904517	1
A5	0.08294	0.11147	−0.0198	−0.0266	0	0.18442	0.16906	0.20331	0.19954	0.904390479	3
A6	−0.1003	−0.072	−0.202	−0.2087	−0.1824	0	−0.0145	0.01889	0.01458	−0.746376122	7
A7	−0.0862	−0.0578	−0.1884	−0.1952	−0.1687	0.01459	0	0.03356	0.02927	−0.618801784	6
A8	−0.1192	−0.0909	−0.2209	−0.2277	−0.2013	−0.0189	−0.0333	0	−0.0044	−0.916517883	9
A9	−0.1145	−0.0863	−0.2159	−0.2226	−0.1963	−0.0145	−0.0289	0.00434	0	−0.874717077	8

Table 8 Comparison with existing traditional MCDM methods

Method	Ranking
SWARA-TOPSIS	$A4 \succ A3 \succ A5 \succ A1 \succ A2 \succ A7 \succ A6 \succ A9 \succ A8$
SWARA-MARCOS	$A4 \succ A3 \succ A5 \succ A1 \succ A2 \succ A7 \succ A6 \succ A9 \succ A8$
SWARA-WASPAS	$A4 \succ A3 \succ A5 \succ A1 \succ A2 \succ A7 \succ A6 \succ A9 \succ A8$
Proposed method	$A4 \succ A3 \succ A5 \succ A1 \succ A2 \succ A7 \succ A6 \succ A9 \succ A8$

Accordingly, Agartala, Tripura is identified as the optimal site for establishing E-Waste Recycling and Upcycling Centres in North-East India, based on its superior aggregated performance across multiple evaluation criteria. This outcome provides a data-driven foundation for informed decision-making aimed at enhancing sustainable waste management infrastructure in the region, mitigating environmental impacts, and supporting circular economy initiatives. To ensure the robustness and validity of the ranking results derived from the proposed AO-based approach, comparative analyse has been conducted.

4.3 Comparative Analysis

Ranking results are compared with other current MCDM approaches described under CLDFS, as demonstrated in Table 8, in order to assess the dependability and longevity of the suggested approach. The key characteristics of the current approaches have been highlighted in order to understand how the suggested approach differs from them. In the proposed integrated framework, Agartala is identified as the best option for establishing E-Waste Recycling and Upcycling Centres in North-East India. A number of current techniques have been taken into consideration in order to confirm the robustness and effectiveness of the rankings produced by the suggested method such as SWARA-TOPSIS [50], SWARA-MARCOS [51], SWARA-WASPAS [52]. Table 8 depicts that all the methods suggest Agartala as the best alternative by giving highest weightage and the ranking of other alternatives are also remain same. As a result, the suggested framework is appropriate for handling any MCGDM problem and has a good correlation.

5 Conclusion

The CLDFAO serves as the foundation for the innovative decision-making paradigm that this study suggests. integrated with SWARA and CODAS, aimed at enhancing the performance of MCGDM methods in addressing the complexity and uncertainty involved in selecting EWRUC-NEI. The core contribution of this study is the introduction and conceptual discussion of the AO within a structured methodological framework. This operator has the potential to support more robust and informed decision-making in complex fuzzy environments.

The methodology is applied to a real-world case study of India's NER, which comprises eight states and is particularly vulnerable to the growing challenge of e-waste generation. A comprehensive set of EWRUC-NEI alternatives and evaluation criteria was developed, incorporating input from three expert decision-makers. Their assessments were synthesized using the proposed operator within an MCGDM framework, enabling the effective aggregation of complex fuzzy information and facilitating precise, adaptable, and informed decision-making.

Following the structured analysis, comprising normalization, transformation, and aggregation of decision matrices, Agartala (A4), Tripura has been identified as the optimal location for establishing EWRUC-NEI, based on its superior aggregated performance across multiple evaluation criteria. This outcome highlights not only Agartala's strategic advantage in supporting regional e-waste management efforts but also validates the strength of the proposed decision-making context in generating rational, context-specific solutions. The model's reliability was further validated through comparative analysis, confirming its adaptability and practical relevance for real-world applications.

In summary, the key points of this study are:

- Proposed a novel decision-making framework using the CLDF.
- The AO in CLDFS has been discussed and integrated with SWARA and CODAS.
- Enhanced a MCGDM methodology for handling complexity and uncertainty in EWRUC-NEI selection.
- Applied to a case study in NEI, a region highly vulnerable to rising e-waste challenges.
- Agartala, Tripura identified as the most suitable EWRUC-NEI alternative.
- Validated through comparative analysis, proving reliability and adaptability.

However, this study has some limitations like: The proposed framework, while robust and theoretically sound, relies on expert judgments which may present subjectivity. The case study was limited to a single geographic region (North-East India), and the specific criteria and alternatives considered may not be universally applicable. Additionally, the effectiveness of the model may vary when scaled to larger datasets or applied to different geographic or socio-economic contexts.

Future research can expand the framework's applicability by testing it across diverse urban and non-urban settings with varying climatic, geographical, and socio-economic conditions, thereby ensuring broader relevance and generalizability. Engaging a diverse set of stakeholders, including urban planners, engineers, policymakers, analysts, researchers, and community representatives, can help minimize subjectivity and contribute to more informed and balanced decision-making. Integrating real-time data such as e-waste generation rates, collection efficiency, and disposal trends can enhance system responsiveness, while advanced technologies such as machine learning, Artificial Intelligence, big data analytics, and fuzzy logic systems can support automated, adaptive decision-making for sustainable e-waste management, particularly in complex and data-scarce regions like North-East India. Beyond e-waste management, the proposed framework demonstrates strong potential across diverse sectors including economy, agriculture, business, healthcare, banking, tourism, and scientific research supporting critical decision-making tasks such as risk analysis, research prioritization, diagnostics, and resource planning, especially in complex and resource-constrained regions like North-East India. Comparative studies involving other fuzzy aggregation and hybrid MCDM models are recommended to benchmark the framework's performance, thereby enhancing its mathematical robustness and practical applicability in sustainable e-waste management.

In conclusion, the CLDFAAO framework represents the key innovation of this research, providing a robust, mathematically grounded tool for decision-making in complex and uncertain scenarios. Its successful application to the North-East India case study has demonstrated its effectiveness in evaluating multiple e-waste management alternatives with precision. Importantly, the identification of Agartala, Tripura as the most suitable EWRUC-NEI alternative highlights the framework's ability to support practical, technically viable, and environmentally sustainable solutions. Overall, this study offers both a meaningful theoretical advancement in fuzzy MCGDM approaches and a practical decision-support tool for addressing the escalating challenges of e-waste management in the region.

AI Use Disclosure Statement The authors did not use any generative AI or AI-assisted technologies in the creation of this manuscript.

Abbreviations

E-waste	Electronic Waste
EWRUC	E-Waste Recycling and Up-cycling Centres
EWSS-NEI	E-Waste Site Selection in North-Eastern India
EWRUC-NEI	E-Waste Recycling and Up-cycling Centres in North-East India
NER	North-Eastern Region
NERI	North-Eastern Region of India
SEWM	Sustainable E-Waste Management
MCDM	Multi-Criteria Decision Making
MCGDM	Multi-Criteria Group Decision Making
FS	Fuzzy Set
IVFS	Interval-Valued Fuzzy Set
MD	Membership Degree
NMD	Non-Membership Degree
CFS	Complex Fuzzy Set
IFS	Intuitionistic Fuzzy Set
CIFS	Complex Intuitionistic Fuzzy Set

PyFS	Pythagorean Fuzzy Set
CPyFS	Complex Pythagorean Fuzzy Set
q-ROFS	Q-Rung Orthopair Fuzzy Set
LDFS	Linear Diophantine Fuzzy Set
CLDFS	Complex Linear Diophantine Fuzzy Set
CLDFN	Complex Linear Diophantine Fuzzy Number
RP	Reference Parameters
TN	Triangular Norm
TCN	Triangular Conorm
AO	Aggregation Operators
WV	Weight Vector
q-ROFNSS	Q-Rung Orthopair Fuzzy N-Soft Sets
CLDFAO	Complex Linear Diophantine Fuzzy Aggregation Operator
LDFWA	Linear Diophantine Fuzzy Weighted Average
CLDFWA	Complex Linear Diophantine Fuzzy Weighted Average
CLDFOWA	Complex Linear Diophantine Fuzzy Ordered Weighted Average
LDFWG	Linear Diophantine Fuzzy Weighted Geometric
CLDFWG	Complex Linear Diophantine Fuzzy Weighted Geometric
CLDFOWG	Complex Linear Diophantine Fuzzy Ordered Weighted Geometric
DM	Decision-Maker
SWARA	Stepwise Weight Assessment Ratio Analysis
CODAS	Combinative Distance-based Assessment
FDMat	Fuzzy Decision Matrix

Appendix

See Tables 9, 10, 11, 12, 13 and 14.

Table 9 Complex linear diophantine fuzzy decision matrix based on DM1

	A1	A2	A3	A4	A5
C1	$\begin{pmatrix}\langle 0.82e^{2\pi i(0.41)}, 0.26e^{2\pi i(0.15)}\rangle, \\ \langle 0.22, 0.34\rangle\end{pmatrix}$	$\begin{pmatrix}\langle 0.70e^{2\pi i(0.56)}, 0.86e^{2\pi i(0.16)}\rangle, \\ \langle 0.14, 0.46\rangle\end{pmatrix}$	$\begin{pmatrix}\langle 0.82e^{2\pi i(0.41)}, 0.26e^{2\pi i(0.15)}\rangle, \\ \langle 0.22, 0.34\rangle\end{pmatrix}$	$\begin{pmatrix}\langle 0.82e^{2\pi i(0.41)}, 0.26e^{2\pi i(0.15)}\rangle, \\ \langle 0.22, 0.34\rangle\end{pmatrix}$	$\begin{pmatrix}\langle 0.82e^{2\pi i(0.41)}, 0.26e^{2\pi i(0.15)}\rangle, \\ \langle 0.22, 0.34\rangle\end{pmatrix}$
C2	$\begin{pmatrix}\langle 0.90e^{2\pi i(0.70)}, 0.70e^{2\pi i(0.90)}\rangle, \\ \langle 0.80, 0.20\rangle\end{pmatrix}$	$\begin{pmatrix}\langle 0.82e^{2\pi i(0.41)}, 0.26e^{2\pi i(0.15)}\rangle, \\ \langle 0.22, 0.34\rangle\end{pmatrix}$	$\begin{pmatrix}\langle 0.90e^{2\pi i(0.70)}, 0.70e^{2\pi i(0.90)}\rangle, \\ \langle 0.80, 0.20\rangle\end{pmatrix}$	$\begin{pmatrix}\langle 0.90e^{2\pi i(0.70)}, 0.70e^{2\pi i(0.90)}\rangle, \\ \langle 0.80, 0.20\rangle\end{pmatrix}$	$\begin{pmatrix}\langle 0.90e^{2\pi i(0.70)}, 0.70e^{2\pi i(0.90)}\rangle, \\ \langle 0.80, 0.20\rangle\end{pmatrix}$
C3	$\begin{pmatrix}\langle 0.98e^{2\pi i(0.38)}, 0.38e^{2\pi i(0.26)}\rangle, \\ \langle 0.30, 0.19\rangle\end{pmatrix}$	$\begin{pmatrix}\langle 0.98e^{2\pi i(0.38)}, 0.38e^{2\pi i(0.26)}\rangle, \\ \langle 0.30, 0.19\rangle\end{pmatrix}$	$\begin{pmatrix}\langle 0.98e^{2\pi i(0.38)}, 0.38e^{2\pi i(0.26)}\rangle, \\ \langle 0.30, 0.19\rangle\end{pmatrix}$	$\begin{pmatrix}\langle 0.98e^{2\pi i(0.38)}, 0.38e^{2\pi i(0.26)}\rangle, \\ \langle 0.30, 0.19\rangle\end{pmatrix}$	$\begin{pmatrix}\langle 0.95e^{2\pi i(0.93)}, 0.48e^{2\pi i(0.40)}\rangle, \\ \langle 0.73, 0.14\rangle\end{pmatrix}$
C4	$\begin{pmatrix}\langle 0.67e^{2\pi i(0.24)}, 0.34e^{2\pi i(0.78)}\rangle, \\ \langle 0.79, 0.10\rangle\end{pmatrix}$	$\begin{pmatrix}\langle 0.90e^{2\pi i(0.70)}, 0.70e^{2\pi i(0.90)}\rangle, \\ \langle 0.80, 0.20\rangle\end{pmatrix}$	$\begin{pmatrix}\langle 0.70e^{2\pi i(0.56)}, 0.86e^{2\pi i(0.16)}\rangle, \\ \langle 0.14, 0.46\rangle\end{pmatrix}$	$\begin{pmatrix}\langle 0.70e^{2\pi i(0.56)}, 0.86e^{2\pi i(0.16)}\rangle, \\ \langle 0.14, 0.46\rangle\end{pmatrix}$	$\begin{pmatrix}\langle 0.78e^{2\pi i(0.23)}, 0.57e^{2\pi i(0.33)}\rangle, \\ \langle 0.30, 0.22\rangle\end{pmatrix}$
C5	$\begin{pmatrix}\langle 0.78e^{2\pi i(0.23)}, 0.57e^{2\pi i(0.33)}\rangle, \\ \langle 0.30, 0.22\rangle\end{pmatrix}$	$\begin{pmatrix}\langle 0.78e^{2\pi i(0.23)}, 0.57e^{2\pi i(0.33)}\rangle, \\ \langle 0.30, 0.22\rangle\end{pmatrix}$	$\begin{pmatrix}\langle 0.78e^{2\pi i(0.23)}, 0.57e^{2\pi i(0.33)}\rangle, \\ \langle 0.30, 0.22\rangle\end{pmatrix}$	$\begin{pmatrix}\langle 0.78e^{2\pi i(0.23)}, 0.57e^{2\pi i(0.33)}\rangle, \\ \langle 0.30, 0.22\rangle\end{pmatrix}$	$\begin{pmatrix}\langle 0.70e^{2\pi i(0.56)}, 0.86e^{2\pi i(0.16)}\rangle, \\ \langle 0.14, 0.46\rangle\end{pmatrix}$
C6	$\begin{pmatrix}\langle 0.95e^{2\pi i(0.93)}, 0.48e^{2\pi i(0.40)}\rangle, \\ \langle 0.73, 0.14\rangle\end{pmatrix}$	$\begin{pmatrix}\langle 0.95e^{2\pi i(0.93)}, 0.48e^{2\pi i(0.40)}\rangle, \\ \langle 0.73, 0.14\rangle\end{pmatrix}$	$\begin{pmatrix}\langle 0.95e^{2\pi i(0.93)}, 0.48e^{2\pi i(0.40)}\rangle, \\ \langle 0.73, 0.14\rangle\end{pmatrix}$	$\begin{pmatrix}\langle 0.95e^{2\pi i(0.93)}, 0.48e^{2\pi i(0.40)}\rangle, \\ \langle 0.73, 0.14\rangle\end{pmatrix}$	$\begin{pmatrix}\langle 0.98e^{2\pi i(0.38)}, 0.38e^{2\pi i(0.26)}\rangle, \\ \langle 0.30, 0.19\rangle\end{pmatrix}$
C7	$\begin{pmatrix}\langle 0.51e^{2\pi i(0.44)}, 0.70e^{2\pi i(0.34)}\rangle, \\ \langle 0.19, 0.80\rangle\end{pmatrix}$	$\begin{pmatrix}\langle 0.51e^{2\pi i(0.44)}, 0.70e^{2\pi i(0.34)}\rangle, \\ \langle 0.19, 0.80\rangle\end{pmatrix}$	$\begin{pmatrix}\langle 0.51e^{2\pi i(0.44)}, 0.70e^{2\pi i(0.34)}\rangle, \\ \langle 0.19, 0.80\rangle\end{pmatrix}$	$\begin{pmatrix}\langle 0.51e^{2\pi i(0.44)}, 0.70e^{2\pi i(0.34)}\rangle, \\ \langle 0.19, 0.80\rangle\end{pmatrix}$	$\begin{pmatrix}\langle 0.63e^{2\pi i(0.45)}, 0.51e^{2\pi i(0.61)}\rangle, \\ \langle 0.14, 0.58\rangle\end{pmatrix}$
C8	$\begin{pmatrix}\langle 0.49e^{2\pi i(0.75)}, 0.19e^{2\pi i(0.60)}\rangle, \\ \langle 0.32, 0.10\rangle\end{pmatrix}$	$\begin{pmatrix}\langle 0.49e^{2\pi i(0.75)}, 0.19e^{2\pi i(0.60)}\rangle, \\ \langle 0.32, 0.10\rangle\end{pmatrix}$	$\begin{pmatrix}\langle 0.49e^{2\pi i(0.75)}, 0.19e^{2\pi i(0.60)}\rangle, \\ \langle 0.32, 0.10\rangle\end{pmatrix}$	$\begin{pmatrix}\langle 0.49e^{2\pi i(0.75)}, 0.19e^{2\pi i(0.60)}\rangle, \\ \langle 0.32, 0.10\rangle\end{pmatrix}$	$\begin{pmatrix}\langle 0.51e^{2\pi i(0.44)}, 0.70e^{2\pi i(0.34)}\rangle, \\ \langle 0.19, 0.80\rangle\end{pmatrix}$
C9	$\begin{pmatrix}\langle 0.63e^{2\pi i(0.45)}, 0.51e^{2\pi i(0.61)}\rangle, \\ \langle 0.14, 0.58\rangle\end{pmatrix}$	$\begin{pmatrix}\langle 0.63e^{2\pi i(0.45)}, 0.51e^{2\pi i(0.61)}\rangle, \\ \langle 0.14, 0.58\rangle\end{pmatrix}$	$\begin{pmatrix}\langle 0.63e^{2\pi i(0.45)}, 0.51e^{2\pi i(0.61)}\rangle, \\ \langle 0.14, 0.58\rangle\end{pmatrix}$	$\begin{pmatrix}\langle 0.63e^{2\pi i(0.45)}, 0.51e^{2\pi i(0.61)}\rangle, \\ \langle 0.14, 0.58\rangle\end{pmatrix}$	$\begin{pmatrix}\langle 0.67e^{2\pi i(0.24)}, 0.34e^{2\pi i(0.78)}\rangle, \\ \langle 0.79, 0.10\rangle\end{pmatrix}$
C10	$\begin{pmatrix}\langle 0.70e^{2\pi i(0.56)}, 0.86e^{2\pi i(0.16)}\rangle, \\ \langle 0.14, 0.46\rangle\end{pmatrix}$	$\begin{pmatrix}\langle 0.67e^{2\pi i(0.24)}, 0.34e^{2\pi i(0.78)}\rangle, \\ \langle 0.79, 0.10\rangle\end{pmatrix}$	$\begin{pmatrix}\langle 0.67e^{2\pi i(0.24)}, 0.34e^{2\pi i(0.78)}\rangle, \\ \langle 0.79, 0.10\rangle\end{pmatrix}$	$\begin{pmatrix}\langle 0.67e^{2\pi i(0.24)}, 0.34e^{2\pi i(0.78)}\rangle, \\ \langle 0.79, 0.10\rangle\end{pmatrix}$	$\begin{pmatrix}\langle 0.49e^{2\pi i(0.75)}, 0.19e^{2\pi i(0.60)}\rangle, \\ \langle 0.32, 0.10\rangle\end{pmatrix}$

(continued)

Table 9 (continued)

	A6	A7	A8	A9
C1	$\left(\begin{array}{c}\langle 0.78e^{2\pi i(0.23)}, 0.57e^{2\pi i(0.33)}\rangle,\\ \langle 0.30, 0.22\rangle\end{array}\right)$	$\left(\begin{array}{c}\langle 0.70e^{2\pi i(0.56)}, 0.86e^{2\pi i(0.16)}\rangle,\\ \langle 0.14, 0.46\rangle\end{array}\right)$	$\left(\begin{array}{c}\langle 0.67e^{2\pi i(0.24)}, 0.34e^{2\pi i(0.78)}\rangle,\\ \langle 0.79, 0.10\rangle\end{array}\right)$	$\left(\begin{array}{c}\langle 0.78e^{2\pi i(0.23)}, 0.57e^{2\pi i(0.33)}\rangle,\\ \langle 0.30, 0.22\rangle\end{array}\right)$
C2	$\left(\begin{array}{c}\langle 0.70e^{2\pi i(0.56)}, 0.86e^{2\pi i(0.16)}\rangle,\\ \langle 0.14, 0.46\rangle\end{array}\right)$	$\left(\begin{array}{c}\langle 0.90e^{2\pi i(0.70)}, 0.70e^{2\pi i(0.90)}\rangle,\\ \langle 0.80, 0.20\rangle\end{array}\right)$	$\left(\begin{array}{c}\langle 0.78e^{2\pi i(0.23)}, 0.57e^{2\pi i(0.33)}\rangle,\\ \langle 0.30, 0.22\rangle\end{array}\right)$	$\left(\begin{array}{c}\langle 0.82e^{2\pi i(0.41)}, 0.26e^{2\pi i(0.15)}\rangle,\\ \langle 0.22, 0.34\rangle\end{array}\right)$
C3	$\left(\begin{array}{c}\langle 0.95e^{2\pi i(0.93)}, 0.48e^{2\pi i(0.40)}\rangle,\\ \langle 0.73, 0.14\rangle\end{array}\right)$	$\left(\begin{array}{c}\langle 0.95e^{2\pi i(0.93)}, 0.48e^{2\pi i(0.40)}\rangle,\\ \langle 0.73, 0.14\rangle\end{array}\right)$	$\left(\begin{array}{c}\langle 0.98e^{2\pi i(0.38)}, 0.38e^{2\pi i(0.26)}\rangle,\\ \langle 0.30, 0.19\rangle\end{array}\right)$	$\left(\begin{array}{c}\langle 0.95e^{2\pi i(0.93)}, 0.48e^{2\pi i(0.40)}\rangle,\\ \langle 0.73, 0.14\rangle\end{array}\right)$
C4	$\left(\begin{array}{c}\langle 0.82e^{2\pi i(0.41)}, 0.26e^{2\pi i(0.15)}\rangle,\\ \langle 0.22, 0.34\rangle\end{array}\right)$	$\left(\begin{array}{c}\langle 0.78e^{2\pi i(0.23)}, 0.57e^{2\pi i(0.33)}\rangle,\\ \langle 0.30, 0.22\rangle\end{array}\right)$	$\left(\begin{array}{c}\langle 0.90e^{2\pi i(0.70)}, 0.70e^{2\pi i(0.90)}\rangle,\\ \langle 0.80, 0.20\rangle\end{array}\right)$	$\left(\begin{array}{c}\langle 0.90e^{2\pi i(0.70)}, 0.70e^{2\pi i(0.90)}\rangle,\\ \langle 0.80, 0.20\rangle\end{array}\right)$
C5	$\left(\begin{array}{c}\langle 0.90e^{2\pi i(0.70)}, 0.70e^{2\pi i(0.90)}\rangle,\\ \langle 0.80, 0.20\rangle\end{array}\right)$	$\left(\begin{array}{c}\langle 0.82e^{2\pi i(0.41)}, 0.26e^{2\pi i(0.15)}\rangle,\\ \langle 0.22, 0.34\rangle\end{array}\right)$	$\left(\begin{array}{c}\langle 0.70e^{2\pi i(0.56)}, 0.86e^{2\pi i(0.16)}\rangle,\\ \langle 0.14, 0.46\rangle\end{array}\right)$	$\left(\begin{array}{c}\langle 0.70e^{2\pi i(0.56)}, 0.86e^{2\pi i(0.16)}\rangle,\\ \langle 0.14, 0.46\rangle\end{array}\right)$
C6	$\left(\begin{array}{c}\langle 0.98e^{2\pi i(0.38)}, 0.38e^{2\pi i(0.26)}\rangle,\\ \langle 0.30, 0.19\rangle\end{array}\right)$	$\left(\begin{array}{c}\langle 0.98e^{2\pi i(0.38)}, 0.38e^{2\pi i(0.26)}\rangle,\\ \langle 0.30, 0.19\rangle\end{array}\right)$	$\left(\begin{array}{c}\langle 0.95e^{2\pi i(0.93)}, 0.48e^{2\pi i(0.40)}\rangle,\\ \langle 0.73, 0.14\rangle\end{array}\right)$	$\left(\begin{array}{c}\langle 0.98e^{2\pi i(0.38)}, 0.38e^{2\pi i(0.26)}\rangle,\\ \langle 0.30, 0.19\rangle\end{array}\right)$
C7	$\left(\begin{array}{c}\langle 0.67e^{2\pi i(0.24)}, 0.34e^{2\pi i(0.78)}\rangle,\\ \langle 0.79, 0.10\rangle\end{array}\right)$	$\left(\begin{array}{c}\langle 0.63e^{2\pi i(0.45)}, 0.51e^{2\pi i(0.61)}\rangle,\\ \langle 0.14, 0.58\rangle\end{array}\right)$	$\left(\begin{array}{c}\langle 0.51e^{2\pi i(0.44)}, 0.70e^{2\pi i(0.34)}\rangle,\\ \langle 0.19, 0.80\rangle\end{array}\right)$	$\left(\begin{array}{c}\langle 0.63e^{2\pi i(0.45)}, 0.51e^{2\pi i(0.61)}\rangle,\\ \langle 0.14, 0.58\rangle\end{array}\right)$
C8	$\left(\begin{array}{c}\langle 0.63e^{2\pi i(0.45)}, 0.51e^{2\pi i(0.61)}\rangle,\\ \langle 0.14, 0.58\rangle\end{array}\right)$	$\left(\begin{array}{c}\langle 0.49e^{2\pi i(0.75)}, 0.19e^{2\pi i(0.60)}\rangle,\\ \langle 0.32, 0.10\rangle\end{array}\right)$	$\left(\begin{array}{c}\langle 0.49e^{2\pi i(0.75)}, 0.19e^{2\pi i(0.60)}\rangle,\\ \langle 0.32, 0.10\rangle\end{array}\right)$	$\left(\begin{array}{c}\langle 0.51e^{2\pi i(0.44)}, 0.70e^{2\pi i(0.34)}\rangle,\\ \langle 0.19, 0.80\rangle\end{array}\right)$
C9	$\left(\begin{array}{c}\langle 0.51e^{2\pi i(0.44)}, 0.70e^{2\pi i(0.34)}\rangle,\\ \langle 0.19, 0.80\rangle\end{array}\right)$	$\left(\begin{array}{c}\langle 0.67e^{2\pi i(0.24)}, 0.34e^{2\pi i(0.78)}\rangle,\\ \langle 0.79, 0.10\rangle\end{array}\right)$	$\left(\begin{array}{c}\langle 0.82e^{2\pi i(0.41)}, 0.26e^{2\pi i(0.15)}\rangle,\\ \langle 0.22, 0.34\rangle\end{array}\right)$	$\left(\begin{array}{c}\langle 0.67e^{2\pi i(0.24)}, 0.34e^{2\pi i(0.78)}\rangle,\\ \langle 0.79, 0.10\rangle\end{array}\right)$
C10	$\left(\begin{array}{c}\langle 0.49e^{2\pi i(0.75)}, 0.19e^{2\pi i(0.60)}\rangle,\\ \langle 0.32, 0.10\rangle\end{array}\right)$	$\left(\begin{array}{c}\langle 0.51e^{2\pi i(0.44)}, 0.70e^{2\pi i(0.34)}\rangle,\\ \langle 0.19, 0.80\rangle\end{array}\right)$	$\left(\begin{array}{c}\langle 0.78e^{2\pi i(0.23)}, 0.57e^{2\pi i(0.33)}\rangle,\\ \langle 0.30, 0.22\rangle\end{array}\right)$	$\left(\begin{array}{c}\langle 0.49e^{2\pi i(0.75)}, 0.19e^{2\pi i(0.60)}\rangle,\\ \langle 0.32, 0.10\rangle\end{array}\right)$

Table 10 Complex linear diophantine fuzzy decision matrix based on DM2

	A1	A2	A3	A4	A5
C1	$\left(\langle 0.91e^{2\pi i(0.71)}, 0.72e^{2\pi i(0.91)}\rangle, \langle 0.78, 0.21\rangle\right)$	$\left(\langle 0.80e^{2\pi i(0.29)}, 0.31e^{2\pi i(0.31)}\rangle, \langle 0.30, 0.23\rangle\right)$	$\left(\langle 0.74e^{2\pi i(0.98)}, 0.64e^{2\pi i(0.96)}\rangle, \langle 0.47, 0.35\rangle\right)$	$\left(\langle 0.87e^{2\pi i(0.94)}, 0.76e^{2\pi i(0.52)}\rangle, \langle 0.75, 0.22\rangle\right)$	$\left(\langle 0.87e^{2\pi i(0.94)}, 0.76e^{2\pi i(0.52)}\rangle, \langle 0.75, 0.22\rangle\right)$
C2	$\left(\langle 0.94e^{2\pi i(0.90)}, 0.72e^{2\pi i(0.65)}\rangle, \langle 0.84, 0.11\rangle\right)$	$\left(\langle 0.91e^{2\pi i(0.71)}, 0.72e^{2\pi i(0.91)}\rangle, \langle 0.78, 0.21\rangle\right)$	$\left(\langle 0.91e^{2\pi i(0.71)}, 0.72e^{2\pi i(0.91)}\rangle, \langle 0.78, 0.21\rangle\right)$	$\left(\langle 0.94e^{2\pi i(0.90)}, 0.72e^{2\pi i(0.65)}\rangle, \langle 0.84, 0.11\rangle\right)$	$\left(\langle 0.94e^{2\pi i(0.90)}, 0.72e^{2\pi i(0.65)}\rangle, \langle 0.84, 0.11\rangle\right)$
C3	$\left(\langle 0.74e^{2\pi i(0.98)}, 0.64e^{2\pi i(0.96)}\rangle, \langle 0.47, 0.35\rangle\right)$	$\left(\langle 0.68e^{2\pi i(0.45)}, 0.80e^{2\pi i(0.77)}\rangle, \langle 0.16, 0.23\rangle\right)$	$\left(\langle 0.80e^{2\pi i(0.29)}, 0.31e^{2\pi i(0.31)}\rangle, \langle 0.30, 0.23\rangle\right)$	$\left(\langle 0.74e^{2\pi i(0.98)}, 0.64e^{2\pi i(0.96)}\rangle, \langle 0.47, 0.35\rangle\right)$	$\left(\langle 0.68e^{2\pi i(0.45)}, 0.80e^{2\pi i(0.77)}\rangle, \langle 0.16, 0.23\rangle\right)$
C4	$\left(\langle 0.87e^{2\pi i(0.94)}, 0.76e^{2\pi i(0.52)}\rangle, \langle 0.75, 0.22\rangle\right)$	$\left(\langle 0.94e^{2\pi i(0.90)}, 0.72e^{2\pi i(0.65)}\rangle, \langle 0.84, 0.11\rangle\right)$	$\left(\langle 0.94e^{2\pi i(0.90)}, 0.72e^{2\pi i(0.65)}\rangle, \langle 0.84, 0.11\rangle\right)$	$\left(\langle 0.91e^{2\pi i(0.71)}, 0.72e^{2\pi i(0.91)}\rangle, \langle 0.78, 0.21\rangle\right)$	$\left(\langle 0.91e^{2\pi i(0.71)}, 0.72e^{2\pi i(0.91)}\rangle, \langle 0.78, 0.21\rangle\right)$
C5	$\left(\langle 0.80e^{2\pi i(0.29)}, 0.31e^{2\pi i(0.31)}\rangle, \langle 0.30, 0.23\rangle\right)$	$\left(\langle 0.74e^{2\pi i(0.98)}, 0.64e^{2\pi i(0.96)}\rangle, \langle 0.47, 0.35\rangle\right)$	$\left(\langle 0.68e^{2\pi i(0.45)}, 0.80e^{2\pi i(0.77)}\rangle, \langle 0.16, 0.23\rangle\right)$	$\left(\langle 0.68e^{2\pi i(0.45)}, 0.80e^{2\pi i(0.77)}\rangle, \langle 0.16, 0.23\rangle\right)$	$\left(\langle 0.80e^{2\pi i(0.29)}, 0.31e^{2\pi i(0.31)}\rangle, \langle 0.30, 0.23\rangle\right)$
C6	$\left(\langle 0.68e^{2\pi i(0.45)}, 0.80e^{2\pi i(0.77)}\rangle, \langle 0.16, 0.23\rangle\right)$	$\left(\langle 0.65e^{2\pi i(0.15)}, 0.30e^{2\pi i(0.63)}\rangle, \langle 0.70, 0.25\rangle\right)$	$\left(\langle 0.65e^{2\pi i(0.15)}, 0.30e^{2\pi i(0.63)}\rangle, \langle 0.70, 0.25\rangle\right)$	$\left(\langle 0.65e^{2\pi i(0.15)}, 0.30e^{2\pi i(0.63)}\rangle, \langle 0.70, 0.25\rangle\right)$	$\left(\langle 0.74e^{2\pi i(0.98)}, 0.64e^{2\pi i(0.96)}\rangle, \langle 0.47, 0.35\rangle\right)$
C7	$\left(\langle 0.65e^{2\pi i(0.15)}, 0.30e^{2\pi i(0.63)}\rangle, \langle 0.70, 0.25\rangle\right)$	$\left(\langle 0.57e^{2\pi i(0.75)}, 0.33e^{2\pi i(0.64)}\rangle, \langle 0.38, 0.54\rangle\right)$	$\left(\langle 0.57e^{2\pi i(0.75)}, 0.33e^{2\pi i(0.64)}\rangle, \langle 0.38, 0.54\rangle\right)$	$\left(\langle 0.57e^{2\pi i(0.75)}, 0.33e^{2\pi i(0.64)}\rangle, \langle 0.38, 0.54\rangle\right)$	$\left(\langle 0.65e^{2\pi i(0.15)}, 0.30e^{2\pi i(0.63)}\rangle, \langle 0.70, 0.25\rangle\right)$
C8	$\left(\langle 0.43e^{2\pi i(0.43)}, 0.15e^{2\pi i(0.14)}\rangle, \langle 0.45, 0.54\rangle\right)$	$\left(\langle 0.43e^{2\pi i(0.43)}, 0.15e^{2\pi i(0.14)}\rangle, \langle 0.45, 0.54\rangle\right)$	$\left(\langle 0.43e^{2\pi i(0.43)}, 0.15e^{2\pi i(0.14)}\rangle, \langle 0.45, 0.54\rangle\right)$	$\left(\langle 0.43e^{2\pi i(0.43)}, 0.15e^{2\pi i(0.14)}\rangle, \langle 0.45, 0.54\rangle\right)$	$\left(\langle 0.43e^{2\pi i(0.46)}, 0.11e^{2\pi i(0.14)}\rangle, \langle 0.48, 0.50\rangle\right)$
C9	$\left(\langle 0.50e^{2\pi i(0.45)}, 0.75e^{2\pi i(0.30)}\rangle, \langle 0.35, 0.70\rangle\right)$	$\left(\langle 0.50e^{2\pi i(0.45)}, 0.75e^{2\pi i(0.30)}\rangle, \langle 0.35, 0.70\rangle\right)$	$\left(\langle 0.50e^{2\pi i(0.45)}, 0.75e^{2\pi i(0.30)}\rangle, \langle 0.35, 0.70\rangle\right)$	$\left(\langle 0.50e^{2\pi i(0.45)}, 0.75e^{2\pi i(0.30)}\rangle, \langle 0.35, 0.70\rangle\right)$	$\left(\langle 0.57e^{2\pi i(0.75)}, 0.33e^{2\pi i(0.64)}\rangle, \langle 0.38, 0.54\rangle\right)$
C10	$\left(\langle 0.57e^{2\pi i(0.75)}, 0.33e^{2\pi i(0.64)}\rangle, \langle 0.38, 0.54\rangle\right)$	$\left(\langle 0.87e^{2\pi i(0.94)}, 0.76e^{2\pi i(0.52)}\rangle, \langle 0.75, 0.22\rangle\right)$	$\left(\langle 0.87e^{2\pi i(0.94)}, 0.76e^{2\pi i(0.52)}\rangle, \langle 0.75, 0.22\rangle\right)$	$\left(\langle 0.80e^{2\pi i(0.29)}, 0.31e^{2\pi i(0.31)}\rangle, \langle 0.30, 0.23\rangle\right)$	$\left(\langle 0.50e^{2\pi i(0.45)}, 0.75e^{2\pi i(0.30)}\rangle, \langle 0.35, 0.70\rangle\right)$

(continued)

Table 10 (continued)

	A6	A7	A8	A9
C1	$\left(\langle 0.74e^{2\pi i(0.98)}, 0.64e^{2\pi i(0.96)}\rangle, \langle 0.47, 0.35\rangle\right)$	$\left(\langle 0.68e^{2\pi i(0.45)}, 0.80e^{2\pi i(0.77)}\rangle, \langle 0.16, 0.23\rangle\right)$	$\left(\langle 0.65e^{2\pi i(0.15)}, 0.30e^{2\pi i(0.63)}\rangle, \langle 0.70, 0.25\rangle\right)$	$\left(\langle 0.68e^{2\pi i(0.45)}, 0.80e^{2\pi i(0.77)}\rangle, \langle 0.16, 0.23\rangle\right)$
C2	$\left(\langle 0.91e^{2\pi i(0.71)}, 0.72e^{2\pi i(0.91)}\rangle, \langle 0.78, 0.21\rangle\right)$	$\left(\langle 0.91e^{2\pi i(0.71)}, 0.72e^{2\pi i(0.91)}\rangle, \langle 0.78, 0.21\rangle\right)$	$\left(\langle 0.87e^{2\pi i(0.94)}, 0.76e^{2\pi i(0.52)}\rangle, \langle 0.75, 0.22\rangle\right)$	$\left(\langle 0.87e^{2\pi i(0.94)}, 0.76e^{2\pi i(0.52)}\rangle, \langle 0.75, 0.22\rangle\right)$
C3	$\left(\langle 0.87e^{2\pi i(0.94)}, 0.76e^{2\pi i(0.52)}\rangle, \langle 0.75, 0.22\rangle\right)$	$\left(\langle 0.87e^{2\pi i(0.94)}, 0.76e^{2\pi i(0.52)}\rangle, \langle 0.75, 0.22\rangle\right)$	$\left(\langle 0.91e^{2\pi i(0.71)}, 0.72e^{2\pi i(0.91)}\rangle, \langle 0.78, 0.21\rangle\right)$	$\left(\langle 0.91e^{2\pi i(0.71)}, 0.72e^{2\pi i(0.91)}\rangle, \langle 0.78, 0.21\rangle\right)$
C4	$\left(\langle 0.94e^{2\pi i(0.90)}, 0.72e^{2\pi i(0.65)}\rangle, \langle 0.84, 0.11\rangle\right)$	$\left(\langle 0.94e^{2\pi i(0.90)}, 0.72e^{2\pi i(0.65)}\rangle, \langle 0.84, 0.11\rangle\right)$	$\left(\langle 0.94e^{2\pi i(0.90)}, 0.72e^{2\pi i(0.65)}\rangle, \langle 0.84, 0.11\rangle\right)$	$\left(\langle 0.94e^{2\pi i(0.90)}, 0.72e^{2\pi i(0.65)}\rangle, \langle 0.84, 0.11\rangle\right)$
C5	$\left(\langle 0.80e^{2\pi i(0.29)}, 0.31e^{2\pi i(0.31)}\rangle, \langle 0.30, 0.23\rangle\right)$	$\left(\langle 0.80e^{2\pi i(0.29)}, 0.31e^{2\pi i(0.31)}\rangle, \langle 0.30, 0.23\rangle\right)$	$\left(\langle 0.68e^{2\pi i(0.45)}, 0.80e^{2\pi i(0.77)}\rangle, \langle 0.16, 0.23\rangle\right)$	$\left(\langle 0.74e^{2\pi i(0.98)}, 0.64e^{2\pi i(0.96)}\rangle, \langle 0.47, 0.35\rangle\right)$
C6	$\left(\langle 0.68e^{2\pi i(0.45)}, 0.80e^{2\pi i(0.77)}\rangle, \langle 0.16, 0.23\rangle\right)$	$\left(\langle 0.74e^{2\pi i(0.98)}, 0.64e^{2\pi i(0.96)}\rangle, \langle 0.47, 0.35\rangle\right)$	$\left(\langle 0.80e^{2\pi i(0.29)}, 0.31e^{2\pi i(0.31)}\rangle, \langle 0.30, 0.23\rangle\right)$	$\left(\langle 0.80e^{2\pi i(0.29)}, 0.31e^{2\pi i(0.31)}\rangle, \langle 0.30, 0.23\rangle\right)$
C7	$\left(\langle 0.65e^{2\pi i(0.15)}, 0.30e^{2\pi i(0.63)}\rangle, \langle 0.70, 0.25\rangle\right)$	$\left(\langle 0.65e^{2\pi i(0.15)}, 0.30e^{2\pi i(0.63)}\rangle, \langle 0.70, 0.25\rangle\right)$	$\left(\langle 0.57e^{2\pi i(0.75)}, 0.33e^{2\pi i(0.64)}\rangle, \langle 0.38, 0.54\rangle\right)$	$\left(\langle 0.65e^{2\pi i(0.15)}, 0.30e^{2\pi i(0.63)}\rangle, \langle 0.70, 0.25\rangle\right)$
C8	$\left(\langle 0.43e^{2\pi i(0.43)}, 0.15e^{2\pi i(0.14)}\rangle, \langle 0.45, 0.54\rangle\right)$	$\left(\langle 0.43e^{2\pi i(0.43)}, 0.15e^{2\pi i(0.14)}\rangle, \langle 0.45, 0.54\rangle\right)$	$\left(\langle 0.43e^{2\pi i(0.43)}, 0.15e^{2\pi i(0.14)}\rangle, \langle 0.45, 0.54\rangle\right)$	$\left(\langle 0.43e^{2\pi i(0.43)}, 0.15e^{2\pi i(0.14)}\rangle, \langle 0.45, 0.54\rangle\right)$
C9	$\left(\langle 0.50e^{2\pi i(0.45)}, 0.75e^{2\pi i(0.30)}\rangle, \langle 0.35, 0.70\rangle\right)$	$\left(\langle 0.57e^{2\pi i(0.75)}, 0.33e^{2\pi i(0.64)}\rangle, \langle 0.38, 0.54\rangle\right)$	$\left(\langle 0.74e^{2\pi i(0.98)}, 0.64e^{2\pi i(0.96)}\rangle, \langle 0.47, 0.35\rangle\right)$	$\left(\langle 0.57e^{2\pi i(0.75)}, 0.33e^{2\pi i(0.64)}\rangle, \langle 0.38, 0.54\rangle\right)$
C10	$\left(\langle 0.57e^{2\pi i(0.75)}, 0.33e^{2\pi i(0.64)}\rangle, \langle 0.38, 0.54\rangle\right)$	$\left(\langle 0.50e^{2\pi i(0.45)}, 0.75e^{2\pi i(0.30)}\rangle, \langle 0.35, 0.70\rangle\right)$	$\left(\langle 0.50e^{2\pi i(0.45)}, 0.75e^{2\pi i(0.30)}\rangle, \langle 0.35, 0.70\rangle\right)$	$\left(\langle 0.50e^{2\pi i(0.45)}, 0.75e^{2\pi i(0.30)}\rangle, \langle 0.35, 0.70\rangle\right)$

Table 11 Complex Linear Diophantine Fuzzy Decision Matrix based on DM3

	A1	A2	A3	A4	A5
C1	$\left(\langle 0.73e^{2\pi i(0.72)}, 0.548e^{2\pi i(0.36)}\rangle, \langle 0.53, 0.11\rangle\right)$	$\left(\langle 0.69e^{2\pi i(0.55)}, 0.85e^{2\pi i(0.19)}\rangle, \langle 0.15, 0.45\rangle\right)$	$\left(\langle 0.64e^{2\pi i(0.12)}, 0.32e^{2\pi i(0.62)}\rangle, \langle 0.75, 0.25\rangle\right)$	$\left(\langle 0.64e^{2\pi i(0.12)}, 0.32e^{2\pi i(0.62)}\rangle, \langle 0.75, 0.25\rangle\right)$	$\left(\langle 0.64e^{2\pi i(0.12)}, 0.32e^{2\pi i(0.62)}\rangle, \langle 0.75, 0.25\rangle\right)$
C2	$\left(\langle 0.64e^{2\pi i(0.12)}, 0.32e^{2\pi i(0.62)}\rangle, \langle 0.75, 0.25\rangle\right)$	$\left(\langle 0.60e^{2\pi i(0.48)}, 0.50e^{2\pi i(0.60)}\rangle, \langle 0.20, 0.61\rangle\right)$	$\left(\langle 0.60e^{2\pi i(0.48)}, 0.50e^{2\pi i(0.60)}\rangle, \langle 0.20, 0.61\rangle\right)$	$\left(\langle 0.60e^{2\pi i(0.48)}, 0.50e^{2\pi i(0.60)}\rangle, \langle 0.20, 0.61\rangle\right)$	$\left(\langle 0.69e^{2\pi i(0.55)}, 0.85e^{2\pi i(0.19)}\rangle, \langle 0.15, 0.45\rangle\right)$
C3	$\left(\langle 0.96e^{2\pi i(0.91)}, 0.46e^{2\pi i(0.41)}\rangle, \langle 0.61, 0.19\rangle\right)$	$\left(\langle 0.96e^{2\pi i(0.91)}, 0.46e^{2\pi i(0.41)}\rangle, \langle 0.61, 0.19\rangle\right)$	$\left(\langle 0.92e^{2\pi i(0.88)}, 0.75e^{2\pi i(0.28)}\rangle, \langle 0.20, 0.29\rangle\right)$	$\left(\langle 0.92e^{2\pi i(0.88)}, 0.75e^{2\pi i(0.28)}\rangle, \langle 0.20, 0.29\rangle\right)$	$\left(\langle 0.92e^{2\pi i(0.88)}, 0.75e^{2\pi i(0.28)}\rangle, \langle 0.20, 0.29\rangle\right)$
C4	$\left(\langle 0.85e^{2\pi i(0.28)}, 0.33e^{2\pi i(0.35)}\rangle, \langle 0.25, 0.20\rangle\right)$	$\left(\langle 0.92e^{2\pi i(0.88)}, 0.75e^{2\pi i(0.28)}\rangle, \langle 0.20, 0.29\rangle\right)$	$\left(\langle 0.85e^{2\pi i(0.28)}, 0.33e^{2\pi i(0.35)}\rangle, \langle 0.25, 0.20\rangle\right)$	$\left(\langle 0.85e^{2\pi i(0.28)}, 0.33e^{2\pi i(0.35)}\rangle, \langle 0.25, 0.20\rangle\right)$	$\left(\langle 0.85e^{2\pi i(0.28)}, 0.33e^{2\pi i(0.35)}\rangle, \langle 0.25, 0.20\rangle\right)$
C5	$\left(\langle 0.60e^{2\pi i(0.48)}, 0.50e^{2\pi i(0.60)}\rangle, \langle 0.20, 0.61\rangle\right)$	$\left(\langle 0.50e^{2\pi i(0.45)}, 0.75e^{2\pi i(0.30)}\rangle, \langle 0.18, 0.78\rangle\right)$	$\left(\langle 0.65e^{2\pi i(0.45)}, 0.55e^{2\pi i(0.30)}\rangle, \langle 0.18, 0.24\rangle\right)$	$\left(\langle 0.65e^{2\pi i(0.45)}, 0.55e^{2\pi i(0.30)}\rangle, \langle 0.18, 0.24\rangle\right)$	$\left(\langle 0.60e^{2\pi i(0.48)}, 0.50e^{2\pi i(0.60)}\rangle, \langle 0.20, 0.61\rangle\right)$
C6	$\left(\langle 0.92e^{2\pi i(0.88)}, 0.75e^{2\pi i(0.28)}\rangle, \langle 0.20, 0.29\rangle\right)$	$\left(\langle 0.85e^{2\pi i(0.28)}, 0.33e^{2\pi i(0.35)}\rangle, \langle 0.25, 0.20\rangle\right)$	$\left(\langle 0.73e^{2\pi i(0.72)}, 0.548e^{2\pi i(0.36)}\rangle, \langle 0.53, 0.11\rangle\right)$	$\left(\langle 0.73e^{2\pi i(0.72)}, 0.548e^{2\pi i(0.36)}\rangle, \langle 0.53, 0.11\rangle\right)$	$\left(\langle 0.73e^{2\pi i(0.72)}, 0.548e^{2\pi i(0.36)}\rangle, \langle 0.53, 0.11\rangle\right)$
C7	$\left(\langle 0.45e^{2\pi i(0.70)}, 0.29e^{2\pi i(0.55)}\rangle, \langle 0.35, 0.15\rangle\right)$	$\left(\langle 0.45e^{2\pi i(0.70)}, 0.29e^{2\pi i(0.55)}\rangle, \langle 0.35, 0.15\rangle\right)$	$\left(\langle 0.45e^{2\pi i(0.70)}, 0.29e^{2\pi i(0.55)}\rangle, \langle 0.35, 0.15\rangle\right)$	$\left(\langle 0.45e^{2\pi i(0.70)}, 0.29e^{2\pi i(0.55)}\rangle, \langle 0.35, 0.15\rangle\right)$	$\left(\langle 0.45e^{2\pi i(0.70)}, 0.29e^{2\pi i(0.55)}\rangle, \langle 0.35, 0.15\rangle\right)$
C8	$\left(\langle 0.69e^{2\pi i(0.55)}, 0.85e^{2\pi i(0.19)}\rangle, \langle 0.15, 0.45\rangle\right)$	$\left(\langle 0.73e^{2\pi i(0.72)}, 0.548e^{2\pi i(0.36)}\rangle, \langle 0.53, 0.11\rangle\right)$	$\left(\langle 0.69e^{2\pi i(0.55)}, 0.85e^{2\pi i(0.19)}\rangle, \langle 0.15, 0.45\rangle\right)$	$\left(\langle 0.69e^{2\pi i(0.55)}, 0.85e^{2\pi i(0.19)}\rangle, \langle 0.15, 0.45\rangle\right)$	$\left(\langle 0.49e^{2\pi i(0.75)}, 0.19e^{2\pi i(0.60)}\rangle, \langle 0.32, 0.10\rangle\right)$
C9	$\left(\langle 0.50e^{2\pi i(0.45)}, 0.75e^{2\pi i(0.30)}\rangle, \langle 0.18, 0.78\rangle\right)$	$\left(\langle 0.64e^{2\pi i(0.12)}, 0.32e^{2\pi i(0.62)}\rangle, \langle 0.75, 0.25\rangle\right)$	$\left(\langle 0.41e^{2\pi i(0.47)}, 0.21e^{2\pi i(0.24)}\rangle, \langle 0.44, 0.48\rangle\right)$	$\left(\langle 0.41e^{2\pi i(0.47)}, 0.21e^{2\pi i(0.24)}\rangle, \langle 0.44, 0.48\rangle\right)$	$\left(\langle 0.41e^{2\pi i(0.47)}, 0.21e^{2\pi i(0.24)}\rangle, \langle 0.44, 0.48\rangle\right)$
C10	$\left(\langle 0.41e^{2\pi i(0.47)}, 0.21e^{2\pi i(0.24)}\rangle, \langle 0.44, 0.48\rangle\right)$	$\left(\langle 0.41e^{2\pi i(0.47)}, 0.21e^{2\pi i(0.24)}\rangle, \langle 0.44, 0.48\rangle\right)$	$\left(\langle 0.96e^{2\pi i(0.91)}, 0.46e^{2\pi i(0.41)}\rangle, \langle 0.61, 0.19\rangle\right)$	$\left(\langle 0.96e^{2\pi i(0.91)}, 0.46e^{2\pi i(0.41)}\rangle, \langle 0.61, 0.19\rangle\right)$	$\left(\langle 0.96e^{2\pi i(0.91)}, 0.46e^{2\pi i(0.41)}\rangle, \langle 0.61, 0.19\rangle\right)$

(continued)

Table 11 (continued)

	A6	A7	A8	A9
C1	$\left(\left\langle 0.69e^{2\pi i(0.55)}, 0.85e^{2\pi i(0.19)}\right\rangle, \langle 0.15, 0.45\rangle\right)$	$\left(\left\langle 0.69e^{2\pi i(0.55)}, 0.85e^{2\pi i(0.19)}\right\rangle, \langle 0.15, 0.45\rangle\right)$	$\left(\left\langle 0.64e^{2\pi i(0.12)}, 0.32e^{2\pi i(0.62)}\right\rangle, \langle 0.75, 0.25\rangle\right)$	$\left(\left\langle 0.69e^{2\pi i(0.55)}, 0.85e^{2\pi i(0.19)}\right\rangle, \langle 0.15, 0.45\rangle\right)$
C2	$\left(\left\langle 0.60e^{2\pi i(0.48)}, 0.50e^{2\pi i(0.60)}\right\rangle, \langle 0.20, 0.61\rangle\right)$	$\left(\left\langle 0.64e^{2\pi i(0.12)}, 0.32e^{2\pi i(0.62)}\right\rangle, \langle 0.75, 0.25\rangle\right)$	$\left(\left\langle 0.60e^{2\pi i(0.48)}, 0.50e^{2\pi i(0.60)}\right\rangle, \langle 0.20, 0.61\rangle\right)$	$\left(\left\langle 0.64e^{2\pi i(0.12)}, 0.32e^{2\pi i(0.62)}\right\rangle, \langle 0.75, 0.25\rangle\right)$
C3	$\left(\left\langle 0.96e^{2\pi i(0.91)}, 0.46e^{2\pi i(0.41)}\right\rangle, \langle 0.61, 0.19\rangle\right)$	$\left(\left\langle 0.96e^{2\pi i(0.91)}, 0.46e^{2\pi i(0.41)}\right\rangle, \langle 0.61, 0.19\rangle\right)$	$\left(\left\langle 0.96e^{2\pi i(0.91)}, 0.46e^{2\pi i(0.41)}\right\rangle, \langle 0.61, 0.19\rangle\right)$	$\left(\left\langle 0.96e^{2\pi i(0.91)}, 0.46e^{2\pi i(0.41)}\right\rangle, \langle 0.61, 0.19\rangle\right)$
C4	$\left(\left\langle 0.92e^{2\pi i(0.88)}, 0.75e^{2\pi i(0.28)}\right\rangle, \langle 0.20, 0.29\rangle\right)$	$\left(\left\langle 0.92e^{2\pi i(0.88)}, 0.75e^{2\pi i(0.28)}\right\rangle, \langle 0.20, 0.29\rangle\right)$	$\left(\left\langle 0.92e^{2\pi i(0.88)}, 0.75e^{2\pi i(0.28)}\right\rangle, \langle 0.20, 0.29\rangle\right)$	$\left(\left\langle 0.92e^{2\pi i(0.88)}, 0.75e^{2\pi i(0.28)}\right\rangle, \langle 0.20, 0.29\rangle\right)$
C5	$\left(\left\langle 0.50e^{2\pi i(0.45)}, 0.75e^{2\pi i(0.30)}\right\rangle, \langle 0.18, 0.78\rangle\right)$	$\left(\left\langle 0.50e^{2\pi i(0.45)}, 0.75e^{2\pi i(0.30)}\right\rangle, \langle 0.18, 0.78\rangle\right)$	$\left(\left\langle 0.50e^{2\pi i(0.45)}, 0.75e^{2\pi i(0.30)}\right\rangle, \langle 0.18, 0.78\rangle\right)$	$\left(\left\langle 0.50e^{2\pi i(0.45)}, 0.75e^{2\pi i(0.30)}\right\rangle, \langle 0.18, 0.78\rangle\right)$
C6	$\left(\left\langle 0.85e^{2\pi i(0.28)}, 0.33e^{2\pi i(0.35)}\right\rangle, \langle 0.25, 0.20\rangle\right)$	$\left(\left\langle 0.85e^{2\pi i(0.28)}, 0.33e^{2\pi i(0.35)}\right\rangle, \langle 0.25, 0.20\rangle\right)$	$\left(\left\langle 0.85e^{2\pi i(0.28)}, 0.33e^{2\pi i(0.35)}\right\rangle, \langle 0.25, 0.20\rangle\right)$	$\left(\left\langle 0.85e^{2\pi i(0.28)}, 0.33e^{2\pi i(0.35)}\right\rangle, \langle 0.25, 0.20\rangle\right)$
C7	$\left(\left\langle 0.45e^{2\pi i(0.70)}, 0.29e^{2\pi i(0.55)}\right\rangle, \langle 0.35, 0.15\rangle\right)$	$\left(\left\langle 0.45e^{2\pi i(0.70)}, 0.29e^{2\pi i(0.55)}\right\rangle, \langle 0.35, 0.15\rangle\right)$	$\left(\left\langle 0.45e^{2\pi i(0.70)}, 0.29e^{2\pi i(0.55)}\right\rangle, \langle 0.35, 0.15\rangle\right)$	$\left(\left\langle 0.45e^{2\pi i(0.70)}, 0.29e^{2\pi i(0.55)}\right\rangle, \langle 0.35, 0.15\rangle\right)$
C8	$\left(\left\langle 0.73e^{2\pi i(0.72)}, 0.548e^{2\pi i(0.36)}\right\rangle, \langle 0.53, 0.11\rangle\right)$	$\left(\left\langle 0.73e^{2\pi i(0.72)}, 0.548e^{2\pi i(0.36)}\right\rangle, \langle 0.53, 0.11\rangle\right)$	$\left(\left\langle 0.73e^{2\pi i(0.72)}, 0.548e^{2\pi i(0.36)}\right\rangle, \langle 0.53, 0.11\rangle\right)$	$\left(\left\langle 0.73e^{2\pi i(0.72)}, 0.548e^{2\pi i(0.36)}\right\rangle, \langle 0.53, 0.11\rangle\right)$
C9	$\left(\left\langle 0.41e^{2\pi i(0.47)}, 0.21e^{2\pi i(0.24)}\right\rangle, \langle 0.44, 0.48\rangle\right)$	$\left(\left\langle 0.41e^{2\pi i(0.47)}, 0.21e^{2\pi i(0.24)}\right\rangle, \langle 0.44, 0.48\rangle\right)$	$\left(\left\langle 0.41e^{2\pi i(0.47)}, 0.21e^{2\pi i(0.24)}\right\rangle, \langle 0.44, 0.48\rangle\right)$	$\left(\left\langle 0.41e^{2\pi i(0.47)}, 0.21e^{2\pi i(0.24)}\right\rangle, \langle 0.44, 0.48\rangle\right)$
C10	$\left(\left\langle 0.64e^{2\pi i(0.12)}, 0.32e^{2\pi i(0.62)}\right\rangle, \langle 0.75, 0.25\rangle\right)$	$\left(\left\langle 0.60e^{2\pi i(0.48)}, 0.50e^{2\pi i(0.60)}\right\rangle, \langle 0.20, 0.61\rangle\right)$	$\left(\left\langle 0.69e^{2\pi i(0.55)}, 0.85e^{2\pi i(0.19)}\right\rangle, \langle 0.15, 0.45\rangle\right)$	$\left(\left\langle 0.60e^{2\pi i(0.48)}, 0.50e^{2\pi i(0.60)}\right\rangle, \langle 0.20, 0.61\rangle\right)$

Table 12 Transformed complex linear diophantine fuzzy decision matrix based on DM1

	A1	A2	A3	A4	A5
C1	$\left(\left\langle 0.82e^{2\pi i(0.41)}, 0.26e^{2\pi i(0.15)}\right\rangle, \langle 0.22, 0.34\rangle\right)$	$\left(\left\langle 0.70e^{2\pi i(0.56)}, 0.86e^{2\pi i(0.16)}\right\rangle, \langle 0.14, 0.46\rangle\right)$	$\left(\left\langle 0.82e^{2\pi i(0.41)}, 0.26e^{2\pi i(0.15)}\right\rangle, \langle 0.22, 0.34\rangle\right)$	$\left(\left\langle 0.82e^{2\pi i(0.41)}, 0.26e^{2\pi i(0.15)}\right\rangle, \langle 0.22, 0.34\rangle\right)$	$\left(\left\langle 0.82e^{2\pi i(0.41)}, 0.26e^{2\pi i(0.15)}\right\rangle, \langle 0.22, 0.34\rangle\right)$
C2	$\left(\left\langle 0.90e^{2\pi i(0.70)}, 0.70e^{2\pi i(0.90)}\right\rangle, \langle 0.80, 0.20\rangle\right)$	$\left(\left\langle 0.82e^{2\pi i(0.41)}, 0.26e^{2\pi i(0.15)}\right\rangle, \langle 0.22, 0.34\rangle\right)$	$\left(\left\langle 0.90e^{2\pi i(0.70)}, 0.70e^{2\pi i(0.90)}\right\rangle, \langle 0.80, 0.20\rangle\right)$	$\left(\left\langle 0.90e^{2\pi i(0.70)}, 0.70e^{2\pi i(0.90)}\right\rangle, \langle 0.80, 0.20\rangle\right)$	$\left(\left\langle 0.90e^{2\pi i(0.70)}, 0.70e^{2\pi i(0.90)}\right\rangle, \langle 0.80, 0.20\rangle\right)$
C3	$\left(\left\langle 0.98e^{2\pi i(0.38)}, 0.38e^{2\pi i(0.26)}\right\rangle, \langle 0.30, 0.19\rangle\right)$	$\left(\left\langle 0.98e^{2\pi i(0.38)}, 0.38e^{2\pi i(0.26)}\right\rangle, \langle 0.30, 0.19\rangle\right)$	$\left(\left\langle 0.98e^{2\pi i(0.38)}, 0.38e^{2\pi i(0.26)}\right\rangle, \langle 0.30, 0.19\rangle\right)$	$\left(\left\langle 0.98e^{2\pi i(0.38)}, 0.38e^{2\pi i(0.26)}\right\rangle, \langle 0.30, 0.19\rangle\right)$	$\left(\left\langle 0.95e^{2\pi i(0.93)}, 0.48e^{2\pi i(0.40)}\right\rangle, \langle 0.73, 0.14\rangle\right)$
C4	$\left(\left\langle 0.67e^{2\pi i(0.24)}, 0.34e^{2\pi i(0.78)}\right\rangle, \langle 0.79, 0.10\rangle\right)$	$\left(\left\langle 0.90e^{2\pi i(0.70)}, 0.70e^{2\pi i(0.90)}\right\rangle, \langle 0.80, 0.20\rangle\right)$	$\left(\left\langle 0.70e^{2\pi i(0.56)}, 0.86e^{2\pi i(0.16)}\right\rangle, \langle 0.14, 0.46\rangle\right)$	$\left(\left\langle 0.70e^{2\pi i(0.56)}, 0.86e^{2\pi i(0.16)}\right\rangle, \langle 0.14, 0.46\rangle\right)$	$\left(\left\langle 0.78e^{2\pi i(0.23)}, 0.57e^{2\pi i(0.33)}\right\rangle, \langle 0.30, 0.22\rangle\right)$
C5	$\left(\left\langle 0.78e^{2\pi i(0.23)}, 0.57e^{2\pi i(0.33)}\right\rangle, \langle 0.30, 0.22\rangle\right)$	$\left(\left\langle 0.78e^{2\pi i(0.23)}, 0.57e^{2\pi i(0.33)}\right\rangle, \langle 0.30, 0.22\rangle\right)$	$\left(\left\langle 0.78e^{2\pi i(0.23)}, 0.57e^{2\pi i(0.33)}\right\rangle, \langle 0.30, 0.22\rangle\right)$	$\left(\left\langle 0.78e^{2\pi i(0.23)}, 0.57e^{2\pi i(0.33)}\right\rangle, \langle 0.30, 0.22\rangle\right)$	$\left(\left\langle 0.70e^{2\pi i(0.56)}, 0.86e^{2\pi i(0.16)}\right\rangle, \langle 0.14, 0.46\rangle\right)$
C6	$\left(\left\langle 0.95e^{2\pi i(0.93)}, 0.48e^{2\pi i(0.40)}\right\rangle, \langle 0.73, 0.14\rangle\right)$	$\left(\left\langle 0.95e^{2\pi i(0.93)}, 0.48e^{2\pi i(0.40)}\right\rangle, \langle 0.73, 0.14\rangle\right)$	$\left(\left\langle 0.95e^{2\pi i(0.93)}, 0.48e^{2\pi i(0.40)}\right\rangle, \langle 0.73, 0.14\rangle\right)$	$\left(\left\langle 0.95e^{2\pi i(0.93)}, 0.48e^{2\pi i(0.40)}\right\rangle, \langle 0.73, 0.14\rangle\right)$	$\left(\left\langle 0.98e^{2\pi i(0.38)}, 0.38e^{2\pi i(0.26)}\right\rangle, \langle 0.30, 0.19\rangle\right)$
C7	$\left(\left\langle 0.70e^{2\pi i(0.34)}, 0.51e^{2\pi i(0.44)}\right\rangle, \langle 0.80, 0.19\rangle\right)$	$\left(\left\langle 0.70e^{2\pi i(0.34)}, 0.51e^{2\pi i(0.44)}\right\rangle, \langle 0.80, 0.19\rangle\right)$	$\left(\left\langle 0.70e^{2\pi i(0.34)}, 0.51e^{2\pi i(0.44)}\right\rangle, \langle 0.80, 0.19\rangle\right)$	$\left(\left\langle 0.70e^{2\pi i(0.34)}, 0.51e^{2\pi i(0.44)}\right\rangle, \langle 0.80, 0.19\rangle\right)$	$\left(\left\langle 0.51e^{2\pi i(0.61)}, 0.63e^{2\pi i(0.45)}\right\rangle, \langle 0.58, 0.14\rangle\right)$
C8	$\left(\left\langle 0.49e^{2\pi i(0.75)}, 0.19e^{2\pi i(0.60)}\right\rangle, \langle 0.32, 0.10\rangle\right)$	$\left(\left\langle 0.49e^{2\pi i(0.75)}, 0.19e^{2\pi i(0.60)}\right\rangle, \langle 0.32, 0.10\rangle\right)$	$\left(\left\langle 0.49e^{2\pi i(0.75)}, 0.19e^{2\pi i(0.60)}\right\rangle, \langle 0.32, 0.10\rangle\right)$	$\left(\left\langle 0.49e^{2\pi i(0.75)}, 0.19e^{2\pi i(0.60)}\right\rangle, \langle 0.32, 0.10\rangle\right)$	$\left(\left\langle 0.51e^{2\pi i(0.44)}, 0.70e^{2\pi i(0.34)}\right\rangle, \langle 0.19, 0.80\rangle\right)$
C9	$\left(\left\langle 0.51e^{2\pi i(0.61)}, 0.63e^{2\pi i(0.45)}\right\rangle, \langle 0.58, 0.14\rangle\right)$	$\left(\left\langle 0.51e^{2\pi i(0.61)}, 0.63e^{2\pi i(0.45)}\right\rangle, \langle 0.58, 0.14\rangle\right)$	$\left(\left\langle 0.51e^{2\pi i(0.61)}, 0.63e^{2\pi i(0.45)}\right\rangle, \langle 0.58, 0.14\rangle\right)$	$\left(\left\langle 0.51e^{2\pi i(0.61)}, 0.63e^{2\pi i(0.45)}\right\rangle, \langle 0.58, 0.14\rangle\right)$	$\left(\left\langle 0.34e^{2\pi i(0.78)}, 0.67e^{2\pi i(0.24)}\right\rangle, \langle 0.10, 0.79\rangle\right)$
C10	$\left(\left\langle 0.70e^{2\pi i(0.56)}, 0.86e^{2\pi i(0.16)}\right\rangle, \langle 0.14, 0.46\rangle\right)$	$\left(\left\langle 0.67e^{2\pi i(0.24)}, 0.34e^{2\pi i(0.78)}\right\rangle, \langle 0.79, 0.10\rangle\right)$	$\left(\left\langle 0.67e^{2\pi i(0.24)}, 0.34e^{2\pi i(0.78)}\right\rangle, \langle 0.79, 0.10\rangle\right)$	$\left(\left\langle 0.67e^{2\pi i(0.24)}, 0.34e^{2\pi i(0.78)}\right\rangle, \langle 0.79, 0.10\rangle\right)$	$\left(\left\langle 0.49e^{2\pi i(0.75)}, 0.19e^{2\pi i(0.60)}\right\rangle, \langle 0.32, 0.10\rangle\right)$

(continued)

Table 12 (continued)

	A6	A7	A8	A9
C1	$\left(\left\langle 0.78e^{2\pi i(0.23)}, 0.57e^{2\pi i(0.33)}\right\rangle, \langle 0.30, 0.22\rangle\right)$	$\left(\left\langle 0.70e^{2\pi i(0.56)}, 0.86e^{2\pi i(0.16)}\right\rangle, \langle 0.14, 0.46\rangle\right)$	$\left(\left\langle 0.67e^{2\pi i(0.24)}, 0.34e^{2\pi i(0.78)}\right\rangle, \langle 0.79, 0.10\rangle\right)$	$\left(\left\langle 0.78e^{2\pi i(0.23)}, 0.57e^{2\pi i(0.33)}\right\rangle, \langle 0.30, 0.22\rangle\right)$
C2	$\left(\left\langle 0.70e^{2\pi i(0.56)}, 0.86e^{2\pi i(0.16)}\right\rangle, \langle 0.14, 0.46\rangle\right)$	$\left(\left\langle 0.90e^{2\pi i(0.70)}, 0.70e^{2\pi i(0.90)}\right\rangle, \langle 0.80, 0.20\rangle\right)$	$\left(\left\langle 0.78e^{2\pi i(0.23)}, 0.57e^{2\pi i(0.33)}\right\rangle, \langle 0.30, 0.22\rangle\right)$	$\left(\left\langle 0.82e^{2\pi i(0.41)}, 0.26e^{2\pi i(0.15)}\right\rangle, \langle 0.22, 0.34\rangle\right)$
C3	$\left(\left\langle 0.95e^{2\pi i(0.93)}, 0.48e^{2\pi i(0.40)}\right\rangle, \langle 0.73, 0.14\rangle\right)$	$\left(\left\langle 0.95e^{2\pi i(0.93)}, 0.48e^{2\pi i(0.40)}\right\rangle, \langle 0.73, 0.14\rangle\right)$	$\left(\left\langle 0.98e^{2\pi i(0.38)}, 0.38e^{2\pi i(0.26)}\right\rangle, \langle 0.30, 0.19\rangle\right)$	$\left(\left\langle 0.95e^{2\pi i(0.93)}, 0.48e^{2\pi i(0.40)}\right\rangle, \langle 0.73, 0.14\rangle\right)$
C4	$\left(\left\langle 0.82e^{2\pi i(0.41)}, 0.26e^{2\pi i(0.15)}\right\rangle, \langle 0.22, 0.34\rangle\right)$	$\left(\left\langle 0.78e^{2\pi i(0.23)}, 0.57e^{2\pi i(0.33)}\right\rangle, \langle 0.30, 0.22\rangle\right)$	$\left(\left\langle 0.90e^{2\pi i(0.70)}, 0.70e^{2\pi i(0.90)}\right\rangle, \langle 0.80, 0.20\rangle\right)$	$\left(\left\langle 0.90e^{2\pi i(0.70)}, 0.70e^{2\pi i(0.90)}\right\rangle, \langle 0.80, 0.20\rangle\right)$
C5	$\left(\left\langle 0.90e^{2\pi i(0.70)}, 0.70e^{2\pi i(0.90)}\right\rangle, \langle 0.80, 0.20\rangle\right)$	$\left(\left\langle 0.82e^{2\pi i(0.41)}, 0.26e^{2\pi i(0.15)}\right\rangle, \langle 0.22, 0.34\rangle\right)$	$\left(\left\langle 0.70e^{2\pi i(0.56)}, 0.86e^{2\pi i(0.16)}\right\rangle, \langle 0.14, 0.46\rangle\right)$	$\left(\left\langle 0.70e^{2\pi i(0.56)}, 0.86e^{2\pi i(0.16)}\right\rangle, \langle 0.14, 0.46\rangle\right)$
C6	$\left(\left\langle 0.98e^{2\pi i(0.38)}, 0.38e^{2\pi i(0.26)}\right\rangle, \langle 0.30, 0.19\rangle\right)$	$\left(\left\langle 0.98e^{2\pi i(0.38)}, 0.38e^{2\pi i(0.26)}\right\rangle, \langle 0.30, 0.19\rangle\right)$	$\left(\left\langle 0.95e^{2\pi i(0.93)}, 0.48e^{2\pi i(0.40)}\right\rangle, \langle 0.73, 0.14\rangle\right)$	$\left(\left\langle 0.98e^{2\pi i(0.38)}, 0.38e^{2\pi i(0.26)}\right\rangle, \langle 0.30, 0.19\rangle\right)$
C7	$\left(\left\langle 0.51e^{2\pi i(0.61)}, 0.63e^{2\pi i(0.45)}\right\rangle, \langle 0.58, 0.14\rangle\right)$	$\left(\left\langle 0.51e^{2\pi i(0.61)}, 0.63e^{2\pi i(0.45)}\right\rangle, \langle 0.58, 0.14\rangle\right)$	$\left(\left\langle 0.51e^{2\pi i(0.61)}, 0.63e^{2\pi i(0.45)}\right\rangle, \langle 0.58, 0.14\rangle\right)$	$\left(\left\langle 0.51e^{2\pi i(0.61)}, 0.63e^{2\pi i(0.45)}\right\rangle, \langle 0.58, 0.14\rangle\right)$
C8	$\left(\left\langle 0.63e^{2\pi i(0.45)}, 0.51e^{2\pi i(0.61)}\right\rangle, \langle 0.14, 0.58\rangle\right)$	$\left(\left\langle 0.49e^{2\pi i(0.75)}, 0.19e^{2\pi i(0.60)}\right\rangle, \langle 0.32, 0.10\rangle\right)$	$\left(\left\langle 0.49e^{2\pi i(0.75)}, 0.19e^{2\pi i(0.60)}\right\rangle, \langle 0.32, 0.10\rangle\right)$	$\left(\left\langle 0.51e^{2\pi i(0.44)}, 0.70e^{2\pi i(0.34)}\right\rangle, \langle 0.19, 0.80\rangle\right)$
C9	$\left(\left\langle 0.70e^{2\pi i(0.34)}, 0.51e^{2\pi i(0.44)}\right\rangle, \langle 0.80, 0.19\rangle\right)$	$\left(\left\langle 0.34e^{2\pi i(0.78)}, 0.67e^{2\pi i(0.24)}\right\rangle, \langle 0.10, 0.79\rangle\right)$	$\left(\left\langle 0.26e^{2\pi i(0.15)}, 0.82e^{2\pi i(0.41)}\right\rangle, \langle 0.34, 0.22\rangle\right)$	$\left(\left\langle 0.34e^{2\pi i(0.78)}, 0.67e^{2\pi i(0.24)}\right\rangle, \langle 0.10, 0.79\rangle\right)$
C10	$\left(\left\langle 0.49e^{2\pi i(0.75)}, 0.19e^{2\pi i(0.60)}\right\rangle, \langle 0.32, 0.10\rangle\right)$	$\left(\left\langle 0.51e^{2\pi i(0.44)}, 0.70e^{2\pi i(0.34)}\right\rangle, \langle 0.19, 0.80\rangle\right)$	$\left(\left\langle 0.78e^{2\pi i(0.23)}, 0.57e^{2\pi i(0.33)}\right\rangle, \langle 0.30, 0.22\rangle\right)$	$\left(\left\langle 0.49e^{2\pi i(0.75)}, 0.19e^{2\pi i(0.60)}\right\rangle, \langle 0.32, 0.10\rangle\right)$

Table 13 Transformed complex linear diophantine fuzzy decision matrix based on DM2

	A1	A2	A3	A4	A5
C1	$\left(\langle 0.91e^{2\pi i(0.71)}, 0.72e^{2\pi i(0.91)}\rangle, \langle 0.78, 0.21\rangle\right)$	$\left(\langle 0.80e^{2\pi i(0.29)}, 0.31e^{2\pi i(0.31)}\rangle, \langle 0.30, 0.23\rangle\right)$	$\left(\langle 0.74e^{2\pi i(0.98)}, 0.64e^{2\pi i(0.96)}\rangle, \langle 0.47, 0.35\rangle\right)$	$\left(\langle 0.87e^{2\pi i(0.94)}, 0.76e^{2\pi i(0.52)}\rangle, \langle 0.75, 0.22\rangle\right)$	$\left(\langle 0.87e^{2\pi i(0.94)}, 0.76e^{2\pi i(0.52)}\rangle, \langle 0.75, 0.22\rangle\right)$
C2	$\left(\langle 0.94e^{2\pi i(0.90)}, 0.72e^{2\pi i(0.65)}\rangle, \langle 0.84, 0.11\rangle\right)$	$\left(\langle 0.91e^{2\pi i(0.71)}, 0.72e^{2\pi i(0.91)}\rangle, \langle 0.78, 0.21\rangle\right)$	$\left(\langle 0.91e^{2\pi i(0.71)}, 0.72e^{2\pi i(0.91)}\rangle, \langle 0.78, 0.21\rangle\right)$	$\left(\langle 0.94e^{2\pi i(0.90)}, 0.72e^{2\pi i(0.65)}\rangle, \langle 0.84, 0.11\rangle\right)$	$\left(\langle 0.94e^{2\pi i(0.90)}, 0.72e^{2\pi i(0.65)}\rangle, \langle 0.84, 0.11\rangle\right)$
C3	$\left(\langle 0.74e^{2\pi i(0.98)}, 0.64e^{2\pi i(0.96)}\rangle, \langle 0.47, 0.35\rangle\right)$	$\left(\langle 0.68e^{2\pi i(0.45)}, 0.80e^{2\pi i(0.77)}\rangle, \langle 0.16, 0.23\rangle\right)$	$\left(\langle 0.80e^{2\pi i(0.29)}, 0.31e^{2\pi i(0.31)}\rangle, \langle 0.30, 0.23\rangle\right)$	$\left(\langle 0.74e^{2\pi i(0.98)}, 0.64e^{2\pi i(0.96)}\rangle, \langle 0.47, 0.35\rangle\right)$	$\left(\langle 0.68e^{2\pi i(0.45)}, 0.80e^{2\pi i(0.77)}\rangle, \langle 0.16, 0.23\rangle\right)$
C4	$\left(\langle 0.87e^{2\pi i(0.94)}, 0.76e^{2\pi i(0.52)}\rangle, \langle 0.75, 0.22\rangle\right)$	$\left(\langle 0.94e^{2\pi i(0.90)}, 0.72e^{2\pi i(0.65)}\rangle, \langle 0.84, 0.11\rangle\right)$	$\left(\langle 0.94e^{2\pi i(0.90)}, 0.72e^{2\pi i(0.65)}\rangle, \langle 0.84, 0.11\rangle\right)$	$\left(\langle 0.91e^{2\pi i(0.71)}, 0.72e^{2\pi i(0.91)}\rangle, \langle 0.78, 0.21\rangle\right)$	$\left(\langle 0.91e^{2\pi i(0.71)}, 0.72e^{2\pi i(0.91)}\rangle, \langle 0.78, 0.21\rangle\right)$
C5	$\left(\langle 0.80e^{2\pi i(0.29)}, 0.31e^{2\pi i(0.31)}\rangle, \langle 0.30, 0.23\rangle\right)$	$\left(\langle 0.74e^{2\pi i(0.98)}, 0.64e^{2\pi i(0.96)}\rangle, \langle 0.47, 0.35\rangle\right)$	$\left(\langle 0.68e^{2\pi i(0.45)}, 0.80e^{2\pi i(0.77)}\rangle, \langle 0.16, 0.23\rangle\right)$	$\left(\langle 0.68e^{2\pi i(0.45)}, 0.80e^{2\pi i(0.77)}\rangle, \langle 0.16, 0.23\rangle\right)$	$\left(\langle 0.80e^{2\pi i(0.29)}, 0.31e^{2\pi i(0.31)}\rangle, \langle 0.30, 0.23\rangle\right)$
C6	$\left(\langle 0.68e^{2\pi i(0.45)}, 0.80e^{2\pi i(0.77)}\rangle, \langle 0.16, 0.23\rangle\right)$	$\left(\langle 0.65e^{2\pi i(0.15)}, 0.30e^{2\pi i(0.63)}\rangle, \langle 0.70, 0.25\rangle\right)$	$\left(\langle 0.65e^{2\pi i(0.15)}, 0.30e^{2\pi i(0.63)}\rangle, \langle 0.70, 0.25\rangle\right)$	$\left(\langle 0.65e^{2\pi i(0.15)}, 0.30e^{2\pi i(0.63)}\rangle, \langle 0.70, 0.25\rangle\right)$	$\left(\langle 0.74e^{2\pi i(0.98)}, 0.64e^{2\pi i(0.96)}\rangle, \langle 0.47, 0.35\rangle\right)$
C7	$\left(\langle 0.30e^{2\pi i(0.63)}, 0.65e^{2\pi i(0.15)}\rangle, \langle 0.25, 0.70\rangle\right)$	$\left(\langle 0.33e^{2\pi i(0.64)}, 0.57e^{2\pi i(0.75)}\rangle, \langle 0.54, 0.38\rangle\right)$	$\left(\langle 0.33e^{2\pi i(0.64)}, 0.57e^{2\pi i(0.75)}\rangle, \langle 0.54, 0.38\rangle\right)$	$\left(\langle 0.33e^{2\pi i(0.64)}, 0.57e^{2\pi i(0.75)}\rangle, \langle 0.54, 0.38\rangle\right)$	$\left(\langle 0.30e^{2\pi i(0.63)}, 0.65e^{2\pi i(0.15)}\rangle, \langle 0.25, 0.70\rangle\right)$
C8	$\left(\langle 0.43e^{2\pi i(0.43)}, 0.15e^{2\pi i(0.14)}\rangle, \langle 0.45, 0.54\rangle\right)$	$\left(\langle 0.43e^{2\pi i(0.43)}, 0.15e^{2\pi i(0.14)}\rangle, \langle 0.45, 0.54\rangle\right)$	$\left(\langle 0.43e^{2\pi i(0.43)}, 0.15e^{2\pi i(0.14)}\rangle, \langle 0.45, 0.54\rangle\right)$	$\left(\langle 0.43e^{2\pi i(0.43)}, 0.15e^{2\pi i(0.14)}\rangle, \langle 0.45, 0.54\rangle\right)$	$\left(\langle 0.43e^{2\pi i(0.46)}, 0.11e^{2\pi i(0.14)}\rangle, \langle 0.48, 0.50\rangle\right)$
C9	$\left(\langle 0.75e^{2\pi i(0.30)}, 0.50e^{2\pi i(0.45)}\rangle, \langle 0.70, 0.35\rangle\right)$	$\left(\langle 0.75e^{2\pi i(0.30)}, 0.50e^{2\pi i(0.45)}\rangle, \langle 0.70, 0.35\rangle\right)$	$\left(\langle 0.75e^{2\pi i(0.30)}, 0.50e^{2\pi i(0.45)}\rangle, \langle 0.70, 0.35\rangle\right)$	$\left(\langle 0.75e^{2\pi i(0.30)}, 0.50e^{2\pi i(0.45)}\rangle, \langle 0.70, 0.35\rangle\right)$	$\left(\langle 0.33e^{2\pi i(0.64)}, 0.57e^{2\pi i(0.75)}\rangle, \langle 0.54, 0.38\rangle\right)$
C10	$\left(\langle 0.57e^{2\pi i(0.75)}, 0.33e^{2\pi i(0.64)}\rangle, \langle 0.38, 0.54\rangle\right)$	$\left(\langle 0.87e^{2\pi i(0.94)}, 0.76e^{2\pi i(0.52)}\rangle, \langle 0.75, 0.22\rangle\right)$	$\left(\langle 0.87e^{2\pi i(0.94)}, 0.76e^{2\pi i(0.52)}\rangle, \langle 0.75, 0.22\rangle\right)$	$\left(\langle 0.80e^{2\pi i(0.29)}, 0.31e^{2\pi i(0.31)}\rangle, \langle 0.30, 0.23\rangle\right)$	$\left(\langle 0.50e^{2\pi i(0.45)}, 0.75e^{2\pi i(0.30)}\rangle, \langle 0.35, 0.70\rangle\right)$

(continued)

Table 13 (continued)

	A6	A7	A8	A9
C1	$\left(\langle 0.74e^{2\pi i(0.98)}, 0.64e^{2\pi i(0.96)}\rangle, \langle 0.47, 0.35\rangle\right)$	$\left(\langle 0.87e^{2\pi i(0.94)}, 0.76e^{2\pi i(0.52)}\rangle, \langle 0.75, 0.22\rangle\right)$	$\left(\langle 0.65e^{2\pi i(0.15)}, 0.30e^{2\pi i(0.63)}\rangle, \langle 0.70, 0.25\rangle\right)$	$\left(\langle 0.68e^{2\pi i(0.45)}, 0.80e^{2\pi i(0.77)}\rangle, \langle 0.16, 0.23\rangle\right)$
C2	$\left(\langle 0.91e^{2\pi i(0.71)}, 0.72e^{2\pi i(0.91)}\rangle, \langle 0.78, 0.21\rangle\right)$	$\left(\langle 0.91e^{2\pi i(0.71)}, 0.72e^{2\pi i(0.91)}\rangle, \langle 0.78, 0.21\rangle\right)$	$\left(\langle 0.87e^{2\pi i(0.94)}, 0.76e^{2\pi i(0.52)}\rangle, \langle 0.75, 0.22\rangle\right)$	$\left(\langle 0.87e^{2\pi i(0.94)}, 0.76e^{2\pi i(0.52)}\rangle, \langle 0.75, 0.22\rangle\right)$
C3	$\left(\langle 0.87e^{2\pi i(0.94)}, 0.76e^{2\pi i(0.52)}\rangle, \langle 0.75, 0.22\rangle\right)$	$\left(\langle 0.87e^{2\pi i(0.94)}, 0.76e^{2\pi i(0.52)}\rangle, \langle 0.75, 0.22\rangle\right)$	$\left(\langle 0.91e^{2\pi i(0.71)}, 0.72e^{2\pi i(0.91)}\rangle, \langle 0.78, 0.21\rangle\right)$	$\left(\langle 0.91e^{2\pi i(0.71)}, 0.72e^{2\pi i(0.91)}\rangle, \langle 0.78, 0.21\rangle\right)$
C4	$\left(\langle 0.94e^{2\pi i(0.90)}, 0.72e^{2\pi i(0.65)}\rangle, \langle 0.84, 0.11\rangle\right)$	$\left(\langle 0.94e^{2\pi i(0.90)}, 0.72e^{2\pi i(0.65)}\rangle, \langle 0.84, 0.11\rangle\right)$	$\left(\langle 0.94e^{2\pi i(0.90)}, 0.72e^{2\pi i(0.65)}\rangle, \langle 0.84, 0.11\rangle\right)$	$\left(\langle 0.94e^{2\pi i(0.90)}, 0.72e^{2\pi i(0.65)}\rangle, \langle 0.84, 0.11\rangle\right)$
C5	$\left(\langle 0.80e^{2\pi i(0.29)}, 0.31e^{2\pi i(0.31)}\rangle, \langle 0.30, 0.23\rangle\right)$	$\left(\langle 0.80e^{2\pi i(0.29)}, 0.31e^{2\pi i(0.31)}\rangle, \langle 0.30, 0.23\rangle\right)$	$\left(\langle 0.68e^{2\pi i(0.45)}, 0.80e^{2\pi i(0.77)}\rangle, \langle 0.16, 0.23\rangle\right)$	$\left(\langle 0.74e^{2\pi i(0.98)}, 0.64e^{2\pi i(0.96)}\rangle, \langle 0.47, 0.35\rangle\right)$
C6	$\left(\langle 0.68e^{2\pi i(0.45)}, 0.80e^{2\pi i(0.77)}\rangle, \langle 0.16, 0.23\rangle\right)$	$\left(\langle 0.74e^{2\pi i(0.98)}, 0.64e^{2\pi i(0.96)}\rangle, \langle 0.47, 0.35\rangle\right)$	$\left(\langle 0.80e^{2\pi i(0.29)}, 0.31e^{2\pi i(0.31)}\rangle, \langle 0.30, 0.23\rangle\right)$	$\left(\langle 0.80e^{2\pi i(0.29)}, 0.31e^{2\pi i(0.31)}\rangle, \langle 0.30, 0.23\rangle\right)$
C7	$\left(\langle 0.30e^{2\pi i(0.63)}, 0.65e^{2\pi i(0.15)}\rangle, \langle 0.25, 0.70\rangle\right)$	$\left(\langle 0.30e^{2\pi i(0.63)}, 0.65e^{2\pi i(0.15)}\rangle, \langle 0.25, 0.70\rangle\right)$	$\left(\langle 0.33e^{2\pi i(0.64)}, 0.57e^{2\pi i(0.75)}\rangle, \langle 0.54, 0.38\rangle\right)$	$\left(\langle 0.30e^{2\pi i(0.63)}, 0.65e^{2\pi i(0.15)}\rangle, \langle 0.25, 0.70\rangle\right)$
C8	$\left(\langle 0.43e^{2\pi i(0.43)}, 0.15e^{2\pi i(0.14)}\rangle, \langle 0.45, 0.54\rangle\right)$	$\left(\langle 0.43e^{2\pi i(0.43)}, 0.15e^{2\pi i(0.14)}\rangle, \langle 0.45, 0.54\rangle\right)$	$\left(\langle 0.43e^{2\pi i(0.43)}, 0.15e^{2\pi i(0.14)}\rangle, \langle 0.45, 0.54\rangle\right)$	$\left(\langle 0.43e^{2\pi i(0.43)}, 0.15e^{2\pi i(0.14)}\rangle, \langle 0.45, 0.54\rangle\right)$
C9	$\left(\langle 0.75e^{2\pi i(0.30)}, 0.50e^{2\pi i(0.45)}\rangle, \langle 0.70, 0.35\rangle\right)$	$\left(\langle 0.33e^{2\pi i(0.64)}, 0.57e^{2\pi i(0.75)}\rangle, \langle 0.54, 0.38\rangle\right)$	$\left(\langle 0.64e^{2\pi i(0.96)}, 0.74e^{2\pi i(0.98)}\rangle, \langle 0.35, 0.47\rangle\right)$	$\left(\langle 0.33e^{2\pi i(0.64)}, 0.57e^{2\pi i(0.75)}\rangle, \langle 0.54, 0.38\rangle\right)$
C10	$\left(\langle 0.57e^{2\pi i(0.75)}, 0.33e^{2\pi i(0.64)}\rangle, \langle 0.38, 0.54\rangle\right)$	$\left(\langle 0.74e^{2\pi i(0.98)}, 0.64e^{2\pi i(0.96)}\rangle, \langle 0.47, 0.35\rangle\right)$	$\left(\langle 0.50e^{2\pi i(0.45)}, 0.75e^{2\pi i(0.30)}\rangle, \langle 0.35, 0.70\rangle\right)$	$\left(\langle 0.50e^{2\pi i(0.45)}, 0.75e^{2\pi i(0.30)}\rangle, \langle 0.35, 0.70\rangle\right)$

Table 14 Transformed complex linear diophantine fuzzy decision matrix based on DM3

	A1	A2	A3	A4	A5
C1	$\left(\langle 0.73e^{2\pi i(0.72)}, 0.54e^{2\pi i(0.36)}\rangle, \langle 0.53, 0.11\rangle\right)$	$\left(\langle 0.69e^{2\pi i(0.55)}, 0.85e^{2\pi i(0.19)}\rangle, \langle 0.15, 0.45\rangle\right)$	$\left(\langle 0.64e^{2\pi i(0.12)}, 0.32e^{2\pi i(0.62)}\rangle, \langle 0.75, 0.25\rangle\right)$	$\left(\langle 0.64e^{2\pi i(0.12)}, 0.32e^{2\pi i(0.62)}\rangle, \langle 0.75, 0.25\rangle\right)$	$\left(\langle 0.64e^{2\pi i(0.12)}, 0.32e^{2\pi i(0.62)}\rangle, \langle 0.75, 0.25\rangle\right)$
C2	$\left(\langle 0.64e^{2\pi i(0.12)}, 0.32e^{2\pi i(0.62)}\rangle, \langle 0.75, 0.25\rangle\right)$	$\left(\langle 0.60e^{2\pi i(0.48)}, 0.50e^{2\pi i(0.60)}\rangle, \langle 0.20, 0.61\rangle\right)$	$\left(\langle 0.60e^{2\pi i(0.48)}, 0.50e^{2\pi i(0.60)}\rangle, \langle 0.20, 0.61\rangle\right)$	$\left(\langle 0.60e^{2\pi i(0.48)}, 0.50e^{2\pi i(0.60)}\rangle, \langle 0.20, 0.61\rangle\right)$	$\left(\langle 0.69e^{2\pi i(0.55)}, 0.85e^{2\pi i(0.19)}\rangle, \langle 0.15, 0.45\rangle\right)$
C3	$\left(\langle 0.96e^{2\pi i(0.91)}, 0.46e^{2\pi i(0.41)}\rangle, \langle 0.61, 0.19\rangle\right)$	$\left(\langle 0.96e^{2\pi i(0.91)}, 0.46e^{2\pi i(0.41)}\rangle, \langle 0.61, 0.19\rangle\right)$	$\left(\langle 0.92e^{2\pi i(0.88)}, 0.75e^{2\pi i(0.28)}\rangle, \langle 0.20, 0.29\rangle\right)$	$\left(\langle 0.92e^{2\pi i(0.88)}, 0.75e^{2\pi i(0.28)}\rangle, \langle 0.20, 0.29\rangle\right)$	$\left(\langle 0.92e^{2\pi i(0.88)}, 0.75e^{2\pi i(0.28)}\rangle, \langle 0.20, 0.29\rangle\right)$
C4	$\left(\langle 0.85e^{2\pi i(0.28)}, 0.33e^{2\pi i(0.35)}\rangle, \langle 0.25, 0.20\rangle\right)$	$\left(\langle 0.92e^{2\pi i(0.88)}, 0.75e^{2\pi i(0.28)}\rangle, \langle 0.20, 0.29\rangle\right)$	$\left(\langle 0.85e^{2\pi i(0.28)}, 0.33e^{2\pi i(0.35)}\rangle, \langle 0.25, 0.20\rangle\right)$	$\left(\langle 0.85e^{2\pi i(0.28)}, 0.33e^{2\pi i(0.35)}\rangle, \langle 0.25, 0.20\rangle\right)$	$\left(\langle 0.85e^{2\pi i(0.28)}, 0.33e^{2\pi i(0.35)}\rangle, \langle 0.25, 0.20\rangle\right)$
C5	$\left(\langle 0.60e^{2\pi i(0.48)}, 0.50e^{2\pi i(0.60)}\rangle, \langle 0.20, 0.61\rangle\right)$	$\left(\langle 0.50e^{2\pi i(0.45)}, 0.75e^{2\pi i(0.30)}\rangle, \langle 0.18, 0.78\rangle\right)$	$\left(\langle 0.65e^{2\pi i(0.45)}, 0.55e^{2\pi i(0.30)}\rangle, \langle 0.18, 0.24\rangle\right)$	$\left(\langle 0.65e^{2\pi i(0.45)}, 0.55e^{2\pi i(0.30)}\rangle, \langle 0.18, 0.24\rangle\right)$	$\left(\langle 0.60e^{2\pi i(0.48)}, 0.50e^{2\pi i(0.60)}\rangle, \langle 0.20, 0.61\rangle\right)$
C6	$\left(\langle 0.92e^{2\pi i(0.88)}, 0.75e^{2\pi i(0.28)}\rangle, \langle 0.20, 0.29\rangle\right)$	$\left(\langle 0.85e^{2\pi i(0.28)}, 0.33e^{2\pi i(0.35)}\rangle, \langle 0.25, 0.20\rangle\right)$	$\left(\langle 0.73e^{2\pi i(0.72)}, 0.54e^{2\pi i(0.36)}\rangle, \langle 0.53, 0.11\rangle\right)$	$\left(\langle 0.73e^{2\pi i(0.72)}, 0.54e^{2\pi i(0.36)}\rangle, \langle 0.53, 0.11\rangle\right)$	$\left(\langle 0.73e^{2\pi i(0.72)}, 0.54e^{2\pi i(0.36)}\rangle, \langle 0.53, 0.11\rangle\right)$
C7	$\left(\langle 0.29e^{2\pi i(0.55)}, 0.45e^{2\pi i(0.70)}\rangle, \langle 0.15, 0.35\rangle\right)$	$\left(\langle 0.29e^{2\pi i(0.55)}, 0.45e^{2\pi i(0.70)}\rangle, \langle 0.15, 0.35\rangle\right)$	$\left(\langle 0.29e^{2\pi i(0.55)}, 0.45e^{2\pi i(0.70)}\rangle, \langle 0.15, 0.35\rangle\right)$	$\left(\langle 0.29e^{2\pi i(0.55)}, 0.45e^{2\pi i(0.70)}\rangle, \langle 0.15, 0.35\rangle\right)$	$\left(\langle 0.29e^{2\pi i(0.55)}, 0.45e^{2\pi i(0.70)}\rangle, \langle 0.15, 0.35\rangle\right)$
C8	$\left(\langle 0.69e^{2\pi i(0.55)}, 0.85e^{2\pi i(0.19)}\rangle, \langle 0.15, 0.45\rangle\right)$	$\left(\langle 0.73e^{2\pi i(0.72)}, 0.54e^{2\pi i(0.36)}\rangle, \langle 0.53, 0.11\rangle\right)$	$\left(\langle 0.69e^{2\pi i(0.55)}, 0.85e^{2\pi i(0.19)}\rangle, \langle 0.15, 0.45\rangle\right)$	$\left(\langle 0.69e^{2\pi i(0.55)}, 0.85e^{2\pi i(0.19)}\rangle, \langle 0.15, 0.45\rangle\right)$	$\left(\langle 0.49e^{2\pi i(0.75)}, 0.19e^{2\pi i(0.60)}\rangle, \langle 0.32, 0.10\rangle\right)$
C9	$\left(\langle 0.75e^{2\pi i(0.30)}, 0.50e^{2\pi i(0.45)}\rangle, \langle 0.78, 0.18\rangle\right)$	$\left(\langle 0.32e^{2\pi i(0.62)}, 0.64e^{2\pi i(0.12)}\rangle, \langle 0.25, 0.75\rangle\right)$	$\left(\langle 0.21e^{2\pi i(0.24)}, 0.41e^{2\pi i(0.47)}\rangle, \langle 0.48, 0.44\rangle\right)$	$\left(\langle 0.21e^{2\pi i(0.24)}, 0.41e^{2\pi i(0.47)}\rangle, \langle 0.48, 0.44\rangle\right)$	$\left(\langle 0.51e^{2\pi i(0.61)}, 0.63e^{2\pi i(0.45)}\rangle, \langle 0.58, 0.14\rangle\right)$
C10	$\left(\langle 0.41e^{2\pi i(0.47)}, 0.21e^{2\pi i(0.24)}\rangle, \langle 0.44, 0.48\rangle\right)$	$\left(\langle 0.41e^{2\pi i(0.47)}, 0.21e^{2\pi i(0.24)}\rangle, \langle 0.44, 0.48\rangle\right)$	$\left(\langle 0.96e^{2\pi i(0.91)}, 0.46e^{2\pi i(0.41)}\rangle, \langle 0.61, 0.19\rangle\right)$	$\left(\langle 0.96e^{2\pi i(0.91)}, 0.46e^{2\pi i(0.41)}\rangle, \langle 0.61, 0.19\rangle\right)$	$\left(\langle 0.96e^{2\pi i(0.91)}, 0.46e^{2\pi i(0.41)}\rangle, \langle 0.61, 0.19\rangle\right)$

(continued)

Table 14 (continued)

	A6	A7	A8	A9
C1	$\left(\begin{array}{c}\langle 0.69e^{2\pi i(0.55)}, 0.85e^{2\pi i(0.19)}\rangle,\\ \langle 0.15, 0.45\rangle\end{array}\right)$	$\left(\begin{array}{c}\langle 0.69e^{2\pi i(0.55)}, 0.85e^{2\pi i(0.19)}\rangle,\\ \langle 0.15, 0.45\rangle\end{array}\right)$	$\left(\begin{array}{c}\langle 0.64e^{2\pi i(0.12)}, 0.32e^{2\pi i(0.62)}\rangle,\\ \langle 0.75, 0.25\rangle\end{array}\right)$	$\left(\begin{array}{c}\langle 0.69e^{2\pi i(0.55)}, 0.85e^{2\pi i(0.19)}\rangle,\\ \langle 0.15, 0.45\rangle\end{array}\right)$
C2	$\left(\begin{array}{c}\langle 0.60e^{2\pi i(0.48)}, 0.50e^{2\pi i(0.60)}\rangle,\\ \langle 0.20, 0.61\rangle\end{array}\right)$	$\left(\begin{array}{c}\langle 0.64e^{2\pi i(0.12)}, 0.32e^{2\pi i(0.62)}\rangle,\\ \langle 0.75, 0.25\rangle\end{array}\right)$	$\left(\begin{array}{c}\langle 0.60e^{2\pi i(0.48)}, 0.50e^{2\pi i(0.60)}\rangle,\\ \langle 0.20, 0.61\rangle\end{array}\right)$	$\left(\begin{array}{c}\langle 0.64e^{2\pi i(0.12)}, 0.32e^{2\pi i(0.62)}\rangle,\\ \langle 0.75, 0.25\rangle\end{array}\right)$
C3	$\left(\begin{array}{c}\langle 0.96e^{2\pi i(0.91)}, 0.46e^{2\pi i(0.41)}\rangle,\\ \langle 0.61, 0.19\rangle\end{array}\right)$	$\left(\begin{array}{c}\langle 0.96e^{2\pi i(0.91)}, 0.46e^{2\pi i(0.41)}\rangle,\\ \langle 0.61, 0.19\rangle\end{array}\right)$	$\left(\begin{array}{c}\langle 0.96e^{2\pi i(0.91)}, 0.46e^{2\pi i(0.41)}\rangle,\\ \langle 0.61, 0.19\rangle\end{array}\right)$	$\left(\begin{array}{c}\langle 0.96e^{2\pi i(0.91)}, 0.46e^{2\pi i(0.41)}\rangle,\\ \langle 0.61, 0.19\rangle\end{array}\right)$
C4	$\left(\begin{array}{c}\langle 0.92e^{2\pi i(0.88)}, 0.75e^{2\pi i(0.28)}\rangle,\\ \langle 0.20, 0.29\rangle\end{array}\right)$	$\left(\begin{array}{c}\langle 0.92e^{2\pi i(0.88)}, 0.75e^{2\pi i(0.28)}\rangle,\\ \langle 0.20, 0.29\rangle\end{array}\right)$	$\left(\begin{array}{c}\langle 0.92e^{2\pi i(0.88)}, 0.75e^{2\pi i(0.28)}\rangle,\\ \langle 0.20, 0.29\rangle\end{array}\right)$	$\left(\begin{array}{c}\langle 0.92e^{2\pi i(0.88)}, 0.75e^{2\pi i(0.28)}\rangle,\\ \langle 0.20, 0.29\rangle\end{array}\right)$
C5	$\left(\begin{array}{c}\langle 0.50e^{2\pi i(0.45)}, 0.75e^{2\pi i(0.30)}\rangle,\\ \langle 0.18, 0.78\rangle\end{array}\right)$	$\left(\begin{array}{c}\langle 0.50e^{2\pi i(0.45)}, 0.75e^{2\pi i(0.30)}\rangle,\\ \langle 0.18, 0.78\rangle\end{array}\right)$	$\left(\begin{array}{c}\langle 0.65e^{2\pi i(0.15)}, 0.30e^{2\pi i(0.63)}\rangle,\\ \langle 0.70, 0.25\rangle\end{array}\right)$	$\left(\begin{array}{c}\langle 0.50e^{2\pi i(0.45)}, 0.75e^{2\pi i(0.30)}\rangle,\\ \langle 0.18, 0.78\rangle\end{array}\right)$
C6	$\left(\begin{array}{c}\langle 0.85e^{2\pi i(0.28)}, 0.33e^{2\pi i(0.35)}\rangle,\\ \langle 0.25, 0.20\rangle\end{array}\right)$	$\left(\begin{array}{c}\langle 0.85e^{2\pi i(0.28)}, 0.33e^{2\pi i(0.35)}\rangle,\\ \langle 0.25, 0.20\rangle\end{array}\right)$	$\left(\begin{array}{c}\langle 0.85e^{2\pi i(0.28)}, 0.33e^{2\pi i(0.35)}\rangle,\\ \langle 0.25, 0.20\rangle\end{array}\right)$	$\left(\begin{array}{c}\langle 0.85e^{2\pi i(0.28)}, 0.33e^{2\pi i(0.35)}\rangle,\\ \langle 0.25, 0.20\rangle\end{array}\right)$
C7	$\left(\begin{array}{c}\langle 0.29e^{2\pi i(0.55)}, 0.45e^{2\pi i(0.70)}\rangle,\\ \langle 0.15, 0.35\rangle\end{array}\right)$	$\left(\begin{array}{c}\langle 0.29e^{2\pi i(0.55)}, 0.45e^{2\pi i(0.70)}\rangle,\\ \langle 0.15, 0.35\rangle\end{array}\right)$	$\left(\begin{array}{c}\langle 0.29e^{2\pi i(0.55)}, 0.45e^{2\pi i(0.70)}\rangle,\\ \langle 0.15, 0.35\rangle\end{array}\right)$	$\left(\begin{array}{c}\langle 0.29e^{2\pi i(0.55)}, 0.45e^{2\pi i(0.70)}\rangle,\\ \langle 0.15, 0.35\rangle\end{array}\right)$
C8	$\left(\begin{array}{c}\langle 0.73e^{2\pi i(0.72)}, 0.548e^{2\pi i(0.36)}\rangle,\\ \langle 0.53, 0.11\rangle\end{array}\right)$	$\left(\begin{array}{c}\langle 0.73e^{2\pi i(0.72)}, 0.54e^{2\pi i(0.36)}\rangle,\\ \langle 0.53, 0.11\rangle\end{array}\right)$	$\left(\begin{array}{c}\langle 0.73e^{2\pi i(0.72)}, 0.54e^{2\pi i(0.36)}\rangle,\\ \langle 0.53, 0.11\rangle\end{array}\right)$	$\left(\begin{array}{c}\langle 0.73e^{2\pi i(0.72)}, 0.54e^{2\pi i(0.36)}\rangle,\\ \langle 0.53, 0.11\rangle\end{array}\right)$
C9	$\left(\begin{array}{c}\langle 0.51e^{2\pi i(0.61)}, 0.63e^{2\pi i(0.45)}\rangle,\\ \langle 0.58, 0.14\rangle\end{array}\right)$	$\left(\begin{array}{c}\langle 0.21e^{2\pi i(0.24)}, 0.41e^{2\pi i(0.47)}\rangle,\\ \langle 0.48, 0.44\rangle\end{array}\right)$	$\left(\begin{array}{c}\langle 0.75e^{2\pi i(0.30)}, 0.50e^{2\pi i(0.45)}\rangle,\\ \langle 0.78, 0.18\rangle\end{array}\right)$	$\left(\begin{array}{c}\langle 0.51e^{2\pi i(0.61)}, 0.63e^{2\pi i(0.45)}\rangle,\\ \langle 0.58, 0.14\rangle\end{array}\right)$
C10	$\left(\begin{array}{c}\langle 0.64e^{2\pi i(0.12)}, 0.32e^{2\pi i(0.62)}\rangle,\\ \langle 0.75, 0.25\rangle\end{array}\right)$	$\left(\begin{array}{c}\langle 0.85e^{2\pi i(0.28)}, 0.33e^{2\pi i(0.35)}\rangle,\\ \langle 0.25, 0.20\rangle\end{array}\right)$	$\left(\begin{array}{c}\langle 0.69e^{2\pi i(0.55)}, 0.85e^{2\pi i(0.19)}\rangle,\\ \langle 0.15, 0.45\rangle\end{array}\right)$	$\left(\begin{array}{c}\langle 0.60e^{2\pi i(0.48)}, 0.50e^{2\pi i(0.60)}\rangle,\\ \langle 0.20, 0.61\rangle\end{array}\right)$

References

1. Abdullah M, Adhikary S, Bhattacharya S, Hazra S, Ganguly A, Nanda S, Rajak P (2024) E-waste in the environment: Unveiling the sources, carcinogenic links, and sustainable management strategies. Toxicology 509:153981
2. Ghulam ST, Abushammala H (2023) Challenges and opportunities in the management of electronic waste and its impact on human health and environment. Sustainability 2023(15):1837
3. Awasthi AK, Awasthi MK, Mishra S, Sarsaiya S, Pandey AK (2022) Evaluation of E-waste materials linked potential consequences to environment in India. Environ Technol Innov 28:102477
4. Barman H, Deb D (2023) Compliance of E-waste (management) rules in the government of Assam offices at Dibrugarh, India. Int J Environ Waste Manage 32(3):315–323
5. Rai S, Gurung A, Sharma HB, Ranjan VP, Cheela VRS (2024) Sustainable solid waste management challenges in hill cities of developing countries: insights from eastern Himalayan smart cities of Sikkim, India. Waste Manag Bull 2(2):1–18
6. Dutta D, Goel S (2021) Understanding the gap between formal and informal e-waste recycling facilities in India. Waste Manage 125:163–171
7. Dasgupta D, Majumder S, Adhikari J, Ghosh P, Purchase D, Garelick H, Chatterjee D (2023) Environmental impact of e-waste management in Indian microscale informal sectors. Environ Sci Pollut Res 30(11):29581–29597
8. Panchal R, Singh A, Diwan H (2021) Economic potential of recycling e-waste in India and its impact on import of materials. Resour Policy 74:102264
9. Sengupta D, Ilankoon IMSK, Kang KD, Chong MN (2022) Circular economy and household e-waste management in India: Integration of formal and informal sectors. Miner Eng 184:107661
10. Xia J, Ghahreman A (2024) Sustainable technologies for the recycling and upcycling of precious metals from e-waste. Sci Total Environ 916:170154
11. Kumar A, Wasan P, Luthra S, Dixit G (2020) Development of a framework for selecting a sustainable location of waste electrical and electronic equipment recycling plant in emerging economies. J Clean Prod 277:122645
12. Majid M, Mir BA (2021) Landfill site selection using GIS based multi criteria evaluation technique: a case study of Srinagar city, India. Environ Chall 3:100031
13. Herva M, Roca E (2013) Review of combined approaches and multi-criteria analysis for corporate environmental evaluation. J Clean Prod 39:355–371
14. Boix-Cots D, Pardo-Bosch F, Pujadas P (2023) A systematic review on multi-criteria group decision-making methods based on weights: analysis and classification scheme. Inf Fusion 96:16–36
15. Luqman A, Siddique S, Shahzadi G, Akram M (2024) Multi-criteria group decision making through full multiplicative form under hesitant fermatean fuzzy environment. Granul Comput 9(2):51
16. Fan J, Jin Z, Wu M (2025) The opinion dynamics model for group decision making with probabilistic uncertain linguistic information. Complex Intell Syst 11(5):1–22
17. Zadeh LA (1965) Fuzzy sets. Inf control 8(3):338–353
18. Gehrke M, Walker C, Walker E (2001) Some basic theory of interval-valued fuzzy sets. In: Proceedings joint 9th IFSA world congress and 20th NAFIPS international conference (Cat. No. 01TH8569) (Vol. 3, pp. 1332–1336). IEEE
19. Atanassov KT (2012) On intuitionistic fuzzy sets theory, vol 283. Springer
20. Garg H (2021) Pythagorean fuzzy sets. Springer, Singapore
21. Basu R, Chakraborty S, Saha AK (2025) Novel q-Rung Orthopair Fuzzy distance based similarity measure and score function in real life decision making. Eng Appl Artif Intell 147:110348
22. Kahraman C (ed) (2008) Fuzzy multi-criteria decision making: theory and applications with recent developments, vol 16. Springer Science & Business Media
23. Kahraman C, Öztayşi B, Çevik Onar S (2016) A comprehensive literature review of 50 years of fuzzy set theory. Int J Comput Intell Syst 9(sup1):3–24

24. Peng X, Yuan H, Yang Y (2017) Pythagorean fuzzy information measures and their applications. Int J Intell Syst 32(10):991–1029
25. Yager RR (2013) Pythagorean membership grades in multicriteria decision making. IEEE Trans Fuzzy Syst 22(4):958–965
26. Liao H, Li Z, Zeng XJ, Liu W (2017) A comparison of distinct consensus measures for group decision making with intuitionistic fuzzy preference relations. Int J Comput Intell Syst 10(1):456–469
27. Tokede O (2025) Application of intuitionistic fuzzy set in social life cycle impact assessment. Int J Life Cycle Assess 30(6):1055–1077
28. Ejegwa PA (2020) New similarity measures for Pythagorean fuzzy sets with applications. Int J Fuzzy Comput Model 3(1):75–94
29. Pinar A, Boran FE (2022) A novel distance measure on q-rung picture fuzzy sets and its application to decision making and classification problems. Artif Intell Rev 55(2):1317–1350
30. Ghoushchi SJ, Vahabzadeh S, Pamucar D (2024) Applying hesitant q-rung orthopair fuzzy sets to evaluate uncertainty in subsidence causes factors. Heliyon 10(8)
31. Riaz M, Hashmi MR (2019) Linear Diophantine fuzzy set and its applications towards multi-attribute decision-making problems. J Intell Fuzzy Syst 37(4):5417–5439
32. Ramot D, Milo R, Friedman M, Kandel A (2002) Complex fuzzy sets. IEEE Trans Fuzzy Syst 10(2):171–186
33. Zeeshan M, Khan M (2022) Complex fuzzy sets with applications in decision making. Iran J Fuzzy Syst 19(4):147–163
34. Jin H, Hussain A, Ullah K, Javed A (2022) Novel complex Pythagorean fuzzy sets under Aczel-Alsina operators and their application in multi-attribute decision making. Symmetry 15(1):68
35. Bi L, Dai S, Hu B (2018) Complex fuzzy geometric aggregation operators. Symmetry 10(7):251
36. Kamacı H (2022) Complex linear Diophantine fuzzy sets and their cosine similarity measures with applications. Complex & Intelligent Systems 8(2):1281–1305
37. Detyniecki, M. (2001). Fundamentals on aggregation operators. *AGOP International Summer School on Aggregation Operators, Asturias.*
38. Ali Z, Mahmood T, Ullah K, Pamucar D, Cirovic G (2021) Power aggregation operators based on t-norm and t-conorm under the complex intuitionistic fuzzy soft settings and their application in multi-attribute decision making. Symmetry 13(11):1986
39. Riaz M, Farid HMA, Karaaslan F (2025) Linear Diophantine fuzzy aggregation operators with multi-criteria decision-making. J Comput Cogn Eng 4(1):24–35
40. Zia MD, Yousafzai F, Abdullah S, Hila K (2024) Complex linear Diophantine fuzzy sets and their applications in multi-attribute decision making. Eng Appl Artif Intell 132:107953
41. Mustafa AW, Bashir Z, Ali J, Syam MI (2025) Complex linear Diophantine fuzzy Dombi prioritized operators-based MULTIMOORA approach with applications to sustainable energy planning. Sci Rep 15(1):29051
42. Maden A (2024) Selection of a sustainable location for waste electrical and electronic equipment recycling plant using entropy and evaluation based on distance from average solution. Waste Manage Res 42(12):1119–1130
43. Garg H, Rani D (2019) Some results on information measures for complex intuitionistic fuzzy sets. Int J Intell Syst 34(10):2319–2363
44. Choudhury S, Saha AK, Bhowmik D, Simic V (2024) A pythagorean fuzzy Einstein weighted averaging operator-based MCDM model for the selection of sustainable urban drainage system. Environ Dev Sustain:1–31
45. Garg H (2022) Q-Rung orthopair fuzzy sets. Springer, New York, NY, USA
46. Klement EP, Mesiar R, Pap E (2004). Triangular norms: position paper I: basic analytical and algebraic properties. Fuzzy Sets Syst 143(1):5–26
47. Keshavarz-Ghorabaee M, Amiri M, Zavadskas EK, Turskis Z, Antucheviciene J (2018) An extended step-wise weight assessment ratio analysis with symmetric interval type-2 fuzzy sets for determining the subjective weights of criteria in multi-criteria decision-making problems. Symmetry 10(4):91

48. Hashemkhani Zolfani S, Yazdani M, Zavadskas EK (2018) An extended stepwise weight assessment ratio analysis (SWARA) method for improving criteria prioritization process. Soft Comput 22(22):7399–7405
49. Ghorabaee MK, Zavadskas EK, Turskis Z, Antucheviciene J (2016) A new combinative distance-based assessment (CODAS) method for multi-criteria decision-making. Econ Comput Econ Cybern Stud Res 50(3):25–44
50. Ayyildiz E, Erdogan M (2024) A fermatean fuzzy SWARA-TOPSIS methodology based on SCOR model for autonomous vehicle parking lot selection. Appl Soft Comput 166:112198
51. Peng JJ, Chen XG, Tan H, Sun JY, Long QQ, Jiang LL (2024) A heterogeneous picture fuzzy SWARA-MARCOS evaluation framework based on a novel cross-entropy measure. Int J Syst Sci 55(8):1528–1552
52. Alrasheedi M, Mardani A, Mishra AR, Rani P, Loganathan N (2022) An extended framework to evaluate sustainable suppliers in manufacturing companies using a new pythagorean fuzzy entropy-SWARA-WASPAS decision-making approach. J Enterp Inf Manag 35(2):333–357

A Multi-view Learning Framework for Obfuscation-Resilient XSS Detection

Mohammad Amin Seifi, Negin Mousavi, Behnam Yousefimehr, Ali Daneshpour, Ayda Soltani, Tohid Asheghimehmandari, and Zhuping Sheng

Abstract XSS is still everywhere and hard to pin down, thanks to layered encodings, hidden control characters, and keywords chopped into pieces. Signature and regex matching tends to miss polymorphic attacks, while many DL approaches can't see both the small details and the long-distance links at the same time. We unveil XSS-Shield, an all-in-one detection framework combining multi-view canonicalization and dual-branch neural networks, together with a public dataset of obfuscated XSS. Our release builds on PayloadAllTheThings, appending systematically created variants together with a curated set of benign payloads. During preprocessing, we step through successive decodings to create multiple canonical views of a payload and compute hand-built obfuscation features. Our study compares a character-level CNN focused on n-grams with a heads-only Transformer encoder aimed at broad, long-range context. Both architectures handle multi-class obfuscation typing, with

M. A. Seifi · B. Yousefimehr · A. Daneshpour
Department of Computer Science, Amirkabir University of Technology, Tehran, Iran
e-mail: seifi@aut.ac.ir

B. Yousefimehr
e-mail: Behnam.y2010@aut.ac.ir

A. Daneshpour
e-mail: alidaneshpour@aut.ac.ir

N. Mousavi
Department of Computer Engineering, Amirkabir University of Technology, Tehran, Iran
e-mail: ngn.m2000@aut.ac.ir

A. Soltani
Department of Civil, Chemical, Environmental and Materials Engineering, University of Bologna, Bologna, Italy
e-mail: ayda.soltani@studio.unibo.it

T. Asheghimehmandari (✉) · Z. Sheng
Civil and Environmental Engineering Department, Morgan State University, Baltimore, United States
e-mail: Tohid.mehmandari@morgan.edu

Z. Sheng
e-mail: Zhuping.sheng@morgan.edu

A. Azizi et al. (eds.), *Advanced Decision-Making Under Uncertainty*, Emerging Trends in Mechatronics, https://doi.org/10.1007/978-981-95-8696-7_9

a second head that gives a simple benign/malicious verdict. Evaluated on our data, CharCNN delivered 0.9958 macro-F1 and the Transformer 0.9905, supporting the case for lightweight models when paired with domain-aware preprocessing. We also introduce a task-specific XAI toolkit that learns execution and hidden patterns, suggesting which characters/tokens are mostly affect output label. We're publishing the dataset, code, and trained models so others can reproduce our results and use them in IPS and WAFs.

Keywords Cross-Site scripting (XSS) · Deep learning · Transformers · CharCNN · Convolutional neural networks · Explainable AI

1 Introduction

As the internet structures day-to-day activity, from commerce and banking to news, web applications are now standard fare. Dependence shows up in the services people use most: online banking, e-commerce, and social networking [1, 2]. However, the expansion of these applications has brought new ways for attackers to get in, making them frequent targets of cyberattacks. In short, we have to secure them, and researchers are spending a lot of time on it [3]. Of the many ways apps can be attacked, Cross-Site Scripting is still very common, which can lead to data exfiltration, session hijacking, and changing what users see without permission.

Despite years of work on XSS, most detection still leans on regular expressions as the main tool. Regex operates as a finite automaton and thus falls short when modeling the complex structure of XSS payloads [4]. Detectors struggle once payloads arrive in sequences and are masked by deep URL/HTML re-encoding, case mangling, null bytes, and zero-width characters. These edits are meant to sidestep static filters, as they proliferate, sustaining manual rules is both costly and ineffective.

In contrast to manual rules, character/byte-level deep learning can capture intricate attack patterns with no feature crafting [5]. Typical character/byte pipelines combine embedding layers [6] with 1-D convolutional filters over characters [7]. Despite their strengths, Char-CNNs mainly detect local n-grams and often fail to model the non-local dependencies required to spot payloads where keywords are broken up by inserted characters [8, 9]. To overcome these limitations and design a detection system resilient to the diverse obfuscation strategies employed in modern XSS attacks, this paper introduces a novel framework that integrates a systematic anti-obfuscation preprocessing pipeline with dual-branch deep learning architectures.

The Primary Contributions of This Study Can Be Summarized as Follows

1. **Obfuscation-Robust Deep Learning Encoders**: We design and evaluate two dual-branch byte-level models, a character-level convolutional neural network (CharCNN) [10] and a Transformer encoder [11], each ingesting both raw and decoded payload representations, augmented with handcrafted features.

2. **Multi-View Canonicalization for Adversarial Resilience**: Our system builds a stack of views from raw to fully decoded view, enriched with handcrafted features, and trains across both obfuscated and canonical versions to handle mixed, multi-layer encodings. Training on both the obfuscated and cleaned versions makes the system tougher to deep nesting and mixed encodings.
3. **Systematic XSS Classification**: We extend beyond binary detection by systematically classifying XSS payloads into distinct attack categories while also maintaining a binary auxiliary head for robust separation between malicious and benign samples.
4. **A Task-Specific XAI Toolkit**: We introduce a task-specific XAI toolkit for XSS that groups obfuscation artifacts into span tokens, using HTML/URL/JS separators, quantifies each token's causal impact via Δ-logit occlusion, and summarizes explanation quality with gradient–occlusion agreement and branch reliance, yielding practitioner-friendly, reproducible insights aligned with the model's dual-view design.

The remainder of this paper is organized as follows. Sections 2 and 3 present the related work and preliminary concepts. In Sect. 4, we present the framework and walk through data collection, multi-view canonicalization, and the model design. Section 5 details our experiments and findings, and Sect. 6 wraps up with takeaways and plans for future work.

2 Related Work

Over time, web application security has progressed from hard-coded rules to flexible deep learning approaches. This section reviews the existing literature, organized by methodological paradigm, in order to provide context and motivation for the proposed framework.

2.1 *Traditional XSS Detection and Its Limitations*

Classic approaches to XSS split into two types of static pattern checks and code analysis.

- **Signature-Based Methods**: Operational defenses commonly hinge on human-written signatures, typically expressed as regex, and shipped with web application firewalls or IDS solutions [12]. By looking for fixed patterns such as **<script>** and **javascript**:, they work well on common, easy-to-recognize attacks. In practice, signature systems react after the fact, offer little protection against tricks such as nested URL/HTML encoding, and are expensive to keep current with continual manual edits.

- **Code Analysis Methods**: It targets the code directly, both at rest and in runtime. SAST tools comb through source files to flag risky validation or sanitization logic [13]. DAST tools hit the deployed app with tailored inputs and monitor its behavior [14]. While handy during development, these tools churn out lots of false alarms and still miss bugs that only show up in messy, real-world runtime. The inability of such traditional approaches to adapt to evolving and obfuscated attack strategies has motivated a transition toward machine learning-based solutions.

2.2 *Machine Learning for Payload Analysis*

Adopting machine learning makes it possible to learn complex, non-linear malicious patterns in a purely data-driven way.

Classical ML and Feature Engineering: The early playbook leaned on classical ML, with Random Forests and SVMs posting promising detection scores [15]. These approaches, however, were heavily dependent on manually engineered lexical features, such as frequency counts of high-risk characters, the presence of HTML tags, and measures of payload entropy. We saw that ML works for XSS, but it only went so far since it depended on human-crafted features.

Deep Learning and Sequential Modeling: Researchers began using deep learning along with sequence modeling to replace brittle, human-crafted features [1, 16]. Working at the character level, Char-CNNs learn from raw payloads and are good at spotting short n-gram signatures of encoders and evasion motifs. Stepping beyond Char-CNNs, bidirectional LSTMs [17] read left-to-right and right-to-left, grabbing context from the whole sequence. Despite the benefits, recurrent nets can't reliably connect distant pieces of a payload, particularly when padding or broken keywords get in the way. Attention fixed a lot of this.

2.3 *Advanced Architecture and Robustness*

Attention-Based Models: Recent advances have focused on attention mechanisms to overcome the limitations of recurrent and convolutional architectures. Self-attention gives the Transformer sequence-wide reach, enabling detection of obfuscation tactics like keyword fragmentation [11]. We use both pure Transformer encoders and a hybrid CNN–Transformer model so we can capture local obfuscation with convolutional layers and global relationships with attention.

Adversarial Defense and Representation Learning: Prior studies show XSS models can be prodded into errors by adversarial edits. By applying contrastive training, we collapse distances among related exploits and expand them from harmless text to resist manipulation. Research on adversarial robustness has specifically examined this threat in the context of XSS detection [18]. In relation to this challenge, our work incorporates contrastive learning [19] to improve resilience by enforcing that embeddings of semantically similar yet syntactically varied malicious payloads remain close in representation space. Put simply, it helps the model generalize both with augmented data and with adversarial tweaks so the whole defense gets tougher.

3 Preliminaries

Cross-Site Scripting is a web security vulnerability in which adversaries inject malicious scripts into web pages, thereby causing the scripts to execute in the browsers of unsuspecting users. Most of the time it's just JavaScript, and it runs right in the victim's browser as if it came from the real site. If input isn't validated and output isn't encoded, bad actors can work around key safeguards such as the same-origin policy [20].

3.1 XSS Types

XSS generally splits into three flavors, based on how the payload gets into the page and when it runs.

- Stored XSS (Persistent): The payload is written to the app's backend (e.g., in a post or profile field). Anyone who opens that page gets the payload—no individual targeting needed. Whenever the infected content is requested, the script rides along and executes in the user's browser. Since it's stored server-side, there's no need for targeted delivery; any visitor to that resource is exposed.
- Reflected XSS (Non-Persistent): Reflected XSS is the bounce-back kind where the app echoes the attacker's input in its next response. You'll most often see them as crafted URLs, circulated via scam emails or planted in hyperlinks. If someone clicks the link, the vulnerable site bounces the payload back in its reply and the browser runs it as if it came from the real domain. In contrast to stored XSS, reflected XSS typically affects a victim only for the duration of a single request.
- DOM-based XSS: In contrast to stored/reflected attacks, DOM-based XSS executes solely in the browser's DOM. The payload doesn't travel through the server, instead it piggybacks on insecure handling of outside input by the browser's scripts. Common entry points are user-controlled values, things like URL parameters, hash fragments, and similar inputs. Insecure processing by the browser's JavaScript triggers payload execution in the DOM, resulting in the same consequences as other XSS, including data loss or page manipulation.

3.2 XSS Payload Structure and Evasion Techniques

The challenge in catching XSS comes from obfuscated request payloads that piggy-back on input normalization by servers and browsers. Put simply, attackers tweak three main pieces of how the payload is built.

1. **Encoding and Obfuscation**: One of the oldest moves is to disguise **< script >** and on error via URL/HTML entity encodings and similar transforms. Standard moves include URL encoding (e.g., < → **%3C**) and HTML entity encoding (e.g., < → **< / <**), sometimes combined or nested so you need multiple decoding passes; the tricky bit is reconstructing layer count and sequence, which rule-based logic often fumbles.
2. **Injected Control Characters**: By inserting control bytes (**\ x 00**, **\n**, **\r**), the keyword gets chopped up and turns into payloads like **ja\x00vascript**:. Different treatment of the characters at the browser, server, and proxy layers undermines defenses based on exact literals or uncomplicated regexes. As a result, slipping in control characters makes life hard for static, pattern-only detectors.
3. **Context and Long-Range Dependencies**: Rather than one obvious payload, the logic is spread across non-adjacent segments, hidden by bulky benign text or by variables that trigger execution later on. Such strategies highlight the necessity of models capable of capturing long-range dependencies and contextual relationships [11]. Instead of focusing just on neighbors like sequential models do, Transformers connect distant parts of the text, catching dispersed or fragmented exploits.

The complications from these evasions make a strong case for byte-level models and multi-view analysis, which maintain robustness against diverse canonicalization and obfuscation.

3.3 Byte-Pair Encoding Tokenization

Byte-Pair Encoding (BPE) [21] helps maintain a manageable vocabulary while still representing low-frequency strings. BPE was originally introduced as a text compression algorithm, however, BPE was later adapted for natural language processing tasks to mitigate the out-of-vocabulary problem inherent in word-level tokenization [22]. Formally, let a corpus C be represented as a sequence over an initial symbol vocabulary V_0 (typically characters or bytes). BPE proceeds iteratively by selecting the most frequent adjacent pair of symbols in the corpus and merging them into a new symbol.

At iteration t, define the vocabulary as V_t and the corpus as C_t. Let the set of all symbol pairs in C_t be denoted by. The frequency of a pair $(a, b) \in P_t$ is given by where count $(a, b) \in w$ is the number of times the adjacent pair (a, b) appears in word w. At each step, the pair with the highest frequency is selected. In other words,. Thereby, when a new symbol such as *ab* is created, the algorithm expands the vocabulary as $V_{t+1} = V_t \bigcup\{ab\}$. The corpus is then updated by replacing every occurrence of the pair (a^*, b^*) with the new merged symbol ab. Merging proceeds iteratively, stopping when the vocab hits $|V_T|$ or the count of merges reaches T. BPE creates a V_T that treats frequent patterns as one token and shards infrequent ones into smaller parts. At the byte level, we cover the entire input space, Unicode bytes, percent-encoded strings, and HTML entities commonly seen in attacks [23].

3.4 Learned Embedding Vectors

Neural models can't do much with raw symbols like words or characters until we turn them into continuous vectors. Think of a learned embedding as a low-dimensional vector assigned to every symbol, and those vector entries are learned with the rest of the network during training [6]. With one-hot, letters don't know about each other, embeddings place similar symbols near each other in a shared vector space. Backprop drives embeddings to group together for symbols that share usage patterns. For example, a model will place **S** near **s** in the embedding space, and < close to its encoded form **%3C**. Day to day, that means these encoders cope with orthographic changes, obfuscation schemes, and unseen strings. With embeddings, tokens live in a concise d-dimensional space instead of an enormous $|V|$ one-hot space, yielding faster computation [24].

3.5 Convolutional Neural Networks

As deep learning models, CNNs [25] were designed with image recognition in mind and shine by learning local patterns across space. The core building block of a CNN is the convolutional layer, which applies learnable filters (or kernels) across the input. Each weight matrix, or filter, shifts across the input and multiplies-and-sums with each neighborhood it meets.

For a sequence matrix $x_{L\times d}$ and a kernel $w_{k\times d}$, running a convolution produces the resulting feature map.

$$c_i = f(w \cdot x_{i:i+k-1} + b)$$

where, $x_{i:i+k-1}$ is the k-long window of x, b is a bias parameter, and f is a nonlinearity like ReLU.

Pooling layers, such as max pooling, are typically applied after convolution to downsample feature maps by retaining only the most salient activations. This confers translation invariance and reduces computational cost. By stacking multiple convolutional and pooling layers, CNNs gradually compose low-level local patterns, for instance, edges and character bigrams, into increasingly abstract features such as shapes and words. While CNNs were first applied to vision, one-dimensional CNNs have proven highly effective in NLP tasks for extracting local lexical and syntactic features [7].

3.6 *Character-Level Convolutional Neural Networks*

With word-level systems, any word that wasn't in the training vocab simply can't be represented. Char-CNNs [10] sidestep this by reading raw character strings instead of words (Fig. 1).

a. A one-hot vector matching the alphabet's size: very sparse but exact.
b. A learned embedding vector in a low-dimensional space that is dense, but generalizable.

This yields an input matrix $X \in \mathbb{R}^{L \times d}$, where L is the maximum character sequence length and d is the embedding dimension. As the sequence is traversed, the filters detect small n-gram patterns, for example affixes, punctuation, and obfuscation tokens.

For example, a filter with kernel width 3 detects character trigrams such as < **sc** or **ipt**, which are highly indicative of injected **<script>** tags. Wider kernels (5–7) capture longer motifs such as **%3Csc**. By stacking filters with multiple widths, Char-CNNs automatically learn multi-scale features that correspond to n-gram patterns of varying size.

After the conv step, global max pooling picks the peak value from each feature map and packs them into a same-length vector for any input. This representation is then passed through one or more fully connected layers for classification.

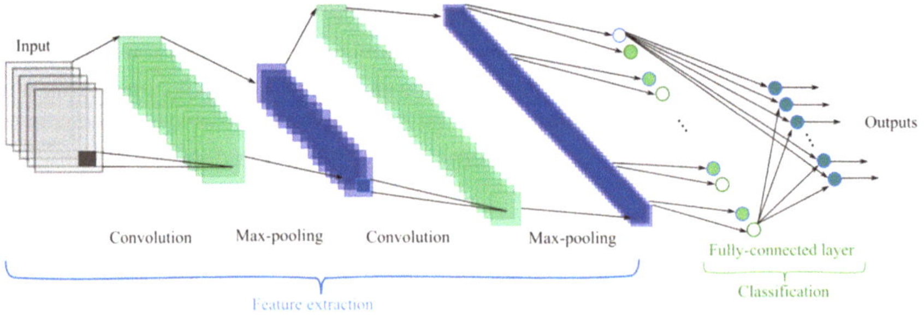

Fig. 1 Architecture of CNNs [26]

3.7 Transformers

Transformers changed sequence modeling by dropping recurrence and convolution and leaning on self-attention [11]. Where RNNs are strictly stepwise, Transformers compute per-token context concurrently across the whole sequence. This enables highly efficient training and makes it possible to capture long-range dependencies that are often inaccessible to CNNs or RNNs.

In essence, Transformer encoders hinge on scaled dot-product self-attention. Given a sequence of input embeddings $X \in \mathbb{R}^{L\times d}$, three linear projections are applied to obtain queries Q, keys K, and V:

$$Q = XW^Q, \;\; K = XW^K, \;\; V = XW^V$$

where $W^Q, W^K, W^V \in R^{d\times d_k}$ are trainable weight matrices. The attention weights are then computed as follows:

$$Attention(Q, K, V) = softmax\left(\frac{QK^T}{\sqrt{d_k}}\right)V$$

In practice, each token can look at all the others, and the attention scores decide how much signal passes between them. To improve expressivity, multi-head attention extends this mechanism by computing attention in parallel across h subspaces.

$$Multi\,Head(Q, K, V) = Concat(head_1, \ldots, head_h)W^O$$

where each $head_i = Attention\left(QW_i^Q, KW_i^K, VW_i^V\right)$.

Each Transformer layer typically consists of multi-head attention followed by a position-wise feed-forward network.

$$FFN(x) = max(0, xW_1 + b_1)W_2 + b_2$$

with residual connections and layer normalization applied after each sublayer. Since self-attention is permutation-invariant, positional encodings are added to the input embeddings to preserve word order information.

From a practical standpoint, Transformers excel at modeling sequences where global context matters. For instance, in adversarial payloads, attackers often split malicious keywords with injected characters or null bytes for instance **s\0elec\0t**, meaning that local n-gram features alone are insufficient. Thanks to self-attention, the model links separated but related symbols, which exposes patterns spanning large contexts.

3.8 Explainability

One of our goals in this paper is explaining why the detector flags an input by locating evidence in the string, measuring which branch (raw or dec) drives the decision, and checking faithfulness.

For input x with length L, the model outputs logits $z \in \mathbb{R}^{C}$. We explain the class logit z_c, which is the output of model before appling softmax. Each position i has embedding $\mathbf{e_i} \in \mathbb{R}^{d}$. Additionally, for two raw and dec view we will use notation $V_0 = \text{RAW}$, $V_1 = \text{DEC}$.

We rely on two character-level, gradient-based signals. Saliency scores output sensitivity using the gradient's norm $S_i^{(c)} = \| \partial z_c / \partial e_{i,k} \|_p$, where p can be 1 or 2, and traces back to early saliency maps for deep nets [27]. Gradient × Input uses the signed interaction $G_i^{(c)} = \sum_k e_{i,k} \, \partial z_c / \partial e_{i,k}$, a simple and widely used baseline (and a first-order step toward path/reference methods such as Integrated Gradients and DeepLIFT) [28].

To handle obfuscated inputs, we convert character attributions into XSS-aware tokens that group **%HH** bytes (e.g., **%3C**), HTML entities (**&name;**, **&#dd**;, **&#xhh;**), tag-like tags (**<... >**), JavaScript escapes (**\xHH**, **\uXXXX**), and also identifiers, numbers, and whitespace. For faithfulness by perturbation, we use token occlusion, masking a token span and measuring the logit change $\Delta(t) = z_c(x) - z_c(x_{\setminus t})$, following the occlusion tradition in model visualization [29]. We then report gradient–occlusion agreement as the Spearman correlation between token Gradient × Input scores and token occlusion Δ-logits, and we quantify view usage in our fused network via a branch-reliance ratio.

$$R_{\text{RAW}} = \frac{\sum_i \left|A_i^{\text{RAW}}\right|}{\sum_i \left|A_i^{\text{RAW}}\right| + \sum_j \left|A_j^{\text{DEC}}\right|}$$

where A denotes either S (Saliency) or G (Gradient × Input).

4 Proposed Framework

We detect XSS payloads by pairing a well-built dataset with multi-view preprocessing and a two-branch model that catches both low-level obfuscations and high-level semantics. The system is meant to catch surface-level disguises and long-range semantic regularities in tandem. The overall pipeline is shown in Fig. 2.

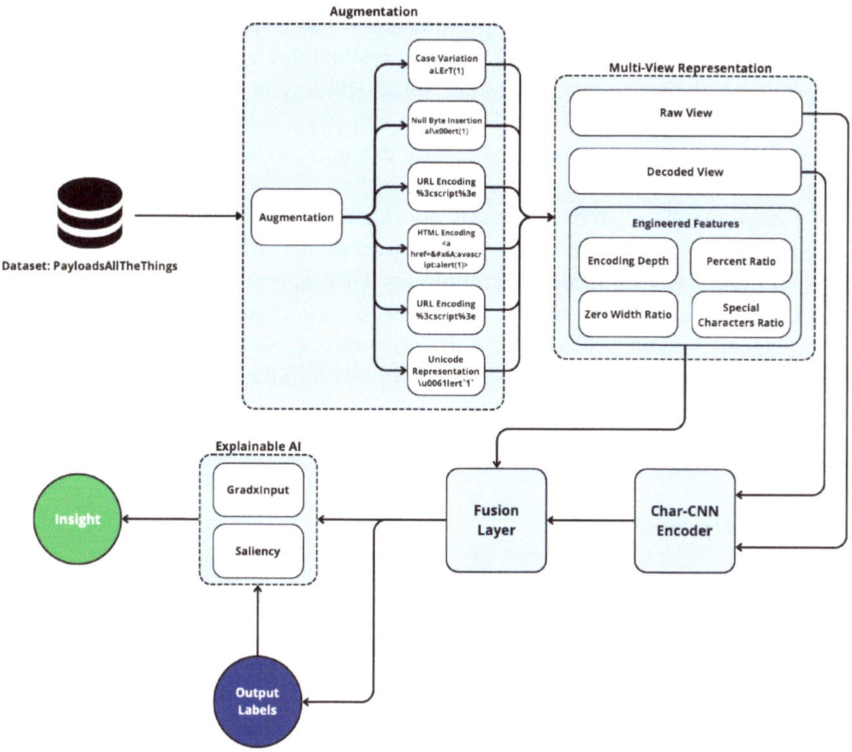

Fig. 2 Overview of our methodology

4.1 Dataset Construction

We used canonical seeds from PayloadAllTheThings, the community's living list of payloads that mirrors real adversarial usage. Starting from the canonical list, we created many obfuscated alternatives, giving the corpus both the classic patterns and the hard-to-parse, in-the-wild variants. To make the augmentation strategy concrete, consider a generic token **PAYLOAD**. Case variation produces **PAYLOAD**, **payload**, or **PaYlOaD**, while null-byte insertion yields **\x00PAYLOAD** or **PAYLOAD\ x 00** (the null marker is recorded but not executed). These simple changes model attempts to evade naive string matching by altering superficial orthography. We applied encoder-style obfuscations extensively. URL percent-encodings and HTML entity encodings were applied partially and repeatedly so the corpus contains once-encoded, repeatedly-encoded, and mixed-encoding forms. The data includes plain once-encoded strings, heavily re-encoded strings, and mixed encodings. Typical cases are percent encodings such as **%50%41%59**…, partial percent mixes like **%50AYLOAD**, and HTML entities in decimal/hex, e.g., **PA**…, **&# x 50;&# x 41;**. We additionally produced Unicode escapes (e.g., **\u0050\u0041**…) so we could handle environment-specific normalization.

Structural obfuscations were also simulated. A token split with an invisible character, such as **P\u200BAYLOAD**, will bypass simple "contains" checks without altering what the string is trying to do. Token-splitting is modeled by surrounding payloads with % runs or dropped-in delimiters (e.g., %%%**PAYLOAD**%%%); we also include control/non-printable byte examples to reflect client/proxy/server parsing quirks.

To help models separate malicious obfuscations from normal text, we curated a parallel set of benign examples, short conversational strings, typical form values (e.g., **contact@example.com**), and ordinary URL/query fragments. Each sample was labeled either normal or one of the obfuscation classes (case-variation, null-byte, percent-encoding, HTML-entity, Unicode-escape, zero-width/percentrun, etc.), resulting in a multi-class scheme that supports both binary decisions and detailed attack-type identification.

4.2 Preprocessing

For every payload, we assemble an information-dense input by stitching together its different text views and a few curated features. Given x as received, we run $D(.)$, a one-shot canonicalization step that performs one step of canonicalization for example: one-pass URL-percent decoding, HTML-entity unescaping, or Unicode unescaping. Iterative application of D generates a sequence of decoded views (Fig. 3).

$$V^{(0)}(x) = x, \quad V^{(k)}(x) = D\big(V^{(k-1)}(x)\big) \quad (k = 1, \ldots, K)$$

where K is a practical upper bound on the number of decoding passes, which is 5 as a conservative default. We cut the loop the moment two consecutive views match. We therefore define the effective decode depth $d(x)$ as the smallest integer k satisfying $V^{(k)}(x) = V^{(k-1)}(x)$, with $d(x) = K$ if no such K exists. The depth based on stability counts the passes that make a difference and offers a clear measure of how obfuscated the input is.

Along with the raw and decoded views, we pull a small set of numerical features $f(x) \in \mathbb{R}^m$ that capture typical obfuscation cues. We summarize each string with decode depth $d(x)$, counts of % and HTML entities **num_entities(x)**, the longest consecutive % run **longest_percent_run(x)**, a zero-width flag **has_zw(x)**, and normalized alpha/digit ratios. Formally we collect these into a single vector.

$$f(x) = [d(x), num\_percent(x), num\_entities(x),$$

$$longest\_percent\_run(x), has\_zw(x),$$

$$ratio\_alpha(x), ratio\_digit(x)]$$

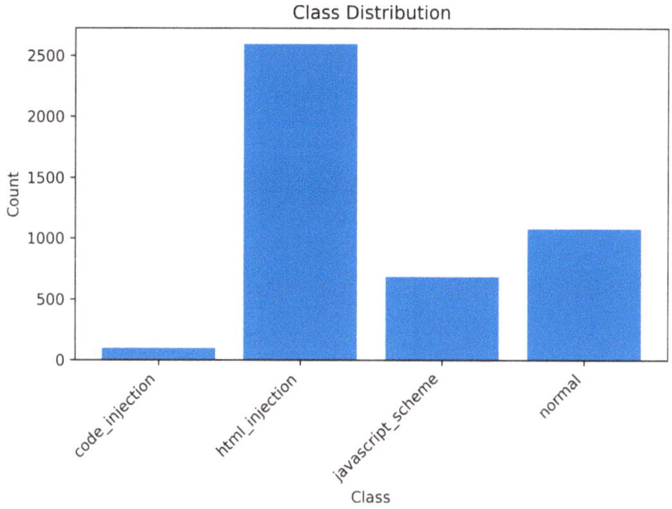

Fig. 3 Bar chart of class distribution in training/validation

Looking at Fig. 4, percent-run length and entity count stand out, concentrating on the malicious side rather than the benign. We also gauge how much each decode changes the string by optionally computing a normalized edit distance between consecutive views. Let $Lev(.,.)$ denote the Levenshtein distance. define the per-step normalized change.

$$\Delta_k(x) = \frac{Lev\left(V^{(k-1)}(x), V^{(k)}(x)\right)}{max\left(1, \left|V^{(k-1)}(x)\right|, \left|V^{(k)}(x)\right|\right)}$$

and summarize cumulative change by $\Delta_{sum}(x) = \sum_1^{d(x)} \Delta_k(x)$. Large values of $\Delta_{sum}(x)$ indicate heavy or semantically obfuscation and can be included as an additional feature when desired.

We then assemble the model input as a multi-view bundle. At minimum, the bundle includes the raw view, the first decoded view, and the feature vector.

$$bundle(x) = \left(V^{(0)}(x), V^{(1)}(x), f(x)\right)$$

More generally, one can include multiple decoded depths up to K.

$$bundle_K(x) = \left(V^{(0)}(x), V^{(1)}(x), \ldots, V^{(K)}(x), f(x)\right)$$

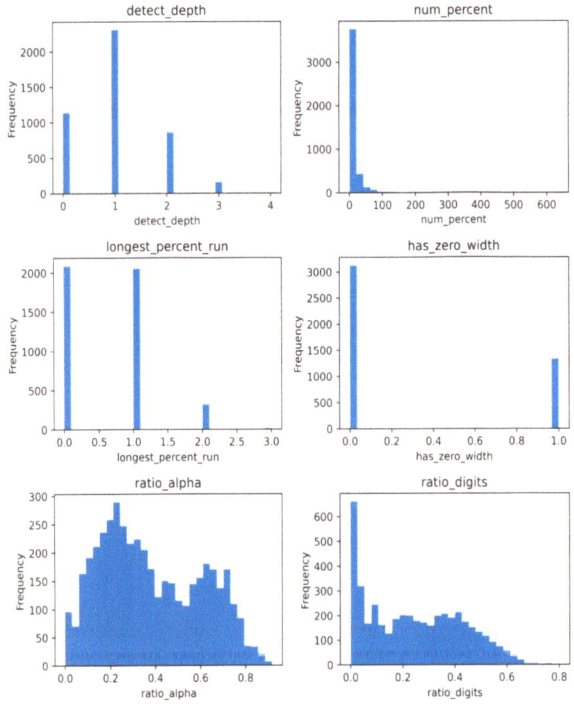

Fig. 4 Distribution of handcrafted features across the dataset

4.3 Model Architecture

We combine a small, dual-branch Character CNN with a dual-branch Transformer that uses only its attention heads. The models operate on the same preprocessed inputs: raw and decoded payload views plus a handful of handcrafted features. The goal of these designs is twofold: first to learn representations that remain stable under obfuscation, and second to combine them with explicit, human-interpretable features that surface low-level encoding patterns.

4.3.1 Dual-Branch CharCNN

CharCNN fits XSS cases where obfuscation happens at the symbol level as it is fine-grained for character-level modeling. Consider, for instance, the difference between **<script>** and **<script>**: semantically identical but obfuscated through character encoding. By operating directly on characters, the model is not restricted by a fixed token vocabulary and can handle unseen or malformed encodings without degradation.

Each input view $V_i(x)$ is read as a sequence of characters $(c_1, \ldots, c_L)$ and then embedded in a continuous vector space.

$$Z^{(i)} = [e(c_1), e(c_2), \ldots, e(c_L)] \in \mathbb{R}^{L \times d},$$

where e is a learned embedding function of dimension d.

The sequence of embeddings is fed into a series of 1-D conv filters. Each filter learns short-range character patterns akin to n-grams. As a concrete case, a kernel of width 3 can light up when it sees the substring **%3C**. The convolutional stack is written as follows:

$$H^{(i)} = Conv\ Stack\left(\left(Z^{(i)}\right)^T\right), \quad H^{(i)} \in \mathbb{R}^{C \times L}$$

where L indexes where in the sequence you are (character position) and C indexes what features the convolutional stack has learned at each position. After convolution, we apply global max pooling.

$$h^{(i)}_{cnn} = MaxH^{(i)}_{:,j}$$

This ensures that the representation emphasizes the most salient features, regardless of their position in the string. Importantly, global pooling introduces invariance to token position, which is desirable since obfuscations can insert junk characters almost anywhere. The representations of both views are concatenated with the handcrafted features.

$$z_{cnn} = h^{(0)}_{cnn} || h^{(1)}_{cnn} || f(x)$$

Fusion occurs through a two-layer MLP with ReLU activations and dropout.

$$u_{cnn} = \phi(W_2 \phi(W_1 z_{cnn} + b_1) + b_2)$$

Predictions are then produced by two heads.

$$\widehat{y_{multi}} = softmax(W_c u_{cnn} + b_c), \quad \widehat{y_{bin}} = \sigma(W_b u_{cnn} + b_b).$$

Then the joint loss will be as follows:

$$\mathcal{L}_{cnn} = \mathcal{L}_{multi} + \lambda \mathcal{L}_{bin}$$

Put simply, the architecture leans toward picking up local attack motifs. It is computationally efficient, with complexity linear in sequence length, and model size in the tens of megabytes. Because it avoids reliance on pretrained resources, the CharCNN can be retrained rapidly when attackers adopt new obfuscation styles.

For real-time detection systems such as intrusion prevention systems or WAFs, this efficiency makes CharCNN an attractive deployment candidate.

4.3.2 Dual-Branch Heads-Only Transformer

CharCNN is great at the short, local patterns, but it struggles to capture long-range dependencies. For instance, obfuscations often split across distant positions in a string, such as breaking < **script** > with null bytes or zero-width characters inserted between every letter. To recognize split-across-positions behavior, we build on a dual-branch Transformer encoder. Each view $V_i(x)$ is tokenized into subword units $(t_1, \ldots, t_L)$ using a byte-level or BPE tokenizer. These tokens are passed to a pretrained Transformer encoder T:

$$H^{(i)} = T\left(V^{(i)}(x)\right) \in \mathbb{R}^{L \times d_T}$$

where L is sequence length in tokens after subword/byte-level tokenization and d_T is hidden dimension of the Transformer encoder which is 512 in this case.

We then compute a pooled embedding using the attention mask.

$$h_{tr}^{(i)} = \frac{1}{\sum_j m_j} \sum_{j=1}^{L} m_j H_j^{(i)}$$

Mean pooling gives a clean, reliable summary and usually beats CLS or attention pooling on efficiency. As before, the two view embeddings are concatenated with features.

$$z_{tr} = h_{tr}^{(0)} \Big\| h_{tr}^{(1)} \Big\| f(x)$$

Fusion proceeds through the same two-layer MLP.

$$u_{tr} = \phi(W_2 \phi(W_1 z_{tr} + b_1) + b_2)$$

The output heads and joint loss are identical.

$$\widehat{y_{multi}} = softmax(W_c u_{tr} + b_c), \quad \widehat{y_{bin}} = \sigma(W_b u_{tr} + b_b).$$

$$\mathcal{L}_{tr} = \mathcal{L}_{multi} + \lambda \mathcal{L}_{bin}$$

The Transformer's advantage lies in contextualization. Even if a malicious sequence is obfuscated by distributing fragments across a long string, self-attention allows the encoder to relate distant tokens. Even though they're trained on natural language, pretrained models capture universal string features (whitespace, punctuation, rare tokens), which lends robustness against obfuscation.

To keep training cheap, we freeze the Transformer and train only the fusion MLP and the output heads. This significantly reduces computational requirements and prevents overfitting, since our dataset is modest in size relative to the billions of parameters in modern Transformers. Complexity is dominated by self-attention $O(L^2 d_T)$, but since XSS payloads rarely exceed a few hundred tokens, inference remains practical.

Deployed models based on the Transformer weather complex obfuscations better, yet maintain light-touch fine-tuning. Compared to CharCNN, it has higher latency but also better resilience to attack variations where local motifs are insufficient.

4.4 Training and Optimization

We chose AdamW for training both models, adding decoupled weight decay to Adam to gain stronger generalization on text classification. We set the learning rate to 2e-3 for CharCNN and 2e-5 for the Transformer, since the former trains from scratch while the latter starts from a pretrained encoder. Early stopping was employed with patience of five epochs, monitoring macro-F1 on a held-out validation split comprising 10% of the dataset. Picking this metric prevents the huge "normal" class from washing out errors on the smaller classes.

Because security data is skewed toward benign traffic, we used two tactics to handle the class imbalance. First, a weighted cross-entropy loss was applied, where each class c is weighted inversely proportional to its frequency π_c.

$$\mathcal{L}_{multi} = -\sum\nolimits_c \alpha_c y_c log\widehat{y_c}, \quad \alpha_c = \frac{1}{\pi_c}$$

Second, balanced sampling was used in the data loader to ensure that minibatches contained a more uniform distribution of attack types. That way, the model didn't just memorize the majority benign class during training.

We trained the CharCNN end to end, updating both the embedding layer and the conv filters as we went. With end-to-end learning, the model picked up little giveaways like **%3C** standing in for < and the repeating **&#x** used for hex entities.

In contrast to CharCNN, the Transformer side used heads-only fine-tuning. We kept the pretrained encoder fixed—DistilRoBERTa as an example—and optimized solely the fusion MLP along with the classification heads. This design choice reflects the limited size of the dataset and the observation that the pretrained layers already provide robust token-level representations.

The final loss function combined a multiclass objective with a binary auxiliary task. Specifically, the primary objective was a weighted multiclass cross-entropy.

$$\mathcal{L}_{multi} = -\sum\nolimits_c \alpha_c y_c log\widehat{y_c}, \quad \alpha_c = \frac{1}{\pi_c}$$

where K is the number of classes (obfuscation strategies plus benign baseline). The auxiliary objective was a binary cross-entropy predicting whether a payload is malicious or benign.

$$\mathcal{L}_{bin} = -\left(y_{bin} log\sigma\left(\hat{z}\right) + (1 - y_{bin}) log\left(1 - \sigma\left(\hat{z}\right)\right)\right)$$

The total training loss was a weighted combination $\mathcal{L} = \mathcal{L}_{multi} + \lambda\mathcal{L}_{bin}$. We purposely use $\lambda < 1$ to down-weight the binary loss, keeping it in a regularizing role alongside the multiclass objective. By introducing a binary head, the multitask approach keeps a clear benign-versus-malicious boundary and thus remains robust when multiclass classes overlap.

Following the definitions in 3.8, we compute post-hoc explanations after training for the predicted class logit $z_{\hat{c}}$ on both views. For each input bundle, we run a single backward pass to obtain Saliency and Gradient × Input at character level per branch, aggregate scores to XSS-aware tokens using the tokenizer described in 3.8, and then measure token occlusion and deletion curves for faithfulness. To make RAW/DEC panels visually comparable we apply joint signed normalization to a shared $[-1,1]$ range. Moreover, overall branch reliance is summarized by the energy ratio R_{RAW}. All XAI runs use the training vocabulary and the same preprocessing as the classifier.

5 Result

5.1 Classification Performance

We evaluate both model families, the dual-branch CharCNN and the heads-only Transformer, on the task of classifying XSS payloads into their respective obfuscation categories and distinguishing them from benign strings. Performance is reported in terms of macro-F1, which treats each class equally and avoids dominance by the majority "normal" class.

Both models achieved near-perfect performance. We saw 0.9958 macro-F1 for CharCNN, versus 0.9905 for the heads-only Transformer, which demonstrate that even lightweight architectures can accurately capture the statistical and structural patterns that differentiate obfuscated payloads from benign inputs. There's a faint edge for CharCNN; nonetheless, the small gap suggests either option is viable with this pipeline.

Figures 5 and 6 plot the training loss over epochs for both models. CharCNN reached a steady state rapidly, well before the tenth epoch. The heads-only Transformer settled in reliably, but its loss fell more slowly and bounced around a bit more from epoch to epoch.

Figures 7, 8 show the validation macro-F1 over epochs. From the curves, we can see that stopping on macro-F1 kept overfitting in check and let the models generalize to both common and rare obfuscations.

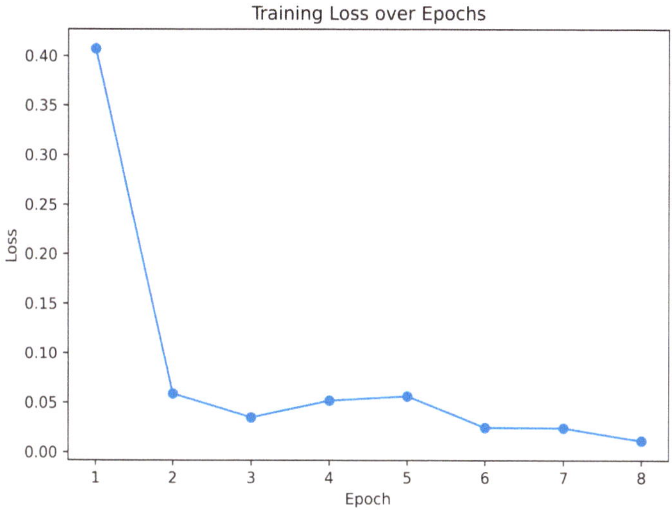

Fig. 5 CharCNN training loss over epochs

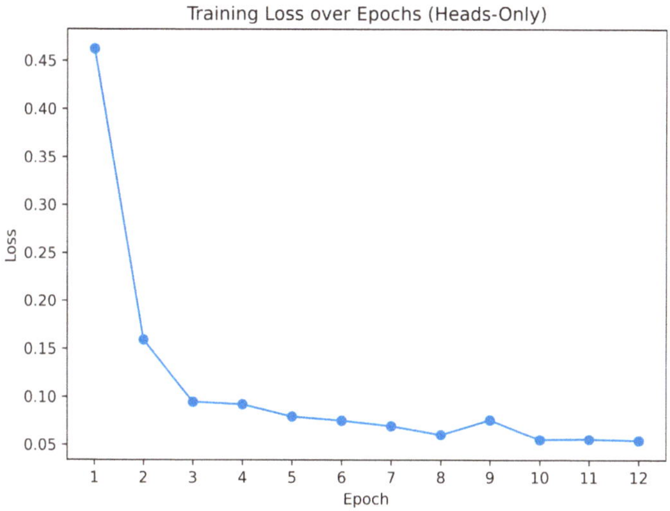

Fig. 6 Transformers training loss over epochs

According to Figs. 9, 10, the confusion matrices show a clear benign–malicious separation for both models, with false positives under 2%.

Within malicious classes, the model at times blurred heavily URL-encoded payloads with mixed-case variants, likely due to similar character statistics.

Nonetheless, the binary auxiliary head reduced such errors by reinforcing a robust separation between malicious and normal samples.

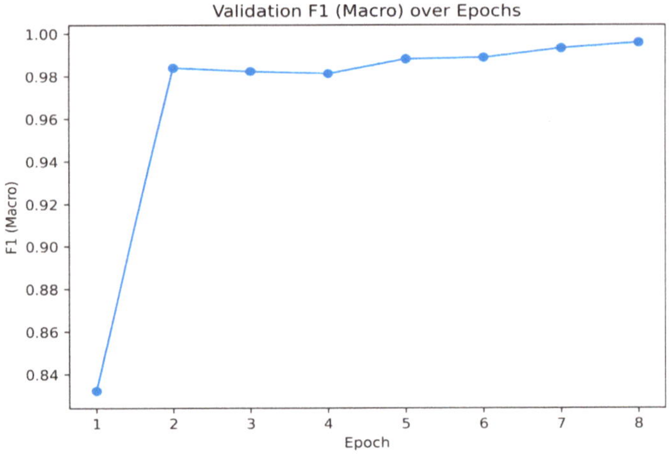

Fig. 7 CharCNN macro F1 validation over epochs

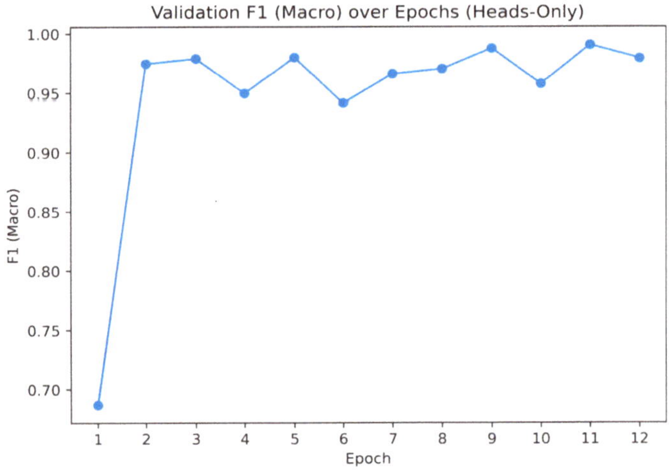

Fig. 8 Transformers macro F1 validation over epochs

You can find the processed dataset and the full set of training notebooks on GitHub to rerun and expand our work: https://github.com/mr-seifi/xss-shield. Included in the repo are the raw/augmented XSS payloads from PayloadAllTheThings, the scripts for multi-view canonicalization, and training code for both the CharCNN and Transformer models. The materials here make it easy to re-create our results, plug in new obfuscation techniques to the dataset, and tune the models for local deployment.

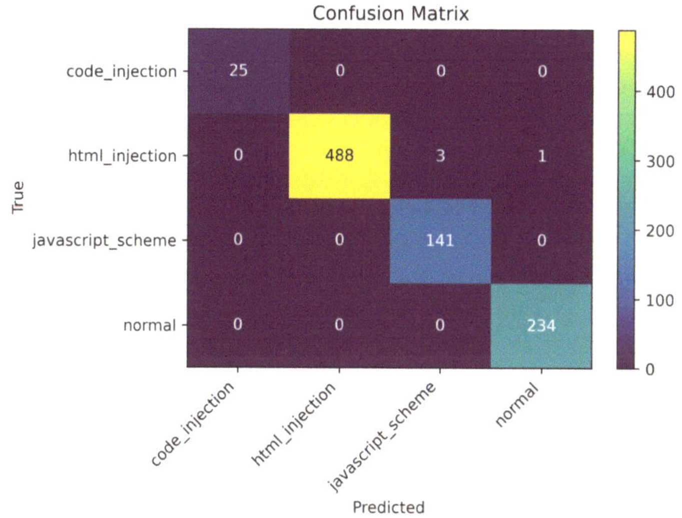

Fig. 9 CharCNN confusion matrix

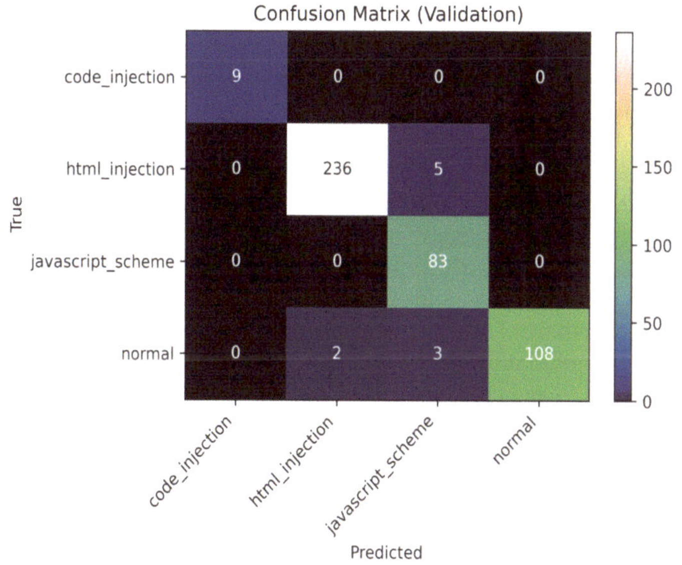

Fig. 10 Transformers confusion matrix

5.2 Explainable Analysis on Curated Payloads

To probe what drives decisions, we analyze 26 handcrafted payloads spanning URL-percent encodings, HTML entities, JavaScript/Unicode escapes, split keywords with zero-width characters, and mixed encodings. For each payload we compute character-level attributions (Saliency, Gradient × Input) on RAW and DEC views, aggregate to XSS-aware tokens, measure token occlusion, and summarize with branch reliance (RAW%), Spearman agreement (token Gradient × Input vs token occlusion), and Occl@1/@3. The full per-payload summary appears in Table 1.

For instance, if you look at the second row of the Table 1, the payload is `<img src = x onerror=alert(1) >` , which is an HTML entity encoded form of `< img src = x onerror = alert(1) >` . The model predicts html_injection with a large margin 38.54, while RAW% is 4.90%, indicating the DEC branch carries the decision after decoding reveals `< img ... onerror = ... >` . On the RAW side, TopTokensByΔ are mostly spacing entities ` ` and a lone 1, and Occl@1/@3 are small (0.052 / 0.107), so masking individual encoded units barely changes the logit once decoded intent is available. The 0.48 Spearman suggests a middling match between token Gradient × Input and occlusion orders, which typically strengthens after merging adjacent entities or switching the metrics to DEC.

We visualize character-level Gradient × Input on the decoded string and the corresponding token-level saliency produced by our XSS-aware tokenizer. Across all exemplars the model consistently assigns large positive importance to canonical triggers, which can be event handlers connectors, the javascript: scheme, the onset of alert, and etc.

For payload #3 `< img/src/onerror = alert(document.origin) >` , which is illustrated in Fig. 11, the strongest Gradient × Input responses occur at the = that binds the event handler **`onerror`** = and at the final > that closes the tag. Security-wise, the = ties onerror to its value so it becomes runnable JavaScript, and the final > commits the tag, triggering the browser to parse and attach the handler. The strong focus on = and > , plus the spike at the start of **alert**, shows the model keys in on the bare-bones structure that makes an inline payload executable, not just long tokens or **document.origin**.

For payload #8, in the char Gradient × Input map (Fig. 12), the top signal lands on the two trailing //, highlighting that in code-injection attacks the first step is to fix the browser's parsing context, so the comment truncates anything that follows so the injected JavaScript executes cleanly. Token saliency (Fig. 13) ranks **document** first, the trailing / next, and **lert** third, highlighting the sensitive source, the comment terminator, and the alert call site.

For payload #11, Gradient × Input rises smoothly across javascript, spikes at alert, and remains positive on **(1)** (Fig. 14). Consistent with expectations, token saliency peaks at **javascript** and **alert**, with only modest weight on the call parentheses and literal (Fig. 15) for scheme-styleXSS.

For payload #15, token saliency likewise highlights the **< img/src/onerror = alert(1) >** token and discounts href and the final **< /a >** (Fig. 16). That's why the margin is so large: the model locks onto the inline handler even when it's tucked inside an otherwise harmless wrapper.

Table 1 Per-payload XAI summary

	Payload	PredClass	Margin	RAW%	Spearman(token $G \times I, \Delta$)	Occl@1	Occl@3
0	%3Cscript%3Ealert(1)%3C%2Fscript%3E	html_ injection	31.45089769	7.664867656	0.827272727	0.228469849	0.307064056
1	<img src = x onerror=a&…	html_ injection	38.54162812	4.902237042	0.481409013	0.051612854	0.107120514
2	< p > Hello < /p >	normal	9.080613136	17.5883232	0.8	2.392601013	3.977284431
3	< img/src/onerror = alert(document.origin) >	html_ injection	37.14737344	10.64054728	–	–7.40215683	–7.40215683
4	< a/href = google.com onfocus = alert(1) autofocus > Hello < /a >	html_ injection	34.3401897	10.94586581	0.5	0.004081726	–2.909454346
5	%3c%61%2f%68%72%65%66%3d%67%6f%6f%67%6c%65%2e%63%6f%6d%20…	html_ injection	40.36363363	3.617986108	0.510258892	0.165512085	0.270755768
6	%25%33%63%25%36%31%25%32%66%25%36%38%25%37%32%25%36%35%25…	html_ injection	2.810449451	13.55367581	0.430020934	0.32113266	0.458646297
7	Hello, this is Amin from space! < p > :) < /p >	normal	22.82054043	18.96296518	0.857585139	1.374446869	3.239189148
8	)\u0061lert‘document.origin‘;//	code_ injection	18.20112026	15.07097698	0.5	0.168716431	0.279094696
9	%48%65%6c%6c%6f%2c%20%74%68%69%73%20%69%73%20%41%6d%69%6e…	normal	0.612280369	5.009922583	0.316307864	0.061138153	0.132707596
10	<a href="https://google.com">Hello, World</a>	html_ injection	19.93334031	10.55847963	1	–0.021190643	–0.668569565
11	jav&# × 61;script;\u0061lert(1)	javascript_ scheme	26.72308135	3.29712162	0.755244755	0.033027649	0.055725098
12	%3c%61%20%68%72%65%66%3d%22%68%74%74%70%73%3a%2f%2f%67%6f…	html_ injection	32.42409849	3.361280962	0.61403991	0.336013794	0.721324921
13	href="https://google.com"><img/src/onerror=alert(1)>">…	html_ injection	30.74810076	10.04749701	0.966666667	8.7738E-05	–0.007171631

(continued)

Table 1 (continued)

	Payload	PredClass	Margin	RAW%	Spearman(token G × I, Δ)	Occl@1	Occl@3
14	%3c%61%20%68%72%65%66%3d%22%68%74%74%70%73%3a%2f%2f%67%6f…	html_ injection	39.2121644	3.182039589	0.492572043	0.292266846	0.635852814
15	<a href="https://google.com"><a onmouseover=alert(documen…	html_ injection	32.27192664	9.329099109	0.972727273	0.006597519	0.006252289
16	< IMG """ > < SCRIPT > alert("HELLO WORLD") < /SCRIPT > " >	html_ injection	31.14593649	8.904027465	0.961538462	0.011651993	0.01729393
17	< a href = "javascript:alert(String.fromCharCode(88,83,83))"…	html_ injection	34.1709547	8.691698916	1	–0.029176712	–0.216238022
18	< IMG onmouseover = "alert('HELLO')" >	html_ injection	35.18642139	8.424244695	–	–6.340719223	–6.340719223
19	< img src = x onerror = "&#0,000,106&#0,000,097&#0,000,118&#0,000,097&…	html_ injection	35.48172522	10.41432327	–	–7.580226898	–7.580226898
20	< a href = "jav ascript:alert('XSS');" > Click Me < /a >	html_ injection	34.8835175	10.56339162	0.202919862	3.8147E-05	9.53674E-05
21	< BODY onload!#$%&()* ~ + -_.,:;?@[/|\]^' = alert("XSS") >	html_ injection	37.00360894	8.830333714	0.958041958	0.02797699	0.034069061
22	%3c%42%4f%44%59%20%6f%6e%6c%6f%61%64%21%23%24%25%26%28%29…	html_ injection	41.90780783	4.041833915	0.646684209	0.193305969	0.389713287
23	%3c%42%4f%44%59%20%6f%6e%6c%6f%61%64%21%23%24%25%26%28%29…	html_ injection	24.60133195	5.315925415	0.73718794	0.171480179	0.298845291
24	<input value="https://google.com"><i&…	html_ injection	36.479491	8.086788744		–6.355262756	–6.355262756
25	%26%23%78%32%32%3b%26%23%78%33%65%3b%26%23%78%33%63%3b%26…	html_ injection	42.66965675	1.777724651	0.232738323	0.124271393	0.271846771

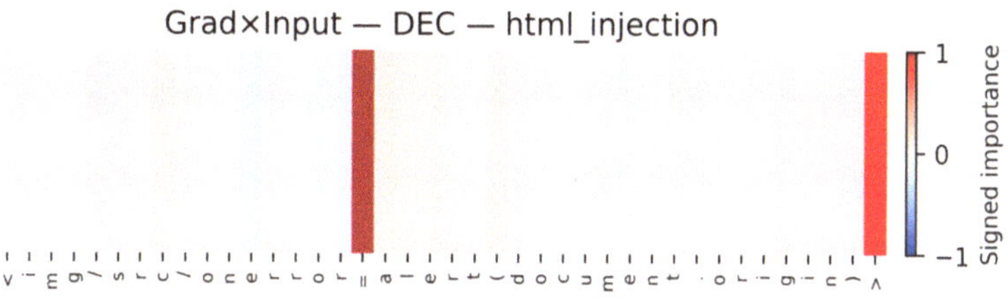

Fig. 11 Heatmap over decoded characters for payload #3, illustrating character's importance on the predicted label using Gradient × Input method

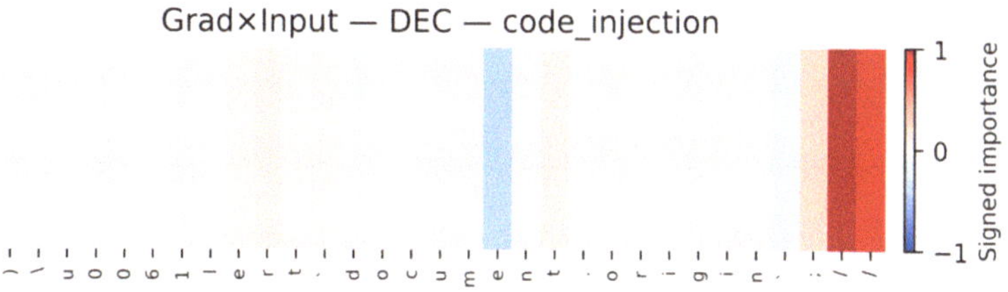

Fig. 12 Heatmap over decoded characters for payload #8, illustrating character's importance on the predicted label using Gradient × Input method

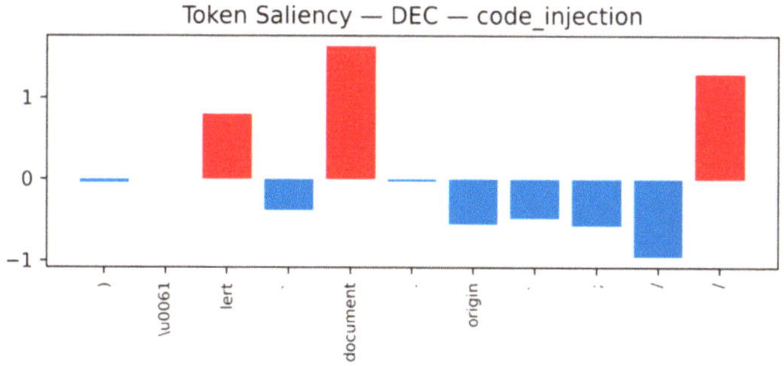

Fig. 13 Token's importance on predicted label using Saliency method for payload #8

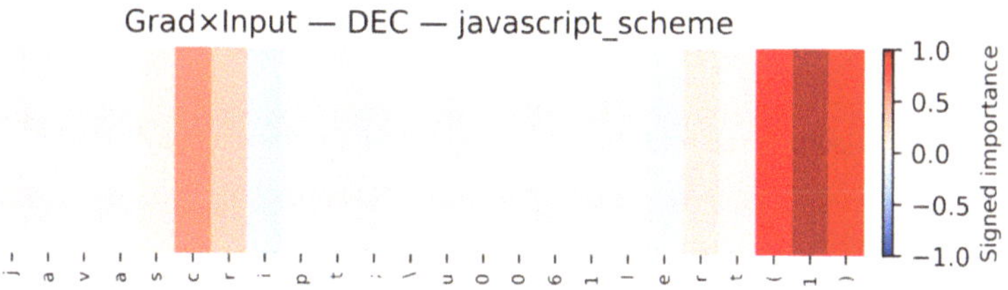

Fig. 14 Heatmap over decoded characters for payload #11, illustrating character's importance on the predicted label using Gradient × Input method

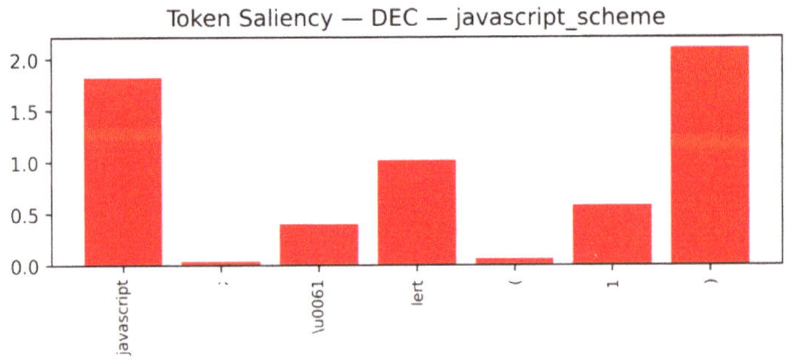

Fig. 15 Token's importance on predicted label using Saliency method for payload #11

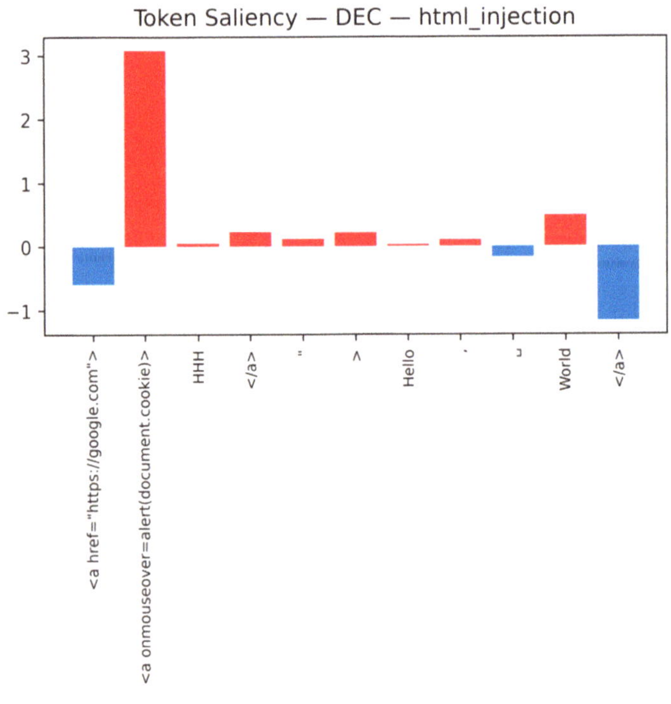

Fig. 16 Token's importance on predicted label using Saliency method for payload #15

The decoded-view explanations show that the model is not memorizing strings. It is learning attack patterns. Character Gradient × Input and token saliency consistently lock onto the structural cues that make code executable, the **javascript**: scheme, event-handler bindings (**onerror =** , **onmouseover =**), boundary symbols like = and the closing > , the **alert** callsite, and context-fixing comment terminators (//, #). Benign scaffolding such as links, surrounding text, and closers receives neutral or negative weight, and obfuscations like percent/HTML/Unicode, and split keywords are discounted once decoded. Overall, it moves past surface form and captures compositional execution patterns, hence the large margins and explanations that square with secure-coding guidance.

6 Conclusion and Future Work

Our setup delivered state-of-the-art performance on the XSS task. We use raw and decoded versions of each payload, along with handcrafted obfuscation features, to represent both visible encodings and the meaning they hide. Of the two models we tried—a dual-branch CharCNN and a dual-branch, heads-only Transformer—both ended up performing almost perfectly, with best validation macro-F1 scores of 0.9958 and 0.9905.

Put simply, the results suggest two ways to deploy the models that work in tandem. The CharCNN, being both character-aware and lightweight, suits always-on systems like intrusion detectors and web application firewalls. Its ability to learn directly from characters ensures resilience against novel or malformed encodings, while its low latency enables integration into production traffic monitoring. On the other hand, the Transformer, even in its lightweight heads-only variant, offers greater robustness to long-range obfuscations and complex token interactions. This makes it a valuable candidate for offline analysis systems or hybrid deployments where latency is less critical.

We integrate XAI that converts character attributions to token summaries, estimates token causality with Δ-logit occlusion, and assesses explanations through branch-wise gradient agreement. The decoded analysis shows that the model focuses on execution signals rather than on filler text. The result is explanations that people can use in practice and that stay true to our fused RAW/DEC setup.

Overall, the results show you don't need huge models or big compute budgets to detect XSS well. Therefore, combining multi-view preprocessing with a modular dual-branch model and you get high accuracy and interpretable outputs.

Several promising directions remain. Begin by adding real-world data and red-team payloads so we can harder-test our invariance to obfuscation. Second, integrate adversarial training and contrastive pretraining across encoding pairs to strengthen alignment between RAW and DEC forms. Thirdly, pursue compression for edge deployment without sacrificing explanation quality. As a last step, let the tokenizer automatically coalesce adjacent encodings and handle additional web contexts, broadening what it can cover.

AI Use Disclosure Statement The authors did not use any generative AI or AI-assisted technologies in the creation of this manuscript. The authors used generative AI technologies only for proofreading. This includes minor corrections to grammar, spelling, and syntax. No AI was used for ideation, data analysis, or writing original content.

References

1. Yousefimehr B, Ghatee M (2025) A distribution-preserving method for resampling combined with LightGBM-LSTM for sequence-wise fraud detection in credit card transactions. Expert Syst Appl 262:125661
2. Ghaffari S, Yousefimehr B, Ghatee M (2024) Generative-AI in e-Commerce: Use-cases and Implementations. In: 2024 20th CSI international symposium on artificial intelligence and signal processing (AISP), 1–5
3. Weamie SJY (2022) Cross-Site scripting attacks and defensive techniques: a comprehensive survey. Int J Commun, Netw Syst Sci, 15(08). https://doi.org/10.4236/ijcns.2022.158010
4. Bates D, Barth A, Jackson C (2010) Regular expressions considered harmful in client-side XSS filters. In: Proceedings of the 19th international conference on world wide web, WWW '10. https://doi.org/10.1145/1772690.1772701
5. Kaur J, Garg U, Bathla G (2023) Detection of cross-site scripting (XSS) attacks using machine learning techniques: a review. Artif Intell Rev, 56(11). https://doi.org/10.1007/s10462-023-10433-3
6. Mikolov T, Chen K, Corrado G, Dean J (2013) Efficient estimation of word representations in vector space. In: 1st international conference on learning representations, ICLR 2013—workshop track proceedings
7. Kim Y (2014) Convolutional neural networks for sentence classification. In: EMNLP 2014 - 2014 conference on empirical methods in natural language processing, proceedings of the conference. https://doi.org/10.3115/v1/d14-1181
8. Lenji AR, Mehmandari TA, Ranjbar S, Tee KF, Fahimifar A (2025) Computational intelligence in rock engineering management systems. In Machine learning in geohazard risk prediction and assessment (pp. 233–285). Elsevier
9. Lenji AR, Mehmandari TA, Afshar D, Imani M, Fahimifar A (2025) Predicting liquefaction resistance of silty sands using machine learning techniques. In Mach Learn Geohazard Risk Predict Assess (pp. 83–105). Elsevier
10. Zhang X, Zhao J, Lecun Y (2015) Character-level convolutional networks for text classification. Adv Neural Inf Process Syst
11. Vaswani A, Shazeer N, Parmar N, Uszkoreit J, Jones L, Gomez AN, Kaiser Ł, Polosukhin I (2017) Attention is all you need. Adv Neural Inf Process Syst,
12. Díaz-Verdejo J, Muñoz-Calle J, Alonso AE, Alonso RE, Madinabeitia G (2022) On the detection capabilities of signature-based intrusion detection systems in the context of web attacks. Appl Sci (Switzerland), 12(2). https://doi.org/10.3390/app12020852
13. Usha G, Kannimuthu S, Mahendiran PD, Shanker AK, Venugopal D (2020) Static analysis method for detecting cross site scripting vulnerabilities. Int J Inf Comput Secur, 13(1). https://doi.org/10.1504/IJICS.2020.108123
14. Singh R, Gupta MK, Patil DR, Patil SM (2024) Analysis of web application vulnerabilities using dynamic application security testing. In: 2024 IEEE 9th international conference for convergence in technology (I2CT), 1–6
15. Hamzah KH, Osman MZ, Anthony T, Ismail MA, Abdullah Z, Alanda A (2024) Comparative analysis of machine learning algorithms for cross-site scripting (XSS) attack detection. JOIV: Int J Inform Vis, 8(3–2), 1678–1685
16. Saronian S, Yousefimehr B, Ghatee M, Bejani MM (2024). An ensemble of deep clustering models with autoencoders to mine travel patterns from smart card data. IEEE Trans Intell Transp Syst
17. Fang Y, Li Y, Liu L, Huang C (2018) DeepXSS: Cross site scripting detection based on deep learning. In: Proceedings of the 2018 international conference on computing and artificial intelligence, 47–51
18. Fang Y, Huang C, Xu Y, Li Y (2019) RLXSS: Optimizing XSS detection model to defend against adversarial attacks based on reinforcement learning. Futur Internet, 11(8). https://doi.org/10.3390/fi11080177

19. Chen T, Kornblith S, Norouzi M, Hinton G (2020) A simple framework for contrastive learning of visual representations. In: 37th international conference on machine learning, ICML 2020, PartF168147–3
20. Nithya V, Lakshmana Pandian S, Malarvizhi C (2015) A survey on detection and prevention of cross-site scripting attack. Int J Secur Its Appl, 9(3). https://doi.org/10.14257/ijsia.2015.9.3.14
21. Gage P (1994) A new algorithm for data compression. The C Users J. doi 10(5555/177910):177914
22. Sennrich R, Haddow B, Birch A (2016) Neural machine translation of rare words with subword units. In: 54th annual meeting of the association for computational linguistics, ACL 2016 - Long Papers, 3. https://doi.org/10.18653/v1/p16-1162
23. Radford Alec, Wu Jeffrey, Child Rewon, Luan David, Amodei Dario, & Sutskever Ilya (2019). Language models are unsupervised multitask learners | enhanced reader. OpenAI Blog, 1(8)
24. Liu Y, Zhang M (2018) Neural network methods for natural language processing. MIT Press One Rogers Street, Cambridge, MA 02142–1209, USA journals-info~
25. LeCun Y, Bengio Y (1995) Convolutional networks for images, speech, and time series. Handb Brain Theory Neural Netw, 3361
26. Khozeimeh F, Sharifrazi D, Izadi NH, Joloudari JH, Shoeibi A, Alizadehsani R, Gorriz JM, Hussain S, Sani ZA, Moosaei H, Khosravi A, Nahavandi S, Islam SMS (2021) Combining a convolutional neural network with autoencoders to predict the survival chance of COVID-19 patients. Sci Rep, 11(1). https://doi.org/10.1038/s41598-021-93543-8
27. Simonyan K, Vedaldi A, Zisserman A (2014) Deep inside convolutional networks: Visualising image classification models and saliency maps. In: 2nd international conference on learning representations, ICLR 2014—workshop track proceedings
28. Ancona M, Ceolini E, Öztireli C, Gross M (2018) Towards better understanding of gradient-based attribution methods for deep neural networks. In: 6th international conference on learning representations, ICLR 2018—conference track proceedings
29. Zeiler MD, Fergus R (2014) Visualizing and understanding convolutional networks. Lect Notes Comput Sci (Incl Subser Lect Notes Artif Intell Lect Notes Bioinform), 8689 LNCS(PART 1). https://doi.org/10.1007/978-3-319-10590-1_53

MIX
Papier aus verantwortungsvollen Quellen
Paper from responsible sources
FSC® C105338

If you have any concerns about our products,
you can contact us on
ProductSafety@springernature.com

In case Publisher is established outside the EU,
the EU authorized representative is:
Springer Nature Customer Service Center GmbH
Europaplatz 3, 69115 Heidelberg, Germany

Printed by Libri Plureos GmbH
in Hamburg, Germany